BIRKHÄUSER

W.D. Wallis

A Beginner's Guide to Graph Theory

Second Edition

Birkhäuser
Boston • Basel • Berlin

W.D. Wallis
Southern Illinois University
Department of Mathematics
Carbondale, IL 62901-4408
U.S.A.
wdwallis@math.siu.edu

Cover design by Alex Gerasev.

Mathematics Subject Classification (2000): 05-01, 05Cxx

Library of Congress Control Number: 2006936809

ISBN-13: 978-0-8176-4484-0 e-ISBN· 13: 978-0-8 176-4580-9

www.birkhausercom (TXQIEB)

Wallis: A Beginner's Guide to Graph Theory, 2nd edition

First Indian Reprint 2011

ISBN 978-81-8489-806-4

This edition is published by Springer (India) Private Limited, (a part of Springer Science+Business Media), Registered Office: 3rd Floor, Gandharva Mahavidyalaya, 212, Deen Dayal Upadhyaya Marg, New Delhi – 110 002, India

Printed in India by Rashtriya Printers, Delhi.

For Jack and Alex

Preface to the Second Edition

In this second edition although I have added two chapters that reflect new directions in graph theory, this presentation still remains relatively compact and concise.

Chapter 9 now focuses on graph labeling problems, which is an area of graph theory that has grown significantly in recent years. It can be approached without a great deal of background, but some of the problems and results presented will provide further insights into graph structure. (This change means that the former Chapters 9 through 13 are now numbered 10 through 14.)

Chapter 15 examines large random graphs, such as the internet and World Wide Web. These graphs have been studied for several years, but the topic received a boost in 1998 when the paper *Collective dynamics of 'small-world' networks* by Watts and Strogatz, appeared in *Nature* [129][1]. Moreover, this field in general has since moved into the public eye with the popularity of the Kevin Bacon game (see [134]) and the publication of popular books by Watts [127, 128], Barabási [6], and others. This chapter is studied in more detail than some others in the book, which I hope will lead to increased discussion rather than just to problem-solving.

There are some other minor changes. More examples are given in the earlier chapters in order to make this book more accessible to less experienced students. In particular, universities in Britain and Australasia often begin graph theory at a lower level than is common in North America. And, of course, I have corrected a number of misprints, hopefully not engendering others.

I received a good deal of feedback on the text of the First Edition; in particular, John George, Gábor Hetyei, Saad El-Zanati and Lane Clark made extensive comments, and Professors George and Hetyei have also reviewed and commented on this Second Edition. I am very grateful for their help.

June 2006 W.D. Wallis

---

[1] Citations refer to the references at the back of the book.

Preface to the First Edition

Many colleges and universities provide a first course in graph theory at about the senior level, intended primarily for mathematics majors but accessible to other students. This book is intended as a text for such a course, which I have given many times .

Over the years my classes have included mainly mathematics and computer science majors, but there have been several engineers and occasional psychologists. Often undergraduate and graduate students are in the same class. Many instructors today will find similar mixed group of students in their classes.

It is to be expected that anybody enrolling in a senior-level mathematics course will be comfortable with mathematical ideas and notations. In particular, I assume that the reader is familiar with the basic concepts of set theory, has seen mathematical induction, and has a passing acquaintance with matrices and algebra. However, one cannot assume that the students in a first graph theory will have a good knowledge of any specific advanced area. My reaction to this is to avoid too many specific prerequisites. The main requirement, namely a little mathematical maturity, may have been acquired in a variety of ways.

My students' reasons for studying graph theory have also been mixed. Some have seen graph theory as an area of pure mathematics to be studied for its own sake, others as an adjunct to such mathematical studies as combinatorics, algebra, or functional analysis, and others as an applied area. Even within a single area of application, there are diverse reasons: one electrical engineer, for example, may use graph theory to study circuits, while another may see it as a foundation for neural networks. Taking this into account, I have concentrated on the topics that appeal to the majority of users, and generally I have omitted those with a smaller readership. I have included a few more specialized chapters dealing with material that students seem to enjoy, and (frankly) ones that I like to teach. Readers can supplement the selection with other topics to meet their specific needs.

Outline of Topics

The first four chapters introduce the main ideas of graph theory and conclude with a short discussion of the minimal spanning tree problem. The idea is to introduce graph-theoretic reasoning, and also introduce an easy algorithm.

The fifth chapter deals with the application of vector space ideas to graphs. This is one of three specialized excursions, and could be omitted or deferred; in particular anybody who has not seen a formal linear algebra course (including at least the general definition of a vector space) should probably skip this chapter. But I have found that students with an algebraic background often like this material, and if it is to be included at all, then this is probably the best place for it.

Chapter 6 explores another special topic, one-factorizations of graphs. All students should read the first section, and most will enjoy the second. The rest is a little specialized, but introduces some good examples of graph-theoretic reasoning.

There follows an exposition of coloring and planarity. A discussion of edge-coloring is included, and should particularly interest those who read all of Chapter 6. Ramsey's Theorem is studied in Chapter 9[2] given later will especially appeal to those with a wider combinatorial background. The later parts of this chapter are quite difficult.

Chapter 10[3] introduces directed graphs. The two following chapters are devoted to two important application areas that will appeal to students of management science, namely critical paths and network flows. Students who do not know a little statistical theory — enough to use the normal distribution, and to look up values in a table of the normal probability function — should skip Section 11.3[4].

A chapter on graph-theoretic algorithms concludes the book. I believe that computer scientists will see more than enough of these topics in other courses, and that graph algorithms are more appropriately studied among other algorithms, not among other aspects of graphs. Moreover, a proper study of algorithms would require some study of computational complexity, which would probably not interest the majority of readers. So my treatment here is intentionally short and quite superficial, but should satisfy the needs of those who are not likely to revisit the topic.

I thought of including several further topics of pure graph theory — covering theorems, line graphs, general problems on cycles, various extremal problems — but rejected them because of their specialized appeal; three specialized topics (graphs and linear spaces, one-factorizations, Ramsey theory) should be enough, and these are my own preferences. A pure-mathematically minded instructor could easily replace Chapters 11 through 13 with other appropriate topics; of course, her/his interpretation of what is "appropriate" could certainly be different from mine.

---

[2] Chapter 10 in this edition

[3] Chapter 11 in this edition

[4] Now Section 12.3

Further Reading

I have made frequent reference to the papers where results first appeared, and to the research literature in general. Those who want to go further into a topic can consult the papers cited. For general reading, the student may wish to consult one of the more advanced volumes on graph theory: one excellent example is West's book [131], an introduction to the subject that goes much deeper than we do here. Volumes of surveys of specific topics include three volumes edited by Beineke and Wilson [10, 11, 12] and two edited by Fulkerson [49, 50]. One very readable book is Tutte's *Connectivity in Graphs* [117], now 35 years old but still an excellent research resource. Yap's collection [135] of short monographs on three topics of graph theory includes an excellent introduction to edge-coloring. Haynes, Hedetneimi and Slater [69] have recently written a first-class introduction to a current hot topic, domination theory. The reader interested in graph matchings and factorization may wish to consult [84] or [125]. Biggs, Lloyd and Wilson provide a good deal of historical information, and some classical papers, in [14].

Exercises

I have tried to include a reasonable number of problems, but not so many that the student is overwhelmed. They range from the easy to the difficult. In a few cases, hints are included, and there are answers and solutions to selected exercises. A hint is indicated by H preceding the exercise number, while box.02inA announces an answer or solution.

Acknowledgments

This book has grown out of graph theory courses that I have taught at the University of Newcastle and Southern Illinois University over the past 30 years. A number of students have made comments and contributions; I hope they will forgive me if I do not mention them by name, but if I tried to do so, I would surely (unintentionally) omit some.

My friend and colleague Roger Eggleton used a draft version of the text for a course at Illinois State University. He made a large number of intelligent and informed comments and corrections, including the discovery of at least two instances where a widely-published, accepted "proof" was in need of amendment. I am very grateful to him and his students for their assistance.

Finally, the book would not exist without the support of George Anastassiou, of the University of Memphis, who recommended it to the publisher, and of Birkhäuser's Ann Kostant, who suggested the title, Elizabeth Loew, and Tom Grasso. Thanks to all of you.

May 2000 W.D. Wallis

Contents

List of Figures

1

Graphs

1.1 Sets, Binary Relations and Graphs

We shall use the standard concepts and notations of set theory. We write $x \in S$ and $y \notin S$ to indicate that x is a member of S, and that y is not a member of S. $|S|$ denotes the number of elements of S, also called the *order* of S. If all elements of S are also elements of T, then S is a subset of T, written $S \subseteq T$. The notation $S \subset T$ means that S is a subset of T but is not identical to T, so that T has at least one element that is not in S.

A set can be specified by writing a description ("the set of positive perfect squares less than 20") or by listing its members ("$\{1, 4, 9, 16\}$"). One can also state a membership law: for example, $\{x^2 \mid x \text{ is an integer}, x^2 < 20\}$. This is called *setbuilder notation*. The | is read as "such that"; it is equally common to use a colon instead of |. Other mathematical abbreviations are often employed in setbuilder notation — for example, $\mathbb{Z}$ usually represents the set of integers, so one might replace "x is an integer" by "$x \in \mathbb{Z}$."

If S and T are any two sets, then $S \cup T$ means the *union* of S and T, the set of everything that is either a member of S or a member of T (or both), and $S \cap T$ is the *intersection*, the set common elements. The *set-theoretic difference* $S \backslash T$, also called the *relative complement* of T in S, consists of all elements of S that are *not* members of T. The cartesian product $S \times T$ is the set of all ordered pairs $\{x, y\}$, where x is a member of S and y is a member of T.

Various identities between sets can be proved. For example, both union and intersection satisfy the commutative laws

$$S \cup T = T \cup S \tag{1.1}$$

$$S \cap T = T \cap S \tag{1.2}$$

for any sets S and T, and the associative laws

$$R \cup (S \cup T) = (R \cup S) \cup T \tag{1.3}$$

$$R \cap (S \cap T) = (R \cap S) \cap T \tag{1.4}$$

for any sets R, S and T.

To prove that two sets are equal, one often proves that every member of one set is an element of the other, and conversely. In other words, to show that $A = B$, first prove $A \subseteq B$ and then prove $B \subseteq A$.

Example. To prove $R \cap (S \cap T) = (R \cap S) \cap T$, first observe that any member x of $R \cap (S \cap T)$ is both a member of R and a member of $S \cap T$, and the latter means that x belongs to both S and T. So all of $x \in R$, $x \in S$ and $x \in T$ are true. From these we see that both $x \in R$ and $x \in S$ are true, so $x \in R \cap S$, and also $x \in T$; therefore $x \in (R \cap S) \cap T$. We have actually shown that

$$R \cap (S \cap T) \subseteq (R \cap S) \cap T.$$

One proves $R \cap (S \cap T) \subseteq (R \cap S) \cap T$ in the same way, and equality has been established.

Example. Prove $R\backslash(S \cup T) = (R\backslash S) \cap (R\backslash T)$for any sets R, S and T.

$R\backslash(S \cup T)$ consists of precisely those members of R that are not members of $S \cup T$, in other words those elements of R that do not belong to S or to T. That is,

$$R\backslash(S \cup T) = \{x \mid x \in R \text{ and } x \notin S \text{ and } x \notin T\}.$$

On the other hand, $(R\backslash S)$ consists of all the things in R that are not in S, and $(R\backslash S) \cap (R\backslash T)$ consists of all the things in R that are not in T; the common elements of these sets are all the things in R but not in S and not in T, which is the same as the description of $R\backslash(S \cup T)$.

Binary relations occur frequently in mathematics and in everyday life. For example, the ordinary mathematical relations $<$, $=$, $>$, $\leq$ and $\geq$ are binary relations on number sets, $\subset$ and $\subseteq$ are binary relations on collections of sets, and so on. If S is the set of all living people, "is the child of" is a typical binary relation on S.

Formally, a *binary relation* $\sim$ on a set S is a rule that stipulates, given any elements x and y of S, whether x bears a certain relationship to y (written $x \sim y$) or not (written $x \not\sim y$). Alternatively, one can define a binary relation $\sim$ on the set S to consist of a set $\sim(S)$ of elements from $S \times S$ (the set of ordered pairs of elements of S), with the notation $x \sim y$ meaning that (x, y) belongs to $\sim(S)$.

One important class of binary relations is arithmetical *congruence*. Two integers a and b are *congruent modulo* n (written $a \equiv b(\bmod n)$) if $a - b$ is divisible by n. The *congruence class* of $a(\bmod n)$ is the set of all integers congruent to a modulo n. There are n different congruence classes modulo n.

One can represent any binary relation by a diagram. The elements of the set S are shown as points (*vertices*), and if $x \sim y$ is true, then a line (*edge*) is shown from x to y, with its direction indicated by an arrow. Provided the set S is finite, all information about any binary relation on S can be shown in the diagram. The diagram is a *directed graph* or *digraph*; if $x \sim x$ is ever true, the diagram is a *looped* digraph.

The binary relation $\sim$ on S is called *reflexive* if $x \sim x$ is true for all x in S, and *antireflexive* if $x \sim x$ is never true (or, equivalently, if $x \not\sim x$ is true for all x). If $y \sim x$ is true whenever $x \sim y$ is true, then $\sim$ is called *symmetric*. If the relation is symmetric, the arrows can be omitted from its diagram. The diagram of a symmetric, antireflexive binary relation on a finite set is called a *graph*.

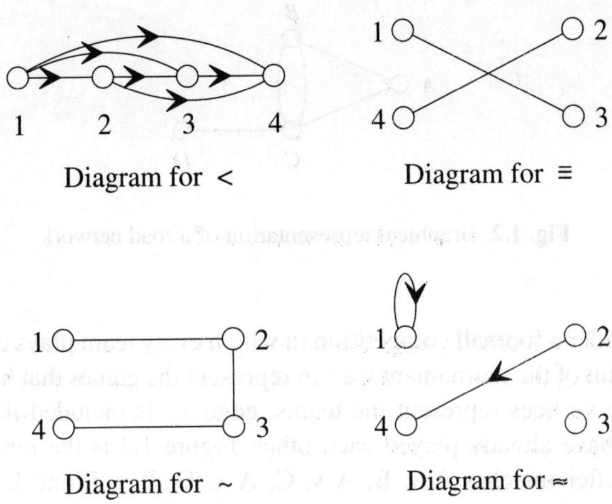

Fig. 1.1. Diagrams of binary relations

Example. Suppose the binary relations $<$, $\equiv$, $\sim$ and $\approx$ are defined on the set $S = \{1, 2, 3, 4\}$ as follows:

$$\begin{aligned}
x < y &\text{ means } x \text{ is less than } y;\\
x \equiv y &\text{ means } x \text{ is congruent to } y \pmod 2 \text{ and } x \neq y;\\
x \sim y &\text{ means } x = y \pm 1;\\
x \approx y &\text{ means } y = x^2.
\end{aligned}$$

Then the corresponding subsets of $S \times S$ are

$$\begin{aligned}
<(S) &= \{(1,2),(1,3),(1,4),(2,3),(2,4),(3,4)\};\\
\equiv(S) &= \{(1,3),(3,1),(2,4),(4,2)\};\\
\sim(S) &= \{(1,2),(2,1),(2,3),(3,2),(3,4),(4,3)\};\\
\approx(S) &= \{(1,1),(2,4)\}.
\end{aligned}$$

The diagrams are shown in Figure 1.1. Relations $\equiv$ and $\sim$ yield graphs, $<$ gives a digraph, and $\approx$ a looped digraph.

In more general situations, it might make sense to use two or more lines to join the same pair of points. For example, suppose we want to describe the roads joining various townships. For many purposes we do not need to know the topography of the region, or whether different roads cross, or various other things. The important information is whether or not there is a road joining two towns. In these cases we could use a complete road map with the exact shapes of the roads and various other details shown, but it would be less confusing to make a diagram, as shown in Figure 1.2, that indicates two roads joining B to C, one road from A to each of B and C and one road from C to D with no direct roads joining A to D or B to D. We say that there is a *multiple edge* (of *multiplicity* 2) joining B to C. If any of the roads were one-way, an arrow could be employed.

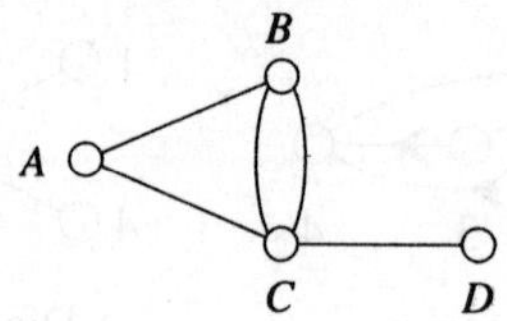

Fig. 1.2. Graphical representation of a road network

Example. Consider a football competition in which every team plays every other team once. At any point of the tournament we can represent the games that have been played by a graph. The vertices represent the teams; edge xy is included if and only if the teams x and y have already played each other. Figure 1.3 is the representation of a 6-team league after matches A v. B, A v. C, A v. D, B v. E and E v. F have been completed.

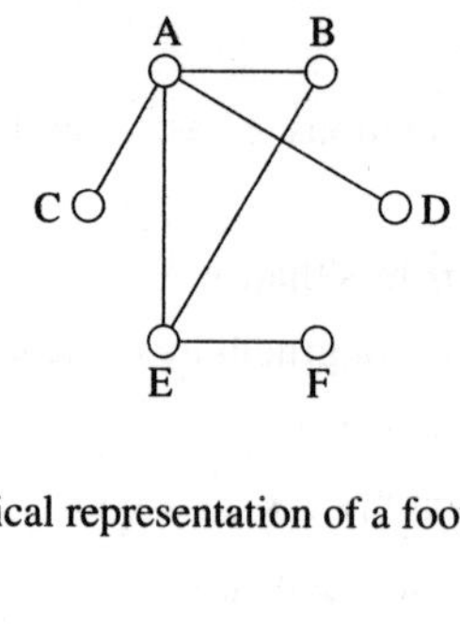

Fig. 1.3. Graphical representation of a football competition

Example. Suppose there are four jobs vacant and five men apply for them. Each man is capable of performing one or more of the jobs. The usual question is whether or not one can allocate jobs to four of the men so that all four jobs are allocated.

This situation can conveniently be represented by a graph. All the applicants and all the jobs are represented by vertices; two vertices are joined if and only if one represents an applicant, the other represents a job, and the applicant is capable of doing the job. Figure 1.4 shows the situation where A, B and C can all handle jobs 1 and 2, D can do 1 and 3 and E can do 2 and 4.

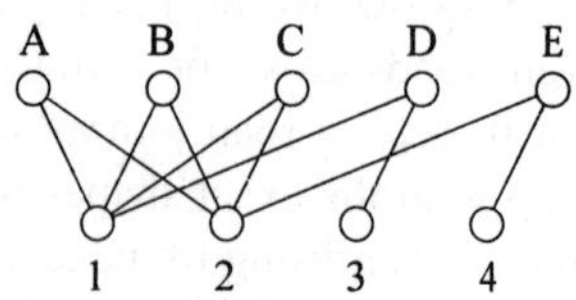

Fig. 1.4. Who is qualified for which job?

Exercises 1.1

1.1.1 Prove that the following identities hold for any sets R, S and T.
(i) $R \cup (S \cup T) = (R \cup S) \cup T$.
(ii) $R \cap (S \cup T) = (R \cap S) \cup (R \cap T)$.
(iii) $R \backslash (S \cap T) = (R \backslash S) \cup (R \backslash T)$.

1.1.2 In each part below, a binary relation $\sim$ is defined on
$$\{-3, -2, -1, 0, 1, 2, 3\}.$$
In each case, is the relation reflexive? Antireflexive? Symmetric?
(i) $x \sim y$ means $x + y \leq 4$.
(ii) $x \sim y$ means $x + y \leq 6$.
(iii) $x \sim y$ means $x = y + 1$.
(iv) $x \sim y$ means $x = \pm y$.

1.1.3 Repeat Exercise 1.1.2 for the following binary relations defined on
$$\{-3, -2, -1, 0, 1, 2, 3\}:$$
(i) $x \sim y$ means $x \leq y^2$.
(ii) $x \sim y$ means $x + y \geq 0$.
(iii) $x \sim y$ means $x + y$ is odd.
(iv) $x \sim y$ means xy is odd.

1.1.4 Draw graphical representations of the relations in Exercise 1.1.2.

1.1.5 Draw graphical representations of the relations in Exercise 1.1.3.

1.1.6 Repeat Exercise 1.1.2 for the following relations defined on the positive integers.
(i) $x \sim y$ means $x + y \leq 4$.
(ii) $x \sim y$ means x divides y.
(iii) $x \sim y$ means x and y have greatest common divisor 1.
(iv) $x \sim y$ means $x + y$ is odd.

1.1.7 Repeat Exercise 1.1.2 for the following relations defined on the positive integers.
(i) $x \sim y$ means x and y are both prime numbers.
(ii) $x \sim y$ means $x = \pm y$.
(iii) $x \sim y$ means xy is odd.

1.1.8 A relation is called *transitive* if every time (x, y) and (y, z) are in the relation, then (x, z) is also.
(i) Describe the graph of a symmetric, transitive relation.
(ii) Which, if any, of the relations in Exercises 1.1.2, 1.1.3, 1.1.6 and 1.1.7 are transitive?

1.1.9 Suppose the binary relation $\sim$ is defined on the set of real numbers as follows: $x \sim y$ means $x \equiv y \pmod 7$. Is $\sim$ reflexive? Antireflexive? Symmetric? Transitive?

1.1.10 A basketball league contains seven teams, denoted by A, B, C, D, E, F, G. Team A has played against each other team once; team B has played against C, E and G; and teams D, E and F have all played. Draw a graph to illustrate this situation.

1.2 Some Definitions

We start by formalizing some of the discussion and definitions from the preceding section. A *graph G* consists of a finite set $V(G)$ of objects called *vertices* together with a set $E(G)$ of unordered pairs of vertices; the elements of $E(G)$ are called *edges*. We write $v(G)$ and $e(G)$ for the orders of $V(G)$ and $E(G)$, respectively; these are often called the *order* and *size* of G. In terms of the more general definitions sometimes used, we can say that "our graphs are finite and contain neither loops nor multiple edges."

Graphs are usually represented by diagrams in which the vertices are points. An edge xy is shown as a line from (the point representing) x to (the point representing) y. To distinguish the vertices from other points in the plane, they are often drawn as small circles or large dots.

Example. A graph has five vertices, a, b, c, d, e, and edges ab, ac, ad, be, de. So its representation is

Often the same graph can give rise to several drawings that look quite dissimilar. For example, the three diagrams in Figure 1.5 all represent the same graph. Although the two diagonal lines cross in the first picture, their point of intersection does not represent a vertex of the graph.

The edge containing x and y is written xy or (x, y); x and y are called its *endpoints*. We say this edge *joins* x to y. If A and B are subsets of $V(G)$, then $[A, B]$ denotes the set of all edges of G with one endpoint in A and the other in B:

$$[A, B] = \{xy : x \in A, y \in B, xy \in E(G)\}. \qquad (1.5)$$

If A consists of the single vertex a, it is usual to write $[a, B]$ instead of $[\{a\}, B]$.

An *isomorphism* of a graph G onto a graph H is a one-to-one map ϕ from $V(G)$ onto $V(H)$ with the property that a and b are adjacent vertices in G if and only if $a\phi$ and $b\phi$ are adjacent vertices in H; G is *isomorphic* to H if and only if there is an isomorphism of G onto H. An isomorphism from a graph G to itself is called an *automorphism* of G.

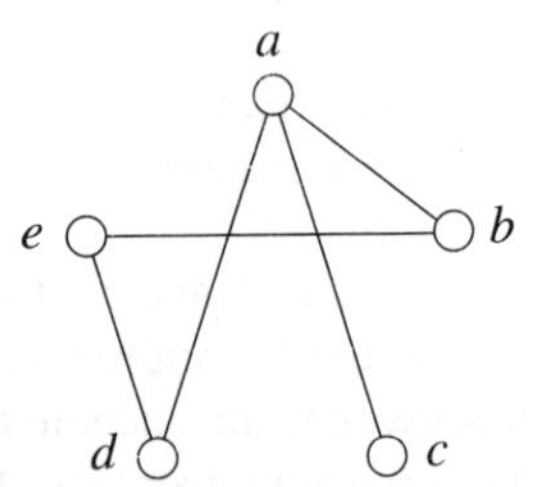

Because graphs are finite, one can prove that a map ϕ is an isomorphism of G onto H by showing that the two graphs have the same numbers of vertices and edges, that $a\phi$ is a vertex of H whenever a is a vertex of G, and that $a\phi$ and $b\phi$ are adjacent vertices in H whenever a and b are adjacent in G.

Given a set S of v vertices, the graph formed by joining each pair of vertices in S is called the *complete* graph on S and denoted by K_S. We also write K_v to mean any complete graph with v vertices. From the definition of isomorphism it follows that all complete graphs on v vertices are isomorphic. The notation K_v can be interpreted as being a generic name for the typical representative of the isomorphism class of all v-vertex complete graphs. The three drawings in Figure 1.5 are all representations of K_4.

A *multigraph* is defined in the same way as a graph except that there may be more than one edge corresponding to the same unordered pair of vertices. The *underlying graph* of a multigraph is formed by replacing all edges corresponding to the unordered pair $\{x, y\}$ by a single edge xy. Unless otherwise mentioned, all definitions and concepts pertaining to graphs will be applied to multigraphs in the obvious way.

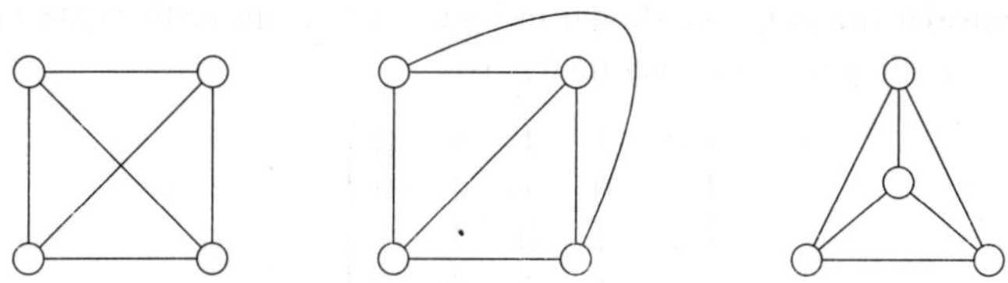

Fig. 1.5. Three representations of K_4

In some cases a direction is imposed on each edge. In this case we call the graph a *directed graph* or *digraph*. Directed edges are usually called *arcs*. An arc is an ordered pair of vertices, the first vertex is the *start* (or *tail* or *origin*) of the arc, and the second is the *finish* (or *head* or *terminus*). Directed graphs can have two arcs with the same endpoints, provided they have opposite directions. The *underlying graph* of a digraph is constructed by ignoring all directions and replacing any resulting multiple edges by single edges.

$G - xy$ denotes the graph produced by deleting edge xy from G. If xy is not an edge of G, then $G + xy$ is the graph constructed from G by adding an edge xy (one often refers to this process as *joining x to y in G*). Figure 1.6 illustrates these ideas. Similarly, $G - x$ means the graph derived from G by deleting one vertex x (and all the edges on which x lies). More generally, $G - S$ is the graph resulting from deleting some set S of vertices.

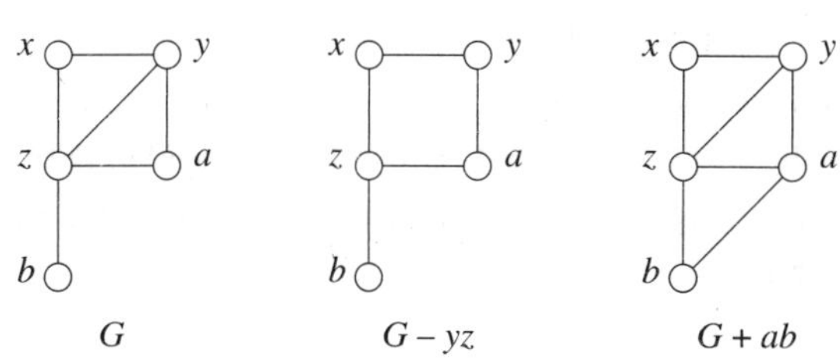

Fig. 1.6. Adding and deleting edges

If vertices x and y are endpoints of one edge in a graph or multigraph, then x and y are said to be *adjacent* to each other, and it is often convenient to write $x \sim y$. Vertices adjacent to x are called *neighbors* of x, and the set of all vertices adjacent to x is called the *neighborhood* of x, and denoted by $N(x)$. If G has v vertices, so that its vertex set is,

$$V(G) = \{x_1, x_2, \ldots, x_v\},$$

then its *adjacency matrix* M_G is the $v \times v$ matrix with entries m_{ij}, such that

$$m_{ij} = \begin{cases} 1 & \text{if } x_i \sim x_j, \\ 0 & \text{otherwise.} \end{cases}$$

The particular matrix will depend on the order in which the vertices are listed.

Example. Consider the graph G shown in Figure 1.6. If its vertices are taken in the order x, y, z, a, b, then its adjacency matrix is

$$\begin{bmatrix} 0 & 1 & 1 & 0 & 0 \\ 1 & 0 & 1 & 1 & 0 \\ 1 & 1 & 0 & 1 & 1 \\ 0 & 1 & 1 & 0 & 0 \\ 0 & 0 & 1 & 0 & 0 \end{bmatrix}.$$

Some authors define the adjacency matrix of a multigraph to be the adjacency matrix of the underlying graph; others set m_{ij} equal to the number of edges joining x_i to x_j. We shall not need to use adjacency matrices of multigraphs in this book.

A vertex and an edge are called *incident* if the vertex is an endpoint of the edge, and two edges are called *adjacent* if they have a common endpoint. A set of edges is called *independent* if no two of its members are adjacent, and a set of vertices is independent if no two of its members are adjacent. The *independence number* $\beta(G)$ of a graph G is the number of elements in the largest independent set in G. For example, the graph G of Figure 1.6 has independence number 3; its largest independent set (which happens to be unique) is $\{a, b, x\}$.

If the edge set of G is

$$E(G) = \{a_1, a_2, \ldots, a_e\},$$

then the *incidence matrix* N_G of G is the $v \times e$ matrix with entries n_{ij}, such that

$$n_{ij} = \begin{cases} 1 & \text{if vertex } x_i \text{ is incident with edge } a_j, \\ 0 & \text{otherwise.} \end{cases}$$

(The adjacency and incidence matrices depend on the orderings chosen for $V(G)$ and $E(G)$; they are not unique, but vary only by row and/or column permutation.)

Example. Here are two copies of the graph of Figure 1.6, with the edges labeled 1, 2, 3, 4, 5, 6 in two different ways.

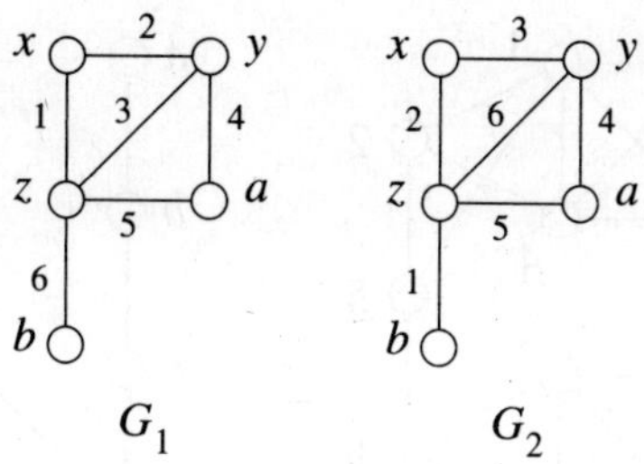

The incidence matrix for the labeling G_1, with edges taken in numerical order and vertices in the order x, y, z, a, b, is

$$\begin{bmatrix} 1 & 1 & 0 & 0 & 0 & 0 \\ 0 & 1 & 1 & 1 & 0 & 0 \\ 1 & 0 & 1 & 0 & 1 & 1 \\ 0 & 0 & 0 & 1 & 1 & 0 \\ 0 & 0 & 0 & 0 & 0 & 1 \end{bmatrix}$$

while the corresponding incidence matrix for the labeling G_2 is

$$\begin{bmatrix} 0 & 1 & 1 & 0 & 0 & 0 \\ 0 & 0 & 1 & 1 & 0 & 1 \\ 1 & 1 & 0 & 0 & 1 & 1 \\ 0 & 0 & 0 & 1 & 1 & 0 \\ 1 & 0 & 0 & 0 & 0 & 0 \end{bmatrix}.$$

If G is a graph, it is possible to choose some of the vertices and some of the edges of G in such a way that these vertices and edges again form a graph, say H. H is then called a *subgraph* of G; one writes $H \leq G$. Clearly every graph G has itself as a subgraph; we say a subgraph H is a *proper* subgraph of G, and write $H < G$, if it does not equal G. The 1-vertex graph (which we shall denote by K_1) is also a subgraph of every graph. If U is any set of vertices of G, then the subgraph consisting of U and all the edges of G that join two vertices of U is called an *induced* subgraph, the *subgraph induced by* U, and is denoted by $\langle U \rangle$ or $G[U]$. A subgraph G of a graph H is called a *spanning* subgraph if $V(G) = V(H)$. Clearly any graph G is a spanning subgraph of $K_{V(G)}$.

In particular, a *clique* in a graph G is a complete subgraph. In other words, it is a subgraph in which every vertex is adjacent to every other. A clique H in G is called *maximal* if no vertex of G outside of H is adjacent to all members of H. The clique structure of G can be illustrated by forming a new graph $C(G)$ called the *clique graph* of G. The vertices of this graph are in one-to-one correspondence with the maximal cliques of the original, and two vertices are adjacent if and only if the corresponding cliques have a common vertex. The size of the largest clique in G is called the *clique number* of G and denoted by $\omega(G)$.

Example. Figure 1.7 shows a graph G and its clique graph $C(G)$. The maximal cliques of G have vertex sets $\{0, 1, 3, 4\}$, $\{1, 2, 4, 5\}$, $\{5, 8\}$, $\{7, 8, 10\}$, $\{7, 9, 10\}$ and $\{6, 7, 9\}$, and are represented in $C(G)$ by a, b, c, d, e and f respectively.

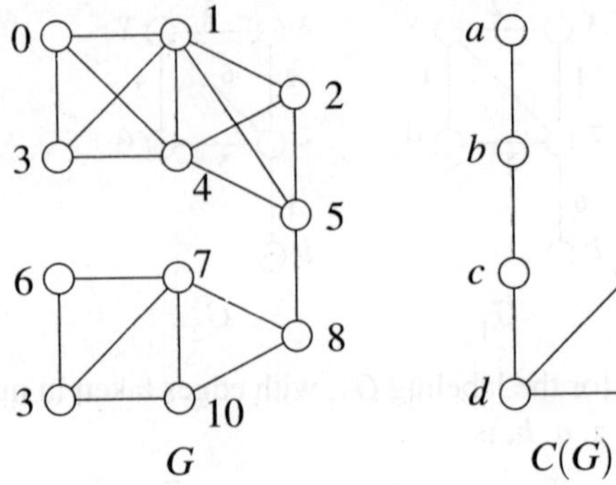

Fig. 1.7. A graph and its clique graph

Given any graph G, the set of all edges of $K_{V(G)}$ that are *not* edges of G will form a graph with $V(G)$ as vertex set; this new graph is called the *complement* of G, and written $\overline{G}$. More generally, if G is a subgraph of H, then the graph formed by deleting all edges of G from H is called the *complement of* G *in* H, denoted by $H - G$. The complement $\overline{K}_S$ of the complete graph K_S on vertex set S is called a *null graph*; we also write $\overline{K}_v$ for a null graph with v vertices.

A graph is called *disconnected* if its vertex set can be partitioned into two subsets, V_1 and V_2, that have no common element, in such a way that there is no edge with one endpoint in V_1 and the other in V_2; if a graph is not disconnected, then it is *connected*. A disconnected graph consists of a number of disjoint subgraphs; a maximal connected subgraph is called a *component*. As an example, instead of three representations of the same graph, Figure 1.5 might show one 12-vertex graph with three 4-vertex components. In a way, connectedness generalizes adjacency. In a connected graph, not all vertices are adjacent, but if x and y are not adjacent, then there must exist vertices $x_1, x_2, \ldots, x_n$ such that x is adjacent to x_1, x_1 is adjacent to $x_2, \ldots$ and x_n is adjacent to y; such a sequence is called an *xy-walk*. Conversely, if every pair of nonadjacent vertices is joined by such a walk, the graph is connected. These ideas will be further explored and generalized in Chapters 2 and 3.

The *complete bipartite graph* on V_1 and V_2 has two disjoint sets of vertices, V_1 and V_2; two vertices are adjacent if and only if they lie in different sets. We write $K_{m,n}$ to mean a complete bipartite graph with m vertices in one set and n in the other. Figure 1.8 shows $K_{4,3}$; $K_{1,n}$ in particular is called an *n-star*. Any subgraph of a complete bipartite graph is called *bipartite*. More generally, the *complete r-partite graph* $K_{n_1,n_2,\ldots,n_r}$ is a graph with vertex set $V_1 \cup V_2 \cup \ldots \cup V_r$, where the V_i are disjoint sets and V_i has order n_i, in which xy is an edge if and only if x and y are in different sets. Any subgraph of this graph is called an *r-partite* graph. If $n_1 = n_2 = \ldots = n_r = n$, we use the abbreviation $K_n^{(r)}$.

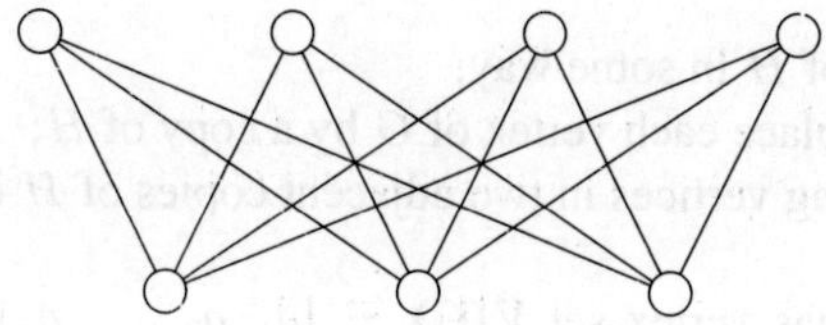

Fig. 1.8. $K_{4,3}$

Example. Prove that the graph G_1, formed by deleting an edge from K_4, and the graph G_2, constructed by adding to $K_{2,2}$ an edge joining two inadjacent vertices, are isomorphic.

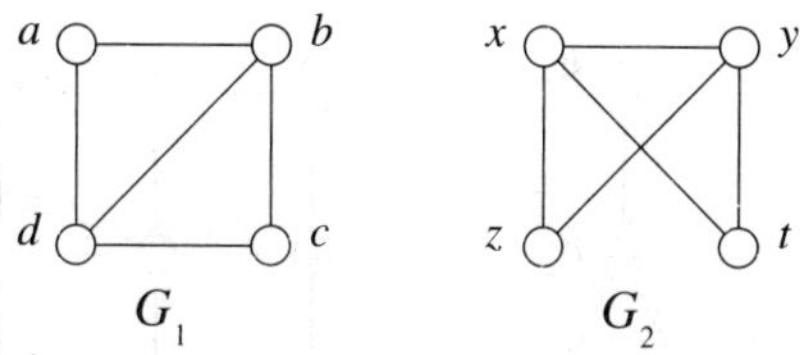

Suppose the graphs are as shown. G_1 is formed by deleting edge ac from the complete graph with vertices a, b, c, d. (If instead edge ab were deleted, for example, the graph formed would be isomorphic to the one shown under the mapping $a \to a$, $b \to c$, $c \to b$, $d \to d$.) G_2 was formed from the complete bipartite graph with vertex-sets $\{x, y\}$ and $\{z, t\}$ by joining one pair — we have joined x to y, but the result of joining z to t is isomorphic. The two have the same numbers of vertices and of edges. On inspection it is seen that $a \to z$, $b \to x$, $c \to t$, $d \to y$ maps the vertices of G_1 to the vertices of G_2 and maps the edges of G_1 to the edges of G_2, so it is an isomorphism.

Several ways of combining two graphs have been studied. The *union* $G \cup H$ of graphs G and H has as vertex set and edge set the unions of the vertex sets and edge sets, respectively, of G and H. The *intersection* $G \cap H$ is defined similarly, using the intersections (but $G \cap H$ is defined only when G and H have a common vertex). If G and H are edge-disjoint graphs on the same vertex set, then their union is often also called their *sum* and written $G \oplus H$. ($A \oplus B$ is often written for the union of disjoint sets A and B; in this notation, the graph $G \oplus H$ has edge set $E(G) \oplus E(H)$.) At the other extreme, disjoint unions can be discussed, and the union of n disjoint graphs all isomorphic to G is denoted by nG.

The notation $G + H$ denotes the *join* of G and H, a graph obtained from G and H by joining every vertex of G to every vertex of H. (This notation is not consistent with the earlier use of the + symbol; $G + xy$ is not the join of G and the K_2 with vertex set $\{x, y\}$. Unfortunately, these two uses of + are common in the graph theory literature.) $G + H$ is also used when G and H represent isomorphism-classes of graphs, with the assumption that G and H are disjoint, so that for example

$$K_{m,n} = \overline{K}_m + \overline{K}_n.$$

The *cartesian product* $G \times H$ of graphs G and H is defined as follows:

(i) label the vertices of H in some way;
(ii) in a copy of G, replace each vertex of G by a copy of H;
(iii) add an edge joining vertices in two adjacent copies of H if and only if they have the same label.

In other words, if G has vertex set $V(G) = \{a_1, a_2, \ldots, a_g\}$ and H has vertex set $V(H) = \{b_1, b_2, \ldots, b_h\}$, then $G \times H$ has vertex set $V(G) \times V(H)$, and (a_i, b_j) is adjacent to (a_k, b_ℓ) if and only if *either* $i = k$ and b_j is adjacent in H to b_ℓ *or* $j = \ell$ and a_i is adjacent in G to a_k. It is clear that $G \times H$ and $H \times G$ are isomorphic. Similarly $(G \times H) \times J$ and $G \times (H \times J)$ are isomorphic, so one can omit parentheses and define cartesian products of three or more graphs in a natural way. An example is shown in Figure 1.9.

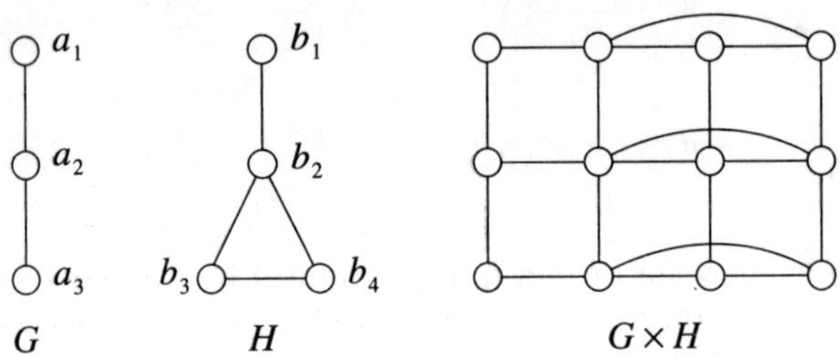

Fig. 1.9. The cartesian product of two graphs

Exercises 1.2

A1.2.1 Write down the incidence and adjacency matrices of the graphs G and H of Figure 1.9.

1.2.2 For each antireflexive symmetric relation in Exercise 1.1.2, write down the incidence and adjacency matrices of the corresponding graph.

1.2.3 Up to isomorphism, there are exactly six connected graphs, and exactly eleven graphs in total, on four vertices. Prove this.

1.2.4 If G is any graph, show that $\overline{\overline{G}} = G$ (that is, the complement of a complement equals the original graph).

A1.2.5 Prove that if G is not a connected graph, then $\overline{G}$ is connected.

1.2.6 Suppose G and H are any two graphs. Show that $\overline{G + H}$ (the complement of the join of G and H) is not connected.

1.2.7 A graph G is *self-complementary* if G and $\overline{G}$ are isomorphic. Prove that the number of vertices in a self-complementary graph must be congruent to 0 or 1 (mod 4).

1.2.8 For any graph G, prove that $\overline{G - x} = \overline{G} - x$, for any vertex x of G.

A1.2.9 G is a bipartite graph with v vertices. Prove that G has at most $\frac{v^2}{4}$ edges.

1.2.10 G is the graph shown in Figure 1.9 and H is the triangle K_3 ; G and H have disjoint vertex sets. Sketch the following graphs:

(i) $3G$
(ii) $2H$
(iii) $G \cup H$
(iv) $G \oplus H$
(v) $G + H$
(vi) $G \times H$
(vii) $G \times G$

2.11 How many edges does the star $K_{1,n}$ have? Write down the adjacency and incidence matrices of $K_{1,5}$.

2.12 The graph W_n, called an *n-wheel*, has $n + 1$ vertices $\{x_0, x_1, \ldots, x_n\}$; x_0 is joined to every other vertex and the other edges are

$$x_1x_2, x_2x_3, \ldots, x_{n-1}x_n, x_nx_1.$$

How many edges does W_n have? Write down the adjacency and incidence matrices of W_5.

2.13 The *n-ladder* L_n has $2n$ vertices $\{x_1, x_2, \ldots, x_n, y_1, y_2, \ldots, y_n\}$; x_i is joined to y_i for every i; x_i is joined to x_{i+1}, and y_i is joined to y_{i+1} for $i = 1, 2, \ldots, n - 1$.
(i) How many edges does L_n have?
(ii) Write down the adjacency and incidence matrices of L_4.

2.14 Suppose G is the graph with vertices x, y, z, t and edges xy, yz, zt; H is the graph with vertices x, z and edge xz; K is a graph K_2 with vertex-set disjoint from G. Show that no two of the graphs $G + xz$, $G + H$ and $G + K$ are isomorphic.

2.15 A *directed multigraph* is a structure similar to a directed graph, in which multiple arcs in the same direction are allowed. Write down a formal definition of a directed multigraph in terms of a vertex set and an edge set. Give two practical examples of situations that are best modeled by directed multigraphs.

1.3 Degree

We define the *degree* or *valency* $d(x)$ of the vertex x to be the number of edges that have x as an endpoint. If $d(x) = 0$, then x is called an *isolated* vertex while a vertex of degree 1 is called *pendant*. The edge incident with a pendant vertex is called a *pendant edge*. A graph is called *regular* if all its vertices have the same degree. If the common degree is r, it is called *r-regular*. In particular, a 3-regular graph is called *cubic*. We write $\delta(G)$ for the smallest of all degrees of vertices of G, and $\Delta(G)$ for the largest. (One also writes either $\Delta(G)$ or $\delta(G)$ for the common degree of a regular graph G.) The degree $d(x)$ of x will equal the sum of the entries in the row of M_G or of N_G corresponding to x.

Theorem 1.1. *In any graph or multigraph, the sum of the degrees of the vertices equals twice the number of edges.*

Proof. It is convenient to work with the incidence matrix: we sum its entries. The sum of the entries in row i is just $d(x_i)$; the sum of the degrees is $\sum_{i=1}^{v} d(x_i)$, which equals the sum of the entries in N. The sum of the entries in column j is 2, since each edge is incident with two vertices; the sum over all columns is thus $2e$, so that

$$\sum_{i=1}^{v} d(x_i) = 2e,$$

giving the result. □

Corollary 1.1.1 *In any graph or multigraph, the number of vertices of odd degree is even. In particular, a regular graph of odd degree has an even number of vertices.*

Example. A graph with the degrees of vertices marked is shown in Figure 1.10. Observe that there are six odd vertices. The isolated vertex has degree 0.

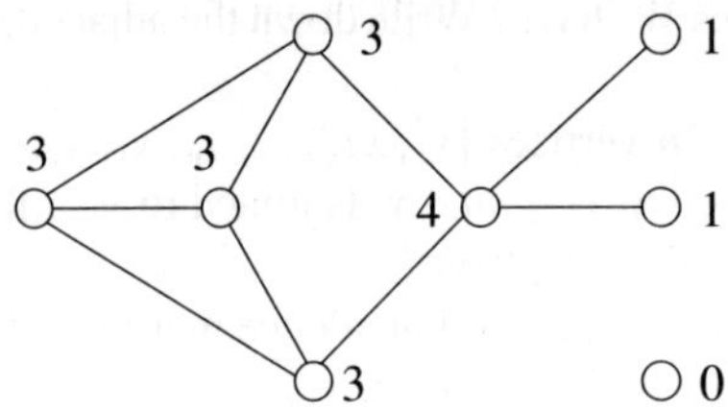

Fig. 1.10. Degrees of vertices

A collection of v nonnegative integers is called *graphical* if and only if there is a graph on v vertices whose degrees are the members of the collection. A graphical collection is called *valid* if and only if there is a *connected* graph with those degrees. (This is one situation where the distinction between graphs and multigraphs is very important: see Exercise 1.3.16.)

Theorem 1.2. [68, 61] *A collection*

$$S = \{d_0, d_1, \ldots, d_{v-1}\}$$

of v integers with $d_0 \geq d_1 \geq \ldots \geq d_{v-1}$, where $d_0 \geq 1$ and $v \geq 2$, is graphical if and only if the collection

$$S' = \{d_1 - 1, \ldots, d_{d_0} - 1, d_{d_0+1}, \ldots, d_{v-1}\}$$

is graphical.

Proof. (i) Suppose S' is graphical. Let H be a graph with vertices $u_1, u_2, \ldots, u_{v-1}$, where

$$\begin{array}{ll} d(u_i) = d_i - 1, & 1 \leq i \leq d_0, \\ d(u_i) = d_i, & d_0 + 1 \leq i \leq v - 1. \end{array}$$

Append a new vertex u_0, and join it to $u_1, u_2, \ldots, u_{d_0}$. The resulting graph has degree sequence S.

(ii) Suppose the collection S is graphical. Let G be a graph with vertex set

$$V(G) = \{x_0, x_1, \dots, x_{v-1}\}$$

such that $d(x_i) = d_i$ for $0 \le i \le v - 1$. Two cases arise:

Case 1. Suppose G contains a vertex y of degree d_0, such that y is adjacent to vertices having degrees $d_1, d_2, \dots, d_{d_0}$. In this case, we remove y and all the edges incident with it. The resulting graph has degree sequence S', whence S' is graphical.

Case 2. Suppose there is no such vertex y. We have the vertex set

$$x_0, x_1, x_2, x_3, x_4, \dots, x_{d_0}, x_{d_0+1}, \dots, x_k, \dots, x_{v-1}$$

with degrees

$$d_0 \ge d_1 \ge d_2 \ge d_3 \ge d_4 \ge \dots \ge d_{d_0} \ge d_{d_0+1} \ge \dots \ge d_k \ge \dots \ge d_{v-1}.$$

Let $X = \{x_{j_1}, x_{j_2}, \dots, x_{j_n}\}$ be the set of all vertices among $x_1, \dots, x_{d_0}$ to which x_0 is *not* adjacent. Then $n \ge 1$. Because x_0 is adjacent to d_0 vertices altogether, there must be exactly n vertices in the set

$$Y = \{x_{k_1}, x_{k_2}, \dots, x_{k_n}\}$$

among $x_{d_0+1}, \dots, x_{v-1}$ to which x_0 *is* adjacent. We show that there is a vertex x_j in X and a vertex x_k in Y such that $d(x_j) > d(x_k)$. Suppose otherwise. Then all the vertices in X and all the vertices in Y have the same degree. Then interchanging x_{j_i} and x_{k_i} in the sequence of vertices for each i, $i = 1, \dots, n$, produces a reordering of S, satisfying the conditions of the theorem, in which x_0 is adjacent to vertices having degrees d_1, $d_2, \dots, d_{d_0}$. So G falls into Case 1, which we assumed was not true.

So there exist vertices x_j and x_k such that x_0 is not adjacent to x_j, x_0 is adjacent to x_k and $d(x_j) > d(x_k)$. Since the degree of x_j is greater than that of x_k, there must be a vertex x_m that is adjacent to x_j but not to x_k.

We delete the edges x_0x_k and x_jx_m from G and add edges x_0x_j and x_kx_m. The result is a graph G' having the same degree sequence s as G. However, the sum of the degrees of the vertices adjacent to x_0 in G' is larger than that in G. If G' falls into Case 1, then S' is graphical. If not, apply the argument to G', obtaining a new graph G'' with the same degree sequence as G, but such that the sum of the degrees of the vertices adjacent to x_0 is larger than the corresponding sum in G'. If G'' falls into Case 1, then S' is graphical; otherwise repeat again. Continuing this procedure must eventually result in a graph satisfying the hypothesis of Case 1, because the total sum of all the degrees remains the same for each new graph, while the sum of the degrees of vertices adjacent to x_0 increases. □

Example. Consider the sequence

$$S = \{6, 3, 3, 3, 3, 2, 2, 2, 2, 1, 1\}.$$

This sequence is graphical if and only if

$$\begin{aligned} S' &= \{2, 2, 2, 2, 1, 1, 2, 2, 1, 1\} \\ &= \{2, 2, 2, 2, 2, 2, 1, 1, 1, 1\} \end{aligned}$$

is graphical; equivalently

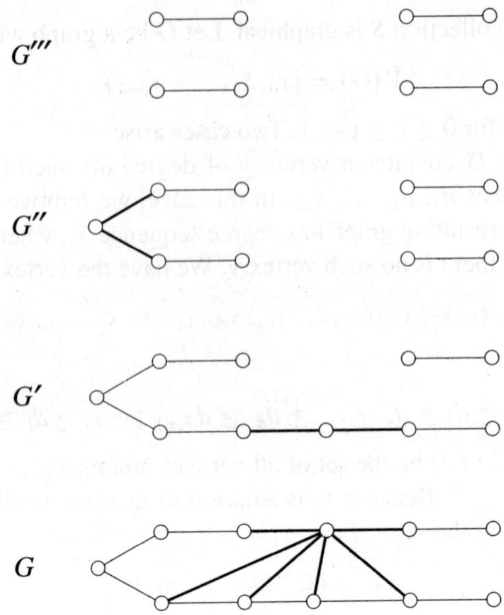

Fig. 1.11. Constructing a graph with given degrees

$$\begin{aligned} S'' &= \{1, 1, 2, 2, 2, 1, 1, 1, 1\} \\ &= \{2, 2, 2, 1, 1, 1, 1, 1, 1\} \end{aligned}$$

must be graphical, as must

$$S''' = \{1, 1, 1, 1, 1, 1, 1, 1\}.$$

Now S''' is easily seen to be graphical: the corresponding graph consists of four independent edges. So S is graphical. The method of constructing a suitable graph is illustrated in Figure 1.11, where G corresponds to S, G' to S', and so on. For example, S' was derived from S by subtracting 1 from six of the degrees, and the six resulting degrees are $\{2, 2, 2, 2, 1, 1\}$, so we select any six vertices in G' whose degrees are $\{2, 2, 2, 2, 1, 1\}$, and join a new vertex to them. In each case the new edges are shown with heavy lines.

As was shown by the above example, application of Theorem 1.2 not only finds whether a sequence is graphical, it also enables us to find a graph with the given degree sequence, if one exists. However, there is no guarantee of uniqueness. In fact, one can often find two graphs with the same degree sequence.

Example. Find three nonisomorphic graphs with degree sequence 3, 3, 2, 2, 2, 2.

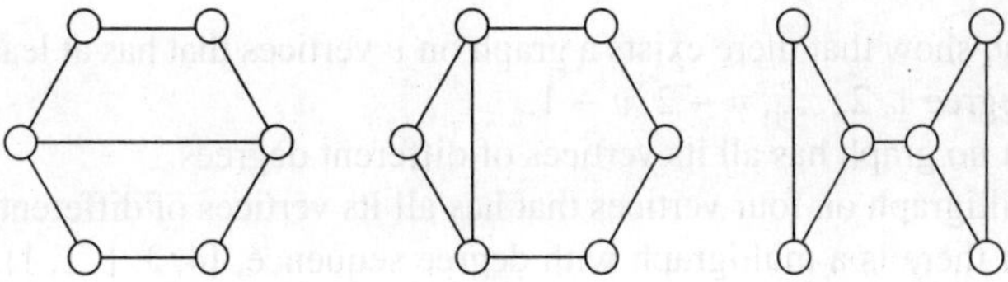

The three graphs are obviously not isomorphic: they have different numbers of subgraphs K_3.

Exercises 1.3

A1.3.1 In each case, find the degrees of the vertices:

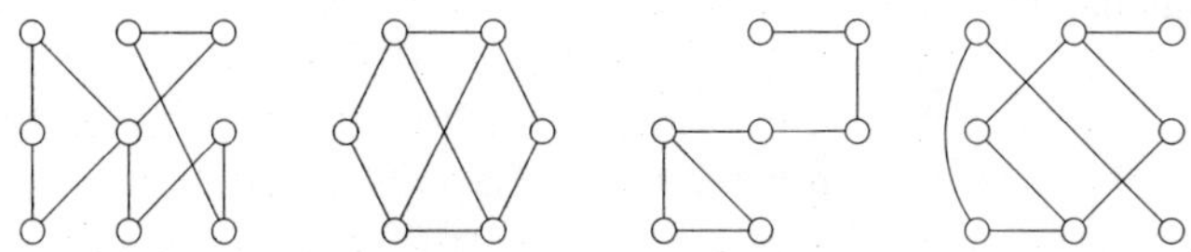

(i) (ii) (iii) (iv)

A1.3.2 Suppose G has v vertices and $\delta(G) \geq (v-1)/2$. Prove that G is connected.

1.3.3 Prove that a regular graph of odd degree can have no component with an odd number of vertices.

A1.3.4 Prove that the collection $\{3, 2, 2, 2, 1\}$ is valid. Find a graph with this collection of degrees.

1.3.5 Prove that the collection $\{3, 3, 2, 1, 1\}$ is valid. Find a graph with this collection of degrees.

H1.3.6 Find two nonisomorphic graphs with degree collection 2, 2, 2, 1, 1.

1.3.7 Find two nonisomorphic graphs with degree collection 3, 3, 3, 3, 3, 3.

H1.3.8 Prove that there do not exist nonisomorphic graphs on four vertices with the same degree collection.

1.3.9 Prove that (up to isomorphism) there is exactly one graph with degree collection $\{5, 5, 4, 4, 3, 3\}$.

1.3.10 Which of the following are graphical?
(i) $\{5, 5, 4, 4, 2, 2\}$;
(ii) $\{5, 4, 4, 3, 3, 3\}$;
(iii) $\{5, 4, 4, 4, 3, 3\}$;
(iv) $\{2, 2, 1, 1, 1, 1\}$?

1.3.11 Which of the following are graphical?
(i) $\{4, 3, 2, 2, 2, 1\}$;
(ii) $\{3, 3, 2, 2, 2, 1\}$;
(iii) $\{5, 4, 4, 4, 2, 1\}$;
(iv) $\{2, 2, 2, 2, 1, 1\}$?

1.3.12 Find a graph on six vertices that has at least one vertex of each degree 1, 2, 3, 4, 5.

1.3.13 For each v, show that there exists a graph on v vertices that has at least one vertex of each degree $1, 2, \ldots, v-2, v-1$.

1.3.14 Prove that no graph has all its vertices of different degrees.

A1.3.15 Find a multigraph on four vertices that has all its vertices of different degrees.

1.3.16 Prove that there is a multigraph with degree sequence, $\{4, 3, 1, 1, 1\}$, but there is no multigraph with degree sequence $\{2, 0, 0, 0\}$. Deduce that Theorem 1.2 does not apply to multigraphs.

A1.3.17 Prove: if d and v are natural numbers, not both odd, with $v > d$, then there there is a regular graph of degree d with exactly v vertices.

A1.3.18 In a looped multigraph, each loop is defined to add 2 to the degree of its vertex.
(i) Do Theorem 1.1 and Corollary 1.3 hold foı looped multigraphs?
(ii) Suppose D is any finite sequence of nonnegative integers such that the sum of all its members is even. Show that there is a looped multigraph with degree sequence D.

2

Walks, Paths and Cycles

2.1 Basic Ideas

A *walk* in a graph G is a finite sequence of vertices $x_0, x_1, \ldots, x_n$ and edges $a_1, a_2, \ldots, a_n$ of G:

$$x_0, a_1, x_1, a_2, \ldots, a_n, x_n,$$

where the endpoints of a_i are x_{i-1} and x_i for each i. A *simple walk* is a walk in which no edge is repeated. If it is desired to specify the terminal vertices, the above walk is called an x_0x_n-*walk*. The *length* of a walk is its number of edges.

A *path* is a walk in which no vertex is repeated. A walk is *closed* when the first and last vertices, x_0 and x_n, are the same. Closed walks are also called *circuits*. A *cycle* of length n is a closed walk of length n, $n \geq 3$, in which the vertices $x_0, x_1, \ldots, x_{n-1}$ are all different. In specifying a path or cycle, it is sufficient to list only the sequence of vertices, because the edges are then uniquely determined. For example, a path consisting of vertices a, b, c, d and edges ab, bc, cd will simply be denoted by $abcd$. The cycle formed by adding edge da to path $abcd$ is often written $(abcd)$. Of course, $(abcd)$, $(bcda)$, $(cdab)$ and $(dabc)$ all represent the same cycle.

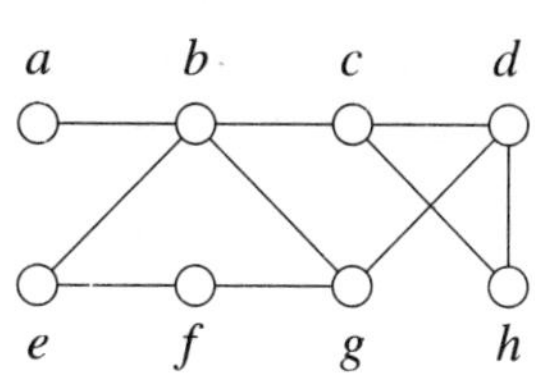

Fig. 2.1. A sample graph for walks and paths

Example. Consider the graph in Figure 2.1. Both $abefgbch$ and $abgdh$ are walks from a to h, but only the latter is a path. $(bcdgfe)$ and $(bcdg)$ are cycles of lengths 6 and 4 respectively.

The following observation, although very easy to prove, will be useful.

Theorem 2.1. *If there is a walk from vertex y to vertex z in the graph G, where y is not equal to z, then there is a path in G with first vertex y and last vertex z.*

Proof. Say the yz-walk is

$$W_1 = x_0, a_1, x_1, a_2, \ldots, a_n, x_n,$$

as above, where $y = x_0$ and $z = x_n$. If the vertices $x_0, x_1, \ldots, x_n$ are all different, W_1 is a path and we are done. If not, select a vertex that appears twice: say $x_i = x_j$, where $i < j$. Write

$$W_2 = x_0, a_1, x_1, \ldots, x_i, a_{j+1}, x_{j+1}, \ldots, a_n, x_n.$$

Then W_2 is a walk from y to z and is shorter than W_1.

If W_2 contains no repeated vertex, then it is the required path. Otherwise, select a repeated vertex, and proceed as above. Again a shorter walk is constructed.

This process must stop at some stage, because each walk is shorter than the preceding one and the length can never be less than 1. So, for some k, W_k cannot be reduced in length. It must be that W_k contains no repeated vertex and is the required path. □

In fact, we have proved a little more than the theorem. We have shown that if there is a yz-walk of length n, there is a yz-path of length at most n.

We say that two vertices are *connected* when there is a walk joining them. (Theorem 2.1 tells us we can replace the word "walk" by "path.") Two vertices of G are connected if and only if they lie in the same component of G; G is a connected graph if and only if all pairs of its vertices are connected. (Observe that this definition of "connected graph" is consistent with the one given in Section 1.2.) If vertices x and y are connected, then their *distance* $D(x, y)$ is the length of the shortest path joining them; by definition $D(x, x) = 0$.

Example. In Figure 2.1, there are several paths from b to h: $bcdh$, bch, $bgdh$, $bgdch$, $befgdh$ and $befgdch$. The shortest is bch, of length 2, so $D(b, h) = 2$.

Cycles give the following useful characterization of bipartite graphs.

Theorem 2.2. *A graph is bipartite if and only if it contains no cycle of odd length.*

Proof. (i) Suppose G is a bipartite graph with disjoint vertex sets U and V. Suppose G contains a cycle of odd length, with $2k + 1$ vertices

$$x_1, x_2, x_3, \ldots, x_{2k+1},$$

where x_i is adjacent to x_{i+1} for $i = 1, 2, \ldots, 2k$, and x_{2k+1} is adjacent to x_1. Suppose x_i belongs to U. Then x_{i+1} must be in V, since otherwise we would have two adjacent vertices in U; and conversely. So, if we assume $x_1 \in U$ we get, successively, $x_2 \in V$, $x_3 \in U, \ldots, x_{2k+1} \in U$. Now $x_{2k+1} \in U$ implies $x_1 \in V$, which contradicts the disjointness of U and V.

(ii) Suppose that G is a graph with no cycle of odd length. Without loss of generality we need only consider the case where G is connected. Choose an arbitrary vertex x in G, and partition the vertex set by defining Y to be the set of vertices whose distance from x is even, and Z to be the set of vertices whose distance from x is odd; x itself belongs to Y.

Now select two vertices y_1 and y_2 in Y. Let P be a shortest path from x to y_1 and Q a shortest path from x to y_2. Denote by u the last vertex common to P and Q. Since P and Q are shortest paths, so are their sections from x to u, which therefore have the same length. Since the lengths of both P and Q are even, the lengths of their sections from u to y_1 and u to y_2 respectively have equal parity, so the path from y_1 to u (in reverse direction along P) to y_2 (along Q) has even length.

If $y_1 \sim y_2$, then this path together with the edge y_1y_2 gives a cycle of odd length, which is a contradiction. Hence no two vertices in Y are adjacent. Similarly no two vertices in Z are adjacent, and G is bipartite. □

A graph that contains no cycles at all is called *acyclic*; a connected acyclic graph is called a *tree*. Trees will be discussed further in Chapter 4. In general, acyclic graphs are called *forests*.

It is clear that the set of vertices and edges that constitute a path in a graph is itself a graph. We define a path P_v to be a graph with v vertices $x_1, x_2, \ldots, x_v$ and $v - 1$ edges $x_1x_2, x_2x_3, \ldots, x_{v-1}x_v$. A cycle C_v is defined similarly, except that the edge x_vx_1 is also included, and (to avoid the triviality of allowing K_2 to be defined as a cycle) v must be at least 3. This convention ensures that every C_v has v edges. Figure 2.2 shows P_4 and C_5.

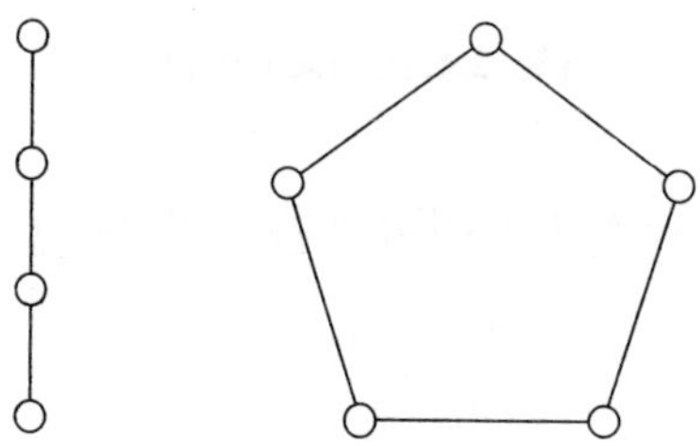

Fig. 2.2. P_4 and C_5

As an extension of the idea of a proper subgraph, we shall define a *proper tree* to be a tree other than K_1, and similarly define a *proper path*. (No definition of a "proper cycle" is necessary.)

Exercises 2.1

A2.1.1 Find all paths from s to t in the graph shown in Figure 2.3(i). What is the length of each? What is the distance from s to t?

A2.1.2 Find the distances between all pairs of vertices in the graph of Figure 2.3(i).

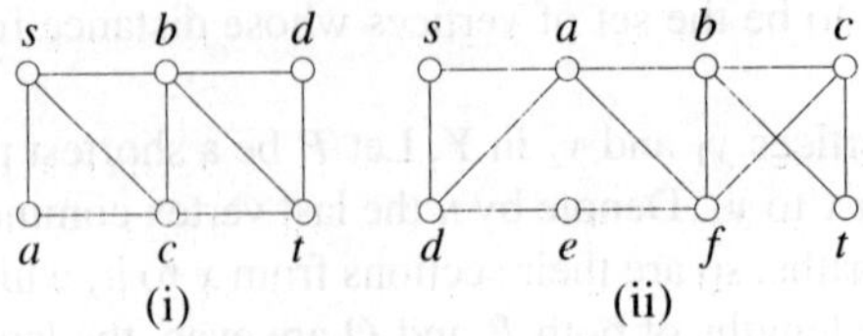

Fig. 2.3. Find all paths from s to t

A2.1.3 Repeat Exercises 2.1.1 and 2.1.2 for the graph of Figure 2.3(ii).

2.1.4 Find the distances between all pairs of vertices in the graph of Figure 2.1.

2.1.5 Figure 2.4 shows the *Petersen graph*, which arises in several contexts in the study of graphs. Find cycles of lengths 5, 6, 8 and 9 in this graph.

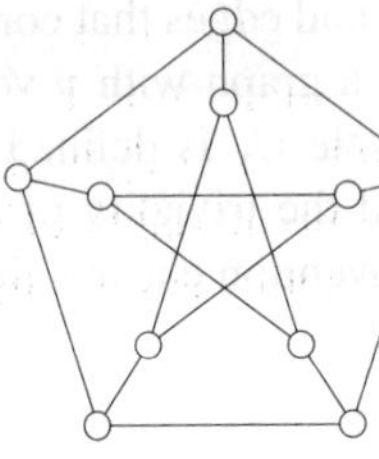

Fig. 2.4. Petersen graph

2.1.6 Show that the graph $C_5 \times K_2$ (called a *pentagonal prism*) contains cycles of lengths 4, 5, 6, 7, 8, 9 and 10.

2.1.7 Find a graph on 10 vertices with 13 edges that contains cycles of all orders 3, 4, 5, 6, 7, 8, 9, 10. Can this be done with fewer edges?

A2.1.8 The graph G contains two different paths from x to y, namely

$$x, t_1, t_2, \ldots, t_m, y$$

and

$$x, u_1, u_2, \ldots, u_m, y.$$

(i) If the vertices t_1 and u_1 are different, prove that G contains a cycle that passes through x.

(ii) Prove that the condition "t_1 and u_1 are different" is necessary.

2.1.9 A graph contains at least two vertices. There is exactly one vertex of degree 1, and every other vertex has degree 2 or greater. Prove that the graph contains a cycle. Would this remain true if we allowed graphs to have infinite vertex sets?

2.1.10 The graph G has adjacency matrix A. Denote the (i, j) entry of the matrix A^k by a_{ij}^k. Prove that there are exactly a_{ij}^k different walks of length k from x_i to x_j.

2.1.11 The *square* G^2 of a graph G has the same vertex set as G; x is adjacent to y in G^2 if and only if $D(x, y) = 1$ or 2 in G. A graph is a *perfect square* if and only if it is the square of some graph.

(i) What are the squares of $K_{1,n}$, P_5 and C_5?

(ii) Find all the connected graphs on four or five vertices that are perfect squares.

2.1.12 If G is a connected graph, show that the distance function $D(x, y)$ has the following properties for all vertices x, y and z:

(i) $D(x, y) = 0$ if and only if $x = y$;
(ii) $D(x, y) = D(y, x)$;
(iii) $D(x, y) + D(y, z) \geq D(x, z)$.

(A mapping to the nonnegative integers that has these properties is called a *metric*.)

2.2 Radius, Diameter and Eccentricity

In this section we introduce some graph parameters related to distance.

Suppose G is a connected graph and x is a vertex of G. The *eccentricity* $\varepsilon(x)$ of x is the largest value of $D(x, y)$, where y ranges through all the vertices. The *diameter* $D = D(G)$ of G is the maximum value of $\varepsilon(x)$ for all vertices x of G, while the *radius* $R = R(G)$ is the smallest value of $\varepsilon(x)$.

Example. Find $\varepsilon(x)$ for every vertex x of the graph G of Figure 2.5. What are the diameter and radius of this graph? The vertices most distant from a are c and e, so $\varepsilon(a) = D(a, c) = D(a, e) = 3$. Similarly $\varepsilon(b) = D(b, c) = D(b, e) = 2$, $\varepsilon(c) = D(c, a) = D(c, d) = 3$, $\varepsilon(d) = D(d, c) = 3$, $\varepsilon(e) = D(e, a) = 3$, $\varepsilon(f) = D(f, a) = D(f, d) = 2$. So $D(G) = 3$ and $R(G) = 2$.

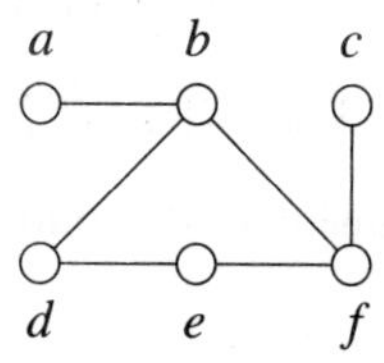

Fig. 2.5. Find eccentricities, diameter and radius

Example. Find the diameter and radius of the cycle C_{2n}.

Every vertex of C_{2n} is distance n from its opposite vertex, and the distance is smaller for all other vertices. So $\varepsilon = n$ for all vertices, and $D = R = n$.

Theorem 2.3. *In any graph G,*

$$R(G) \leq D(G) \leq 2R(G).$$

Proof. By definition $D(G)$ equals the maximum distance between any two vertices in G. Say x and y attain this maximum distance so that $D(x, y) = D(G)$, and say z is a vertex for which $\varepsilon(z) = R(G)$. Since R is a distance between two vertices, necessarily $R \le D$. But by definition $D(z,t) \le \varepsilon(z) = R(G)$ for every vertex t, so $D(G) = D(x,y) \le D(x,z) + D(z,y) \le 2R(G)$. □

The *center* $C(G)$ of a finite graph G of radius R is the set of all vertices x that have eccentricity $\varepsilon(x) = R$.

Example. The graph shown in Figure 2.5 has center $\{f\}$. The center of the cycle C_{2n} is the whole graph.

The *girth* $g(G)$ of a graph G is the length of the shortest cycle contained in G; the *circumference* $c(G)$ of G is the length of the longest cycle on G. If G is acyclic, then $g(G)$ and $c(G)$ are not defined.

Example. Find a graph G with $D(G) = g(G) = c(G)$.

Consider a cycle C_{2n}. To one vertex, attach a path P_{n+1} containing n new vertices. The resulting graph has diameter, girth and circumference $2n$.

Theorem 2.4. *Suppose G is a graph which contains a cycle. Then*

$$g(G) \le 2D(G) + 1.$$

Proof. Suppose $g(G) \ge 2D(G) + 2$. Consider a cycle C of length $g(G)$ in G; label the vertices of C as $x_1, x_2, \ldots, x_{g(G)}$ around the cycle.

Suppose $P = x_1 y_2 y_3 \ldots y_t x_{D(G)+2}$ is a shortest path from x_1 to $x_{D(G)+2}$. P must contain fewer than $D(G)+1$ edges, by the definition of diameter. If P does not contain any of the vertices $x_{D(G)+3}, x_{D(G)+4}, \ldots, x_{g(G)}$, then the cycle

$$(x_1 y_2 y_3 \ldots y_t x_{D(G)+2} x_{D(G)+3} x_{D(G)+4} \ldots x_{g(G)})$$

has length less than $g(G)$. Otherwise, suppose x_k is the highest-numbered member of C that belongs to P; say $x_k = y_s$. Again, $(x_1 y_2 y_3 \ldots y_s x_{h+1} x_{k+2} \ldots x_{g(G)})$ is a cycle shorter than $g(G)$. In either case, we have contradicted the definition of $g(G)$. □

Exercises 2.2

2.2.1 Find the girth, radius and diameter of the following graphs.
(i) The n-vertex path P_n.
(ii) The complete bipartite graph $K_{m,n}$.
(iii) The ladder L_n (see Exercise 1.2.13).
(iv) The Petersen graph (see Exercise 2.1.4).
(v) The pentagonal prism (see Exercise 2.1.5).

A2.2.2 Find the girth, circumference, radius and diameter of the following graphs.

(i)
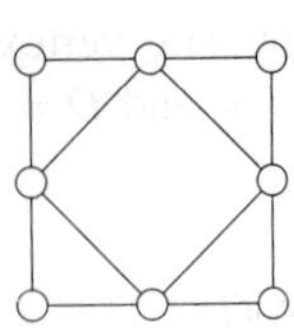
(ii)
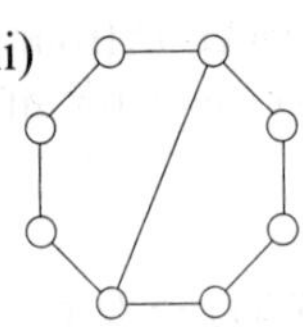
(iii)
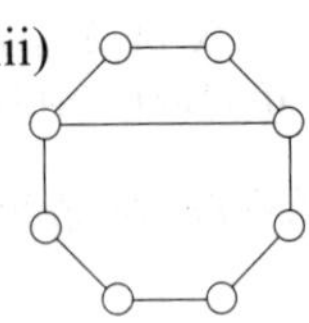
(iv)
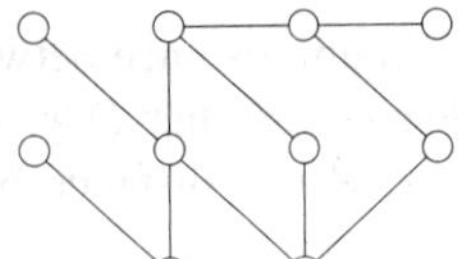
(v)
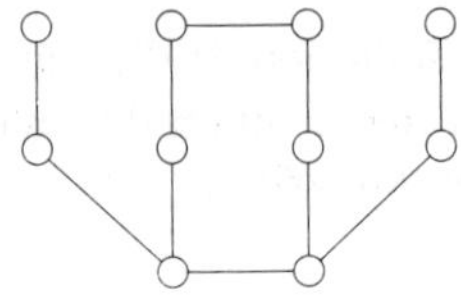

2.2.3 Find the girth, circumference, radius and diameter of the following graphs.

(i) 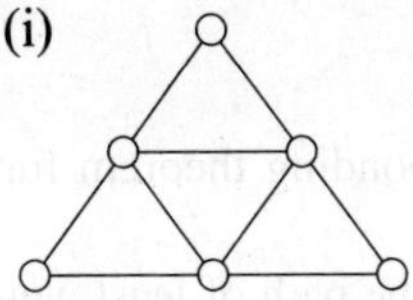 (ii) 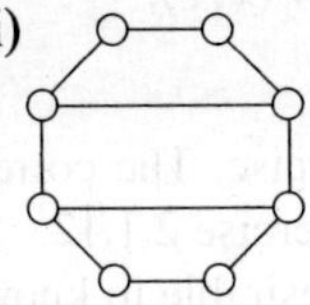 (iii) 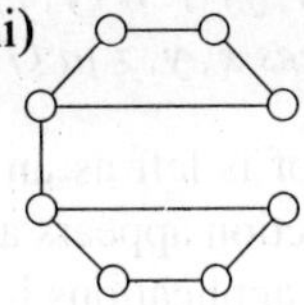

A2.2.4 Find centers of the graphs in Exercise 2.2.2.

2.2.5 Find centers of the graphs in Exercise 2.2.3.

2.2.6 Give an example of a graph other than K_3 with a center of size 3.

2.2.7 Suppose G is connected and suppose every $x \in V(G)$ satisfies $\delta(x) \geq \frac{v-1}{2}$. Show that G has diameter no greater than 2.

2.2.8 A graph G is called *self-centered* if $C(G) = G$. (We observed that C_{2n} is self-centered.)

(i) Prove that a complete graph is self-centered.

(ii) Give an example of a self-centered graph with four vertices other than K_4.

H(iii) Prove that if G is a self-centered graph, then so is the Cartesian product $K_2 \times G$.

A2.2.9 Suppose G is a connected graph, not K_1, which has no induced subgraph isomorphic to $K_{1,3}$. Show that G contains a pair of adjacent vertices x and y such that $G - \{x, y\}$ is connected. [112]

2.2.10 Suppose G has an automorphism f (as defined in Section 1.2) that takes vertex x of G to vertex y. Show that if x is in the center of G, then so is y.

2.3 Weighted Distance

In many applications it is appropriate to define a positive function called a *weight* $w(x, y)$ associated with each edge (x, y). For example, if a graph represents a road system, a common weight is the length of the corresponding stretch of road. In the underlying graph of a multigraph, $w(x, y)$ might be the multiplicity of (x, y) in the multigraph. Weights also often represent costs or durations. The weight of a *path* P is the sum of the weights of the edges in P. Similarly, one can define the weights of walks, cycles, subgraphs and graphs. Weights can also be defined on the vertices; vertex weight will be used in Section 12.2.

Example. The routes traveled by an airline can conveniently be shown in a graph, with vertices representing cities and edges representing services. Several different weights might be used: the distance, the flying time and the airfare are all possibilities.

If x and y are connected vertices, then the *weighted distance* from x to y is the minimum among the weights of all the paths from x to y, and is denoted by $W(x, y)$.

Theorem 2.5. *Suppose G is a connected graph. The weighted distance function $W(x, y)$ has the following properties:*

(i) $W(x, y) = 0$ *if and only if* $x = y$;

(ii) $W(x, y) = W(y, x)$;

(iii) $W(x, y) + W(y, z) \geq W(x, z)$;

for all vertices x, y, z *in* G.

The proof is left as an exercise. The corresponding theorem for the unweighted distance function appears as Exercise 2.1.12.

In many applications it is desirable to know the path of least weight between two vertices. This is usually called the *shortest path* problem, primarily because a common application is one in which weights represent physical distances. We shall describe an algorithm due to Dijkstra [34] that finds the shortest path from vertex s to vertex t in a finite connected graph G. Informally stated, the algorithm arranges the vertices of G in order of increasing weighted distance from s. This algorithm is most easily described if w is defined for all pairs of vertices, so we write $w(x, y) = \infty$ if x and y are not adjacent. (In a computer implementation, ∞ can be replaced by some very large number.)

The algorithm actually orders the vertices of G as $s_0 (= s), s_1, s_2, \ldots$ so that the weighted distances $W(s, s_1),\ W(s, s_2),\ \ldots$ are in nondecreasing order. To do so, it attaches a temporary label $\ell(x)$ to each member x of $V(G)$. $\ell(x)$ is an upper bound on the weighted length of the shortest path from s to x. To start the algorithm, write $s_0 = s$ and $S_0 = \{s\}$. Define $\ell(s_0) = 0$ and for every other vertex y, $\ell(y) = \infty$. Call this step 0. After step k the set

$$S_k = \{s_0, s_1, \ldots, s_k\}$$

has been defined. In the next step, for each x not in S_k the algorithm changes the value $\ell(x)$ to the weighted length of the shortest (s, x) path that has only one edge not in S_k, if this is an improvement. Then s_{k+1} is chosen to be a vertex x such that $\ell(x)$ is minimized. (In practice, it does not matter whether the new value of $\ell(x)$ is used or the old value is retained for those vertices x other than s_{k+1}. Moreover, for each member s_i of S_k, it is only necessary to consider one new vertex, a vertex x such that $w(s_i, x)$ is minimal.)

- s_0 is s;
- s_1 is the vertex x for which $w(s, x)$ is smallest, the vertex "closest" to s;
- to find s_2 one selects, for every vertex x other than s or s_1, either the lowest weight edge sx or the lowest weight path ss_1x, whichever is shorter, and then chooses the shortest of all these to be s_2;

and so on. After a vertex has been labeled s_i, its ℓ-value never changes, and this final value of $\ell(x)$ equals $W(s, x)$. Eventually the process must stop, because a new vertex is added at each step and the vertex set is finite. So $W(s, t)$ must eventually be found. Finally, whenever a vertex is labeled s_{k+1}, one can define a unique vertex s_i that "preceded" it in the process — the vertex before s_{k+1} in the ss_{k+1} path that had length $\ell(s_{k+1})$. By working back from t to the vertex before it, then to the one before that and so on, one eventually reaches s, and reversing the process gives a shortest st path. For convenience, the predecessor of each new vertex s_k is recorded at the step when s_k is selected.

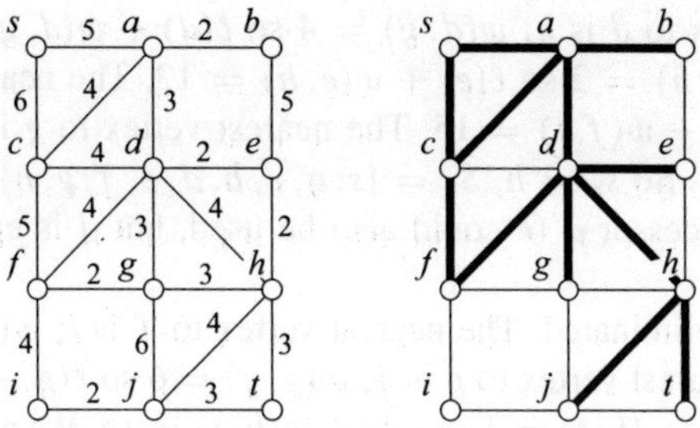

(a) (b)

Fig. 2.6. Find the path of minimum weight from s to t

Example. The left-hand part of Figure 2.6 shows a graph with weighted edges. We shall find the minimum weight path from s to t.

Initially set $s_0 = s$, $S_0 = \{s\}$ and $\ell(s) = 0$. s has no predecessor.

The nearest vertex to s is a ($w(s, a) = 5$, $w(s, c) = 6$). So $s_1 = a$, $S_1 = \{s, a\}$ and $\ell(a) = 5$. a has predecessor s.

We consider each member of S_1. The nearest vertex in $V \backslash S_1$ to s (in fact, the only one) is c, and $w(s, c) = 6$. The nearest vertex to a is b ($w(a, b) = 2$, $w(a, c) = 4$, $w(a, d) = 3$). The candidate values are $\ell(c) = 6$ (reached through s) and $\ell(b) = 7$ (through a). The smaller is chosen. So $s_2 = c$, $S_2 = \{s, a, c\}$ and $\ell(c) = 6$. c has predecessor s.

We now process S_2. There is no vertex in $V \backslash S_1$ adjacent to s, so s can be ignored in this and later iterations. The nearest vertex to a is b; $w(a, b) = 2$ so $\ell(a) + w(a, b) = 7$. The nearest vertex to c is d; $w(c, d) = 4$ so $\ell(c) + w(c, d) = 10$. So $s_3 = b$, $S_3 = \{s, a, c, b\}$ and $\ell(b) = 7$. b has predecessor a.

The nearest vertex to a is d; $w(a, d) = 3$ so $\ell(a) + w(a, d) = 8$. The nearest vertex to c is d; $w(c, d) = 4$ so $\ell(c) + w(c, d) = 10$. The nearest vertex to b is e; $w(b, e) = 5$ so $\ell(b) + w(b, e) = 12$. So $s_4 = d$, $S_4 = \{s, a, c, b, d\}$ and $\ell(d) = 8$. d has predecessor a.

a need not be considered, as all its neighbors are in S_4. The nearest vertex to c is f; $w(c, f) = 5$ so $\ell(c) + w(c, f) = 11$. The nearest vertex to b is e; $w(b, e) = 5$ so $\ell(b) + w(b, e) = 12$. The nearest vertex to d is e; $w(d, e) = 2$ so $\ell(d) + w(d, e) = 10$. So $s_5 = e$, $S_5 = \{s, a, c, b, d, e\}$ and $\ell(e) = 10$. e has predecessor d.

From now on b need not be considered. The nearest vertex to c is f; $w(c, f) = 5$ so $\ell(c) + w(c, f) = 11$. The nearest vertex to d is g; $w(d, g) = 3$ so $\ell(d) + w(d, g) = 11$. The nearest vertex to e is h; $w(e, h) = 2$ so $\ell(e) + w(e, h) = 12$. Either f or g could be chosen. For convenience, suppose the earlier member of the alphabet is always chosen when equal ℓ-values occur. Then $s_6 = f$, $S_6 = \{s, a, c, b, d, e, f\}$ and $\ell(f) = 11$. f has predecessor c.

Now c can be ignored. The nearest vertex to d is g; $w(d, g) = 3$ so $\ell(d) + w(d, g) = 11$. The nearest vertex to e is h; $w(e, h) = 2$ so $\ell(e) + w(e, h) = 12$. The nearest vertex to f is g; $w(f, g) = 2$ so $\ell(f) + w(f, g) = 13$. So $s_7 = g$, $S_7 = \{s, a, c, b, d, e, f, g\}$ and $\ell(g) = 11$. g has predecessor d.

The nearest vertex to d is h; $w(d, g) = 4$ so $\ell(d) + w(d, g) = 12$. The nearest vertex to e is h; $w(e, h) = 2$ so $\ell(e) + w(e, h) = 12$. The nearest vertex to f is i; $w(f, i) = 4$ so $\ell(f) + w(f, i) = 15$. The nearest vertex to g is h; $w(g, h) = 3$ so $\ell(g) + w(g, h) = 14$. So $s_8 = h$, $S_8 = \{s, a, c, b, d, e, f, g, h\}$ and $\ell(h) = 12$. We shall say h has predecessor d (e could also be used, but d is earlier in alphabetical order).

d and e are now eliminated. The nearest vertex to f is i; $w(f, i) = 4$ so $\ell(f) + w(f, i) = 15$. The nearest vertex to g is j; $w(g, j) = 6$ so $\ell(g) + w(g, j) = 17$. The nearest vertex to h is t; $w(h, t) = 3$ so $\ell(h) + w(h, t) = 15$. We always choose t when it has the equal-smallest ℓ-value. So $s_9 = t$, and the algorithm stops with $\ell(t) = 15$. Since t has predecessor h, the minimum weight path is $sadht$, with weight 15. On the right-hand side of Figure 2.6, all links to predecessors are emphasized. It is easy to read off the minimum weight path from this figure, as well as the minimum weight paths from s to a, b, c, d, e, f, g and h.

Exercises 2.3

2.3.1 Prove Theorem 2.5.

A2.3.2 In the weighted graphs in Figure 2.7, find the minimum weight paths from s to t.

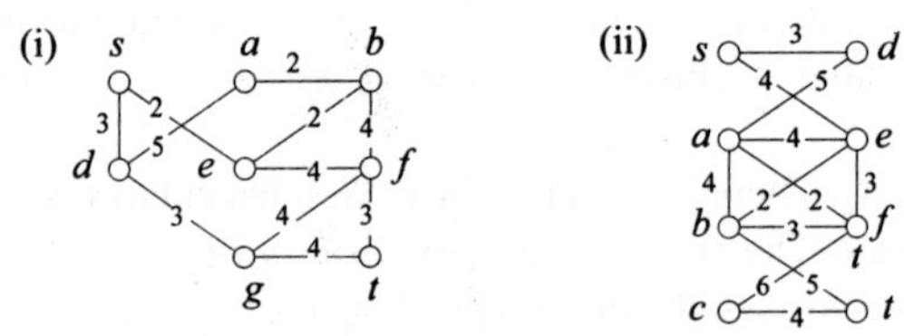

Fig. 2.7. Graphs for Exercise 2.3.2

2.3.3 Repeat the preceding exercise for the weighted graphs in Figure 2.8.

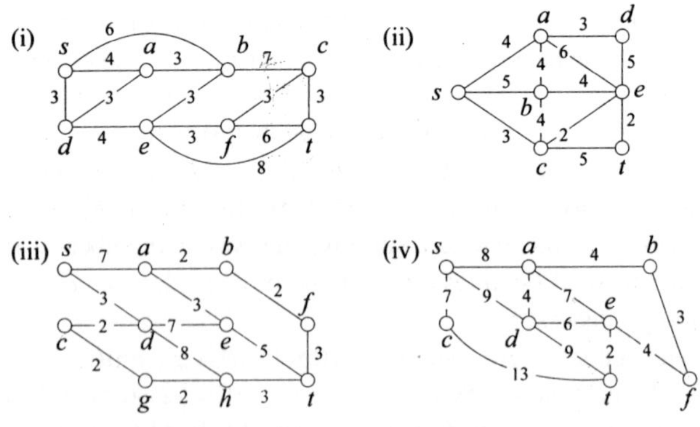

Fig. 2.8. Graphs for Exercise 2.3.3

2.4 Euler Walks

Unlike many parts of mathematics, the theory of graphs has a definite birthdate. The first paper on graphs was published by Euler in 1736, and had been delivered by him to the St. Petersburg Academy one year earlier.

Euler's paper grew out of a famous old problem. The town of Königsberg in Prussia is built near the mouth of the river Pregel. The river divides the town into four parts, including an island called The Kneiphof, and in the eighteenth century the town had seven bridges; the layout is shown in Figure 2.9. The question under discussion was whether it is possible from any point on Königsberg to take a walk in such a way as to cross each bridge exactly once.

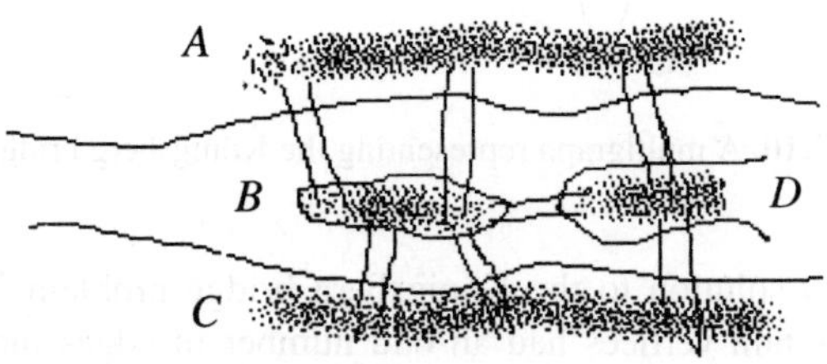

Fig. 2.9. Königsberg bridges

Euler set himself the more general problem: given any configuration of river, islands and bridges, find a general rule for deciding whether there is a walk that covers each bridge precisely once.

We first show that it is impossible to walk over the bridges of Königsberg. For suppose there was such a walk. There are three bridges leading to the area C: you can traverse two of these, one leading into C and the other leading out, at one time in your tour. There is only one bridge left: if you cross it going into C, then you cannot leave C again, unless you use one of the bridges twice, so C must be the finish of the walk; if you cross it in the other direction, C must have been the start of the walk. In either event, C is either the place where you started or the place where you finished.

A similar analysis can be applied to A, B and D, since each has an odd number of bridges. But the walk starts at one place and finishes at one place. Therefore it is impossible for A, B, C and D all to be either the start or the finish.

The ideas we have just used can be applied to more general configurations of bridges and islands, and to other problems. We start by finding a graphical model — in fact, a multigraph — that contains the essential facts of the Königsberg bridge problem. We observe that the topography of C is really irrelevant. When considering C we talked only about entering or leaving the area. For this purpose it would be the same if C were shrunk to a point connecting the three bridges, and the same could be done to A, B and D. The bridges themselves do not have any physical significance, and we are concerned with them only as connections between the points. So we can discuss the question just as well by constructing a multigraph with vertices A, B, C and D

corresponding to the parts A, B, C and D of the town, and with an edge representing each bridge, as shown in Figure 2.10. In terms of this model, the original problem becomes: can a simple walk be found that contains every edge of the multigraph? A simple walk with this property is called an *Euler walk*, and a graph containing such a walk is *Eulerian*.

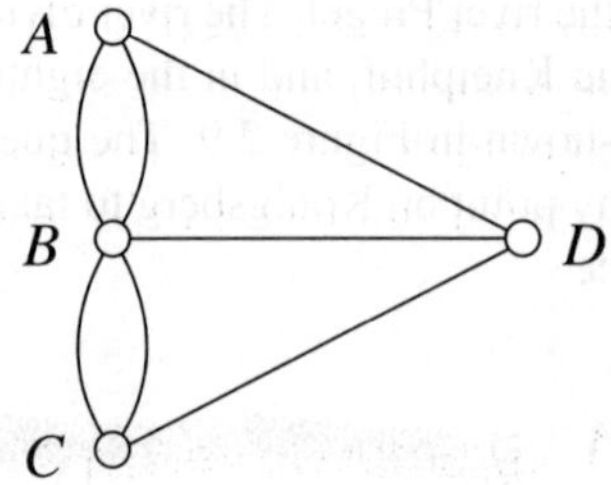

Fig. 2.10. A multigraph representing the Königsberg bridges

In proving that a solution to the Königsberg bridge problem is impossible, we used the fact that certain vertices had an odd number of edges incident with them. (The precise number was not important; oddness was the significant feature.) Let us call a vertex *even* if its degree is even, and *odd* otherwise. It was observed that an odd vertex must be either the first or the last point in the walk. In fact, if a multigraph has an Euler walk, then either the multigraph has two odd vertices, the start and finish of the Euler walk, or the multigraph has no odd vertices, and the Euler walk starts and finishes at the same point. Another obvious necessary condition is that the multigraph must be connected. These two conditions are together sufficient.

Theorem 2.6. *If a connected multigraph has no odd vertices, then it has an Euler walk starting from any given point and finishing at that point. If a connected multigraph has two odd vertices, then it has an Euler walk whose start and finish are the odd vertices.*

Proof. Consider any simple walk in a multigraph that starts and finishes at the same vertex. If one erases every edge in that walk, one deletes two edges touching any vertex that was crossed once in the walk, four edges touching any vertex that was crossed twice, and so on. (For this purpose, count "start" and "finish" combined as one crossing.) In every case an *even* number of edges is deleted.

First, consider a multigraph with no odd vertex. Select any vertex x, and select any edge incident with x. Go along this edge to its other endpoint, say y. Then choose any other edge incident with y. In general, on arriving at a vertex, select any edge incident with it that has not yet been used, and go along the edge to its other endpoint. At the moment when this walk has led into the vertex z, where z is not x, an odd number of edges touching z has been used up (the last edge to be followed, and an even number previously). Since z is even, there is at least one edge incident with z that is still available. Therefore, if the walk is continued until a further edge is impossible, the last vertex must be x — that is, the walk is closed. It will necessarily be a simple walk and it must contain every edge incident with x.

Now assume that a connected multigraph with every vertex even is given, and a simple closed walk has been found in it by the method just described. Delete all the edges in the walk, forming a new multigraph. From the first paragraph of the proof it follows that every vertex of the new multigraph is even. It may be that we have erased every edge in the original multigraph; in that case we have already found an Euler walk. If there are edges still left, there must be at least one vertex, say c, that was in the original walk and that is still on an edge in the new multigraph — if there were no such vertex, then there could be no connection between the edges of the walk and the edges left in the new multigraph, and the original multigraph must have been disconnected. Select such a vertex c, and find a closed simple walk starting from c. Then unite the two walks as follows: at one place where the original walk contained c, insert the new walk. For example, if the two walks are

$$x, y, \ldots, z, c, u, \ldots, x$$

and

$$c, s, \ldots, t, c,$$

then the resulting walk will be

$$x, y, \ldots, z, c, s, \ldots, t, c, u, \ldots, x.$$

(There may be more than one possible answer, if c occurred more than once in the first walk. Any of the possibilities may be chosen.) The new walk is a closed simple walk in the original multigraph. Repeat the process of deletion, this time deleting the newly formed walk. Continue in this way. Each walk contains more edges than the preceding one, so the process cannot go on indefinitely. It must stop: this will only happen when one of the walks contains all edges of the original multigraph, and that walk is an Euler walk.

Finally, consider the case where there are two odd vertices p and q and every other vertex is even. Form a new multigraph by adding an edge pq to the original. This new multigraph has every vertex even. Find a closed Euler walk in it, choosing p as the first vertex and the new edge pq as the first edge. Then delete this first edge; the result is an Euler walk from q to p. □

It is clear that loops make no difference in whether or not a graph has an Euler walk. If there is a loop at vertex x, it can be added when the walk crosses x.

We can adapt the above proof of Theorem 2.6 to make an algorithm for finding a closed Euler walk. All that is needed is to order the vertices in some way; whenever a vertex is to be chosen, select the one earliest in the ordering from among those meeting the requirements.

Example. Find an Euler walk in the road network represented by Figure 2.11.

We order the vertices by standard alphabetical ordering. Starting from A, we find the walk $ACBEA$. When these edges are deleted (see Figure 2.12(a)) there are no edges remaining through A, so a new walk is started. A and B are not available, so we choose C, the first vertex from the first walk that still has edges adjacent to it, and trace the walk $CEDGEIFC$, after which there are no edges available at C (see Figure 2.12(a)). G is available, yielding walk $GHILHJG$. As is clear from Figure 2.12(c), the remaining edges form a walk $JKLJ$.

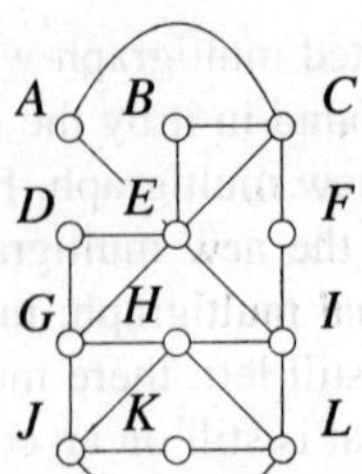

Fig. 2.11. Construct an Euler walk

We start with $ACBEA$. We replace C by $CEDGEIFC$, yielding

$$ACEDGEIFCBEA,$$

then replace the first E by $EGJLIE$, with the result

$$ACEDGHILHJGEIFCBEA.$$

Finally J is replaced by $JKLJ$, and the Euler walk is

$$ACEDGHILHJKLJGEIFCBEA.$$

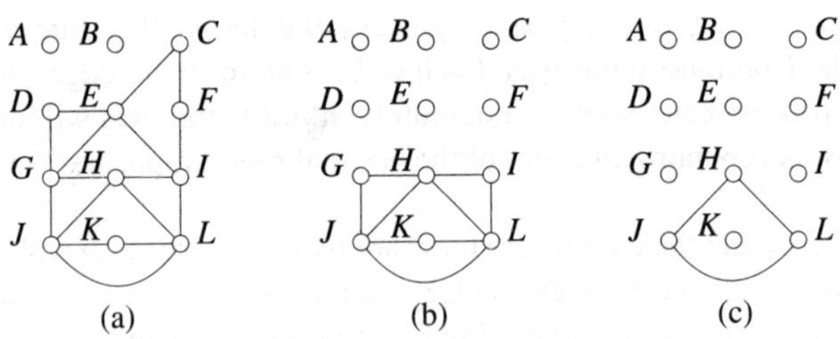

Fig. 2.12. Constructing an Euler walk

A good application of Euler walks is planning the route of a highway inspector or mail contractor, who must travel over all the roads in a highway system. Suppose the system is represented as a multigraph G, as was done in Section 1.1. Then the most efficient route will correspond to an Euler walk in G.

If G contains no Euler walk, the highway inspector must repeat some edges of the graph in order to return to his starting point. Let us define an *Eulerization* of G to be a multigraph with a closed Euler walk, that is formed from G by duplicating some edges. A *good* Eulerization is one that contains the minimum number of new edges, and this minimum number is the *Eulerization number* $eu(G)$ of G.

Example. Consider the multigraph G of Figure 2.13. What is $eu(G)$? Find an Eulerization of G that uses the minimum number of edges.

Look at the representation of the multigraph on the left of Figure 2.14. The black vertices have odd degree, so they need additional edges. As there are four black ver-

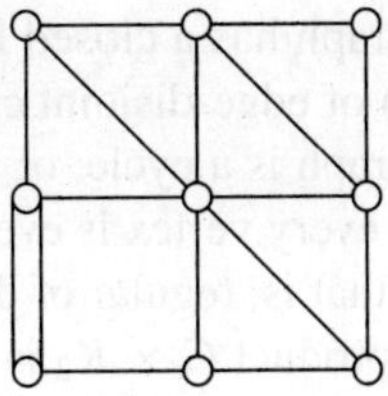

Fig. 2.13. Find an Eulerization

tices, at least two new edges are needed; but obviously no two edges will suffice. However, there are solutions with three added edges — two examples are shown in Figure 2.14 — so $eu(G) = 3$.

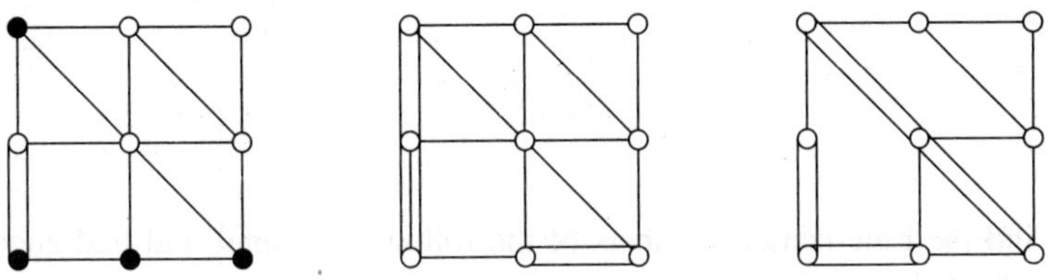

Fig. 2.14. Finding an Eulerization

Exercises 2.4

2.4.1 Which of the following graphs contain Euler walks? If the graph contains an Euler walk, find one. If not, how many edges are required for an Eulerization?

(i) 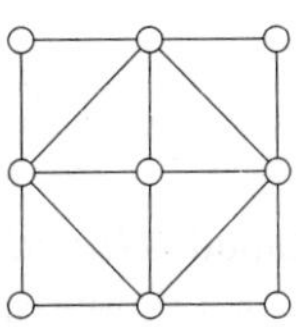 (ii) 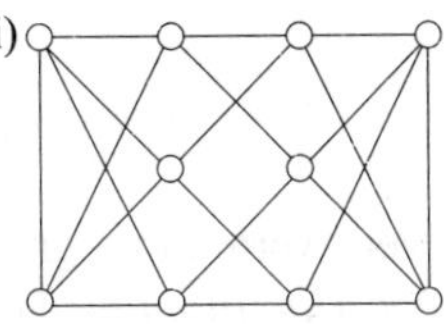 (iii) 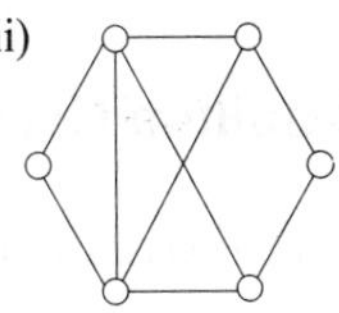

2.4.2 Repeat the previous exercise for these graphs:

(i) 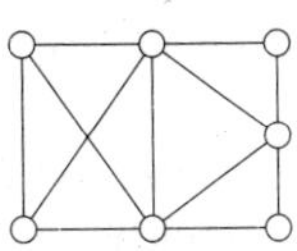 (ii) 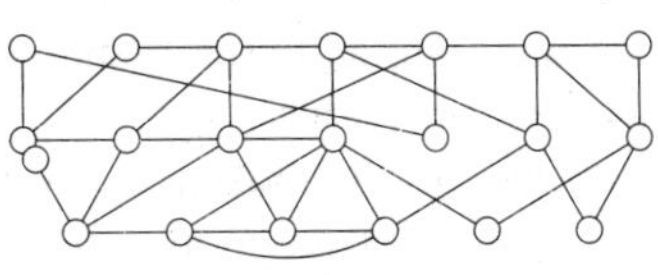

(iii) 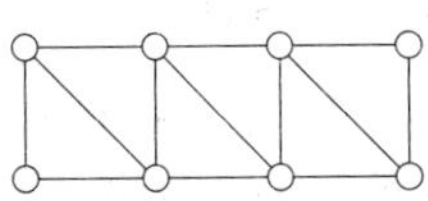 (iv) 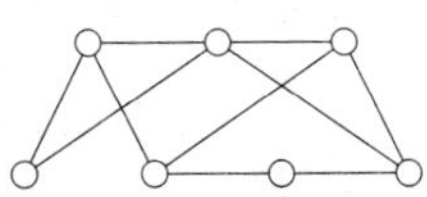

2.4.3 Suppose G is a connected graph with k vertices of odd degree. Show that G can be decomposed into $k/2$ edge-disjoint simple walks.

2.4.4 (i) Show that a connected graph has a closed Euler walk if and only if it can be decomposed into a union of edge-disjoint cycles.
(ii) Show that a connected graph is a cycle, or a union of edge-disjoint cycles, if and only if the degree of every vertex is even.

H2.4.5 Suppose G is a cubic graph (that is, regular of degree 3).
(i) Prove that the Cartesian product $G \times K_2$ is Eulerian.
(ii) Prove that $G \times G$ is Eulerian.

2.4.6 Suppose G is a cubic graph.
(i) Prove that the join $G + K_3$ is Eulerian;
(ii) Prove that $G + G$ is Eulerian.

2.4.7 Find the Eulerization numbers of the following graphs, and find good Eulerizations of them.

(i)
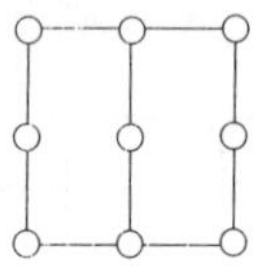
(ii)
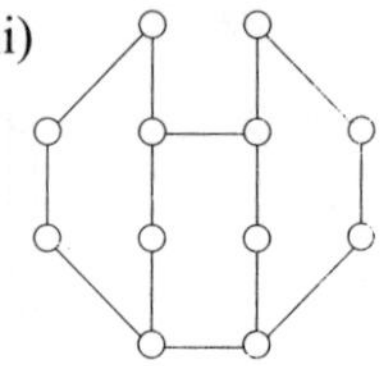

2.4.8 Find the Eulerization numbers of the following graphs, and find good Eulerizations of them.

(i)
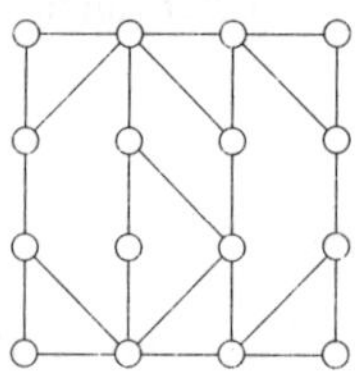
(ii)
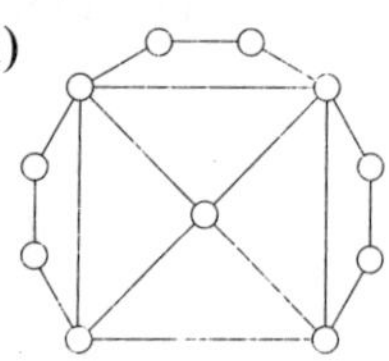

2.5 Hamilton Cycles

A cycle that passes through every vertex in a graph is called a *Hamilton cycle* and a graph with such a cycle is called *Hamiltonian*. Typically one thinks of a Hamiltonian graph as a cycle with a number of other edges (called *chords* of the cycle). The idea of such a spanning cycle was simultaneously developed by Hamilton [62] in the special case of the icosahedron, and more generally by Kirkman [76]. A *Hamilton path* is a path that contains every vertex.

It is easy to discuss Hamiltonicity in particular cases, and there are a number of small theorems. However, no good necessary and sufficient conditions are known for the existence of Hamilton cycles. The following result is a useful sufficient condition.

Theorem 2.7. *If G is a graph with v vertices, $v \geq 3$, and $d(x) + d(y) \geq v$ whenever x and y are nonadjacent vertices of G, then G is Hamiltonian.*

Proof. Suppose the theorem is false. Choose a v such that there is a v-vertex counterexample, and select a graph G on v vertices that has the maximum number of edges among counterexamples. Choose two nonadjacent vertices p and q: because of the maximality of G, $G + pq$ must be Hamiltonian. Moreover, pq must be an edge in every Hamilton cycle of $G + pq$, because otherwise the cycle would be Hamiltonian in G. By hypothesis, $d(p) + d(q) \geq v$.

Consider any Hamilton cycle in $G + pq$:

$$p, x_1, x_2, \ldots, x_{v-2}, q.$$

If x_i is any member of $N(p)$, then x_{i-1} cannot be a member of $N(q)$, because if it were, then

$$p, x_1, x_2, \ldots, x_{i-1}, q, x_{v-2}, x_{v-3}, \ldots, x_i$$

would be a Hamilton cycle in G. So each of the $d(p)$ vertices adjacent to p in G must be preceded in the cycle by vertices not adjacent to q, and none of these vertices can be q itself. So there are at least $d(p) + 1$ vertices in G that are not adjacent to q. So there are at least $d(q) + d(p) + 1$ vertices in G, whence

$$d(p) + d(q) \leq v - 1,$$

a contradiction. □

Corollary 2.8. *If G is a graph with v vertices, $v \geq 3$, and every vertex has degree at least $\frac{v}{2}$, then G is Hamiltonian.*

Theorem 2.7 was first proved by Ore [90] and Corollary 2.8 was proved some years earlier by Dirac [35]. Both can in fact be generalized into the following result of Pósa [97]: a graph with v vertices, $v \geq 3$, has a Hamilton cycle provided the number of vertices of degree less than or equal to k does not exceed k, for each k satisfying $1 \leq k \leq \frac{v-1}{2}$.

Suppose a graph G contains a Hamilton cycle

$$x_1, x_2, \ldots, x_v, x_1.$$

Since x_i occurs only once in the cycle, only two of the edges touching x_i can be in the cycle. One can sometimes use this fact to prove that a graph contains no Hamilton cycle. For example, consider the graph of Figure 2.15.

Suppose the graph contains a Hamilton cycle.

The vertices on the outer circuit are each of degree 3, and only two of the edges touching any given vertex can be in a Hamilton cycle. Figure 2.16(a) shows as dotted lines all the edges touching three of those vertices; of the nine edges, three are not in the cycle. Similarly, Figure 2.16(b) shows the fifteen edges touching the three vertices of degree 5; nine of these are out of the cycle. These sets of edges are disjoint, so there are at least twelve edges not in the cycle. (If the sets were not disjoint, but had k common elements, only $12-k$ edges would definitely be eliminated.) Similarly, one of the edges touching the central vertex must be deleted in forming the cycle. So thirteen edges are barred, and the Hamilton cycle must be chosen from the remaining fourteen edges. Since the graph has sixteen vertices, a Hamilton cycle in it must contain sixteen edges, which is impossible.

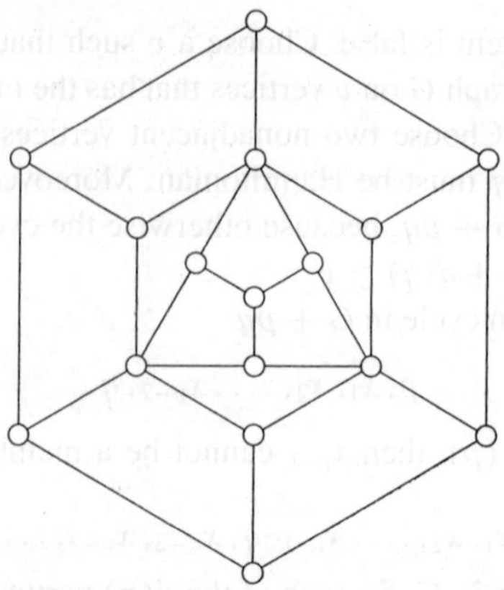

Fig. 2.15. A graph with no Hamilton cycle

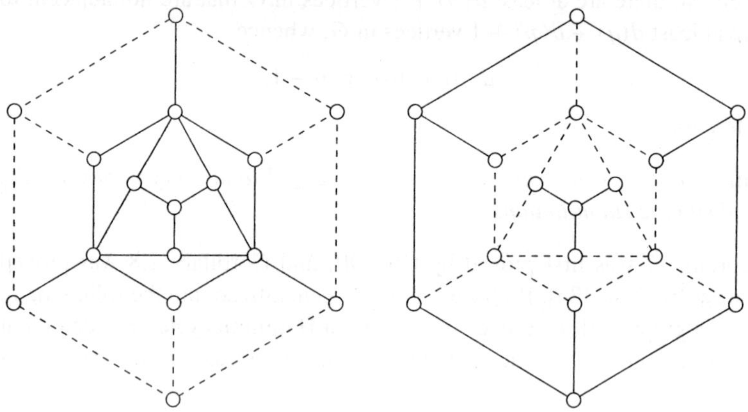

(a) (b)

Fig. 2.16. Steps in proving there is no Hamilton cycle

A similar argument could be used to prove the impossibility of a Hamilton path.

Another test is applicable only to bipartite graphs. As a bipartite graph is a subgraph of some $K_{m,n}$, its vertices can be partitioned into two subsets, of sizes m and n, such that the graph contains no edge that joins two vertices in the same subset.

Theorem 2.9. *A bipartite graph with vertex sets of sizes m and n can contain a Hamilton cycle only if $m = n$, and can contain a Hamilton path only if m and n differ by at most* 1.

Proof. Suppose a bipartite graph G has vertex sets V_1 and V_2, and suppose it contains a Hamilton path:

$$x_1, x_2, \ldots, x_v.$$

Suppose that x_1 belongs to V_1. Then x_2 must be in V_2, x_3 in V_1, and so on. Since the path contains every vertex, it follows that

$$V_1 = \{x_1, x_3, \ldots\},$$
$$V_2 = \{x_2, x_4, \ldots\}.$$

If v is even, then V_1 and V_2 each contain $v/2$ elements; if v is odd, then $|V_1| = (v+1)/2$ and $|V_2| = (v-1)/2$. In either case, the difference in orders is at most 1. If G contains a Hamilton cycle

$$x_1, x_2, \ldots, x_v, x_1,$$

and x_1 is in V_1, then x_v must belong to V_2; so $|V_1| = |V_2| = v/2$. □

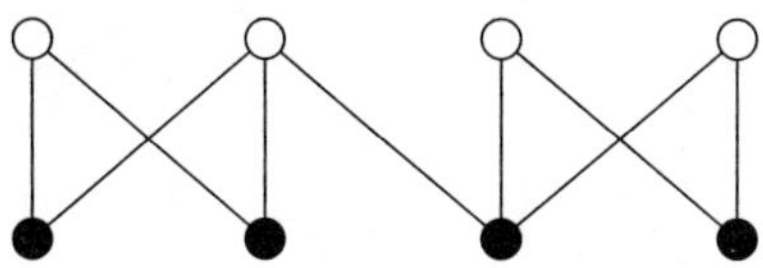

Fig. 2.17. A bipartite graph with no Hamilton cycle

It should be realized that neither of these necessary conditions is sufficient; in particular, Figure 2.17 shows that a bipartite graph with four vertices in each subset contains no Hamilton cycle; this cannot be proven using the above methods.

Example. The result of Theorem 2.7 is good in the following sense: $K_{d,d+1}$ has no Hamilton cycle (by Theorem 2.9), but of its $2d+1$ vertices, $d+1$ have degree d each and d have degree $d+1$ each. Theorem 2.7 shows that $K_{d,d}$ has a Hamilton cycle.

Exercises 2.5

2.5.1 Prove that the following graphs contain no Hamilton paths.

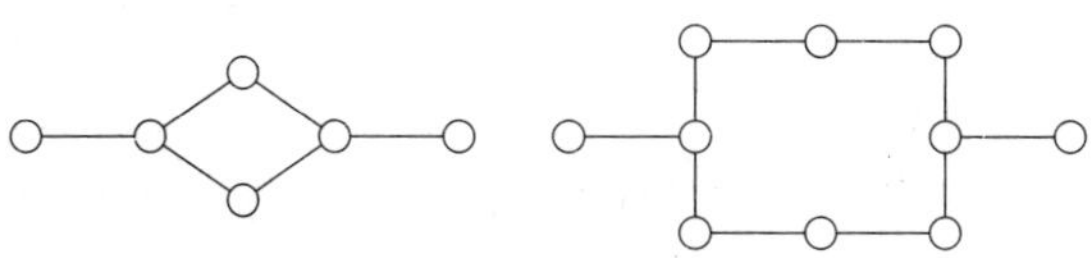

2.5.2 Prove that the following graphs contain no Hamilton cycles.

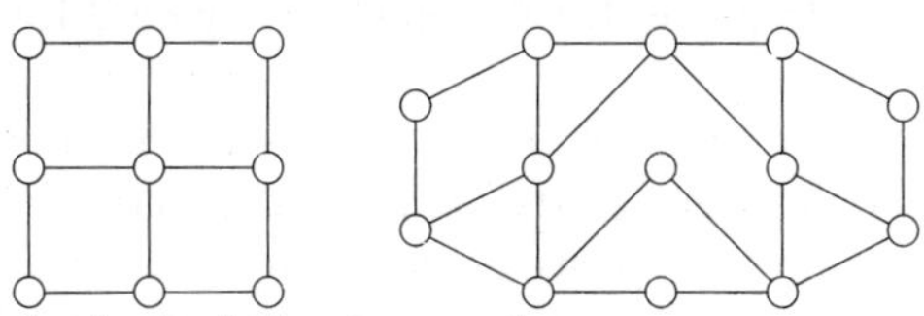

2.5.3 Find Hamilton cycles in the following graphs.

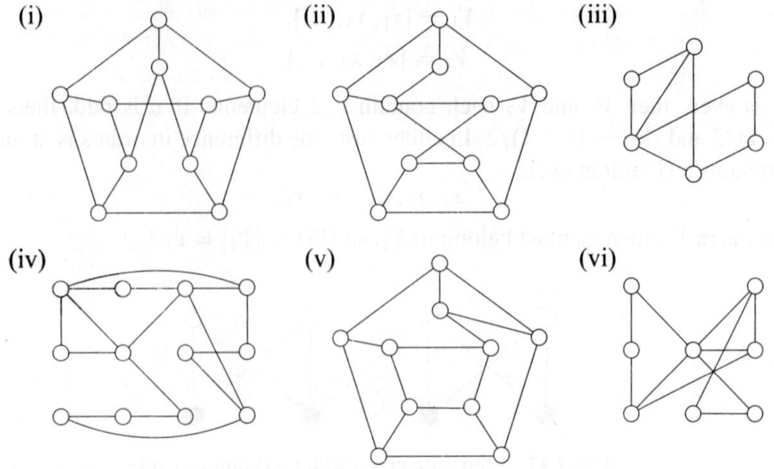

2.5.4 Find Hamilton cycles in the following graphs.

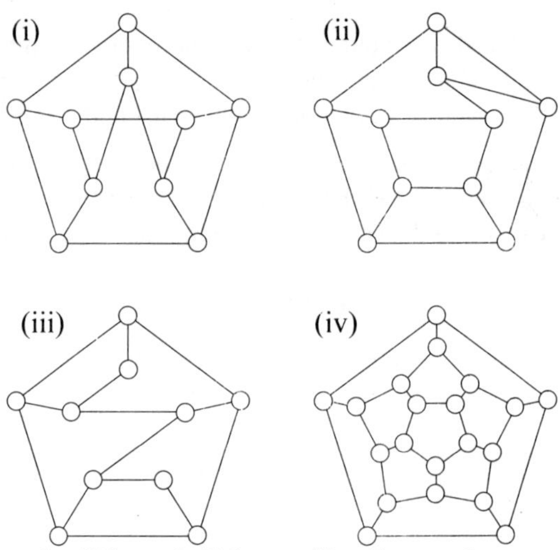

2.5.5 Verify that the graph of Figure 2.17 has no Hamilton cycle.

2.5.6 Use Theorem 2.9 to prove that the graph of Figure 2.15 contains no Hamilton path.

2.5.7 Prove that a spanning subgraph of $K_{m,n}$ can have a Hamilton cycle only if $m = n$

H2.5.8 Prove that for every v there exists a graph (other than K_v) on v vertices such tha $d(x) + d(y) \geq v - 1$ for any two nonadjacent vertices x and y.

2.5.9 G is a graph with v vertices; x and y are nonadjacent vertices of G satisfyin $d(x)+d(y) \geq v$. Prove that $G+xy$ is Hamiltonian if and only if G is Hamiltonian [16]

HA2.5.10 G is a graph with v vertices; $v \geq 3$.

(i) Prove that if G has at least $\frac{v^2-3v+6}{2}$ edges, then G is Hamiltonian.

(ii) Find a non-Hamiltonian graph with $\frac{v^2-3v+6}{2} - 1$ edges (thus proving that th preceding result is best-possible).

11 For which graphs is an Euler walk also a Hamilton path or cycle?

12 Suppose G contains a Hamilton path. Prove that both $G \times K_2$ and $G + K_1$ contain Hamilton cycles.

13 Suppose G is the graph

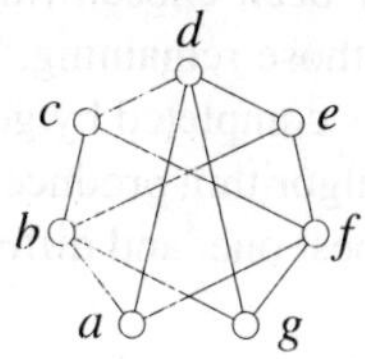

Find Hamiltonian cycles in the following graphs.

(i) $G + ac$	(ii) $G + ae$.
(iii) $G + ag$.	(iv) $G + ce$.
(v) $G + cg$.	(vi) $G + eg$.

Show that G itself is not Hamiltonian.

14 Eleven people plan to have dinner together on a number of different occasions. They sit at a round table. No person has the same neighbor at any two different dinners.

A(i) Show that this can be done for 5 days.

(ii) Generalize to the case of any prime number $2n + 1$ of people, and n days.

(iii) Find a solution for 9 people and 4 days.

(See Theorem 6.3, later, for a general solution.)

2.6 The Traveling Salesman Problem

Suppose a traveling saleman wishes to visit several cities. If the cities are represented as vertices and the possible routes between them as edges, then the salesman's itinerary is a Hamilton cycle in the graph.

In most cases one can associate a cost with every edge. Depending on the salesman's priorities, the cost might be a dollar cost such as airfare, a number of miles, or a number of hours. The most desirable itinerary will be the one for which the sum of costs is a minimum. The problem of finding this cheapest Hamilton cycle is called the Traveling Salesman Problem. Without loss of generality it can be assumed that the graph is complete (if there is no direct route from x to y, associate with xy the cheapest path from x to y).

Not surprisingly, the Traveling Salesman Problem is computationally difficult. There is no algorithm for solving the Traveling Salesman Problem that is substantially better than listing all Hamilton cycles. In K_{24}, for example, there are about 10^{23} such cycles (see Exercise 2.6.1). So some fast methods of solution have been developed. Although they are not guaranteed, these methods have been found to be better than random on average.

The *nearest neighbor* method works as follows. Starting at some vertex x, one first chooses the edge incident with x whose cost is least. Say that edge is xy. Then an edge incident with y is chosen in accordance with the following rule: if y is the vertex most recently reached, then eliminate from consideration all edges incident with y that lead to vertices that have already been chosen (including x), and then select an edge of minimum cost from among those remaining. This rule is followed until every vertex has been chosen. The cycle is completed by going from the last vertex chosen back to the starting position x. This algorithm produces a directed cycle in the complete graph, but not necessarily the cheapest one, and different solutions may come from different choices of initial vertex x.

The *sorted edges* method does not depend on the choice of an initial vertex. One first produces a list of all the edges in ascending order of cost. At each stage, the cheapest edge is chosen with the restriction that no vertex can have degree 3 among the chosen edges, and the collection of edges contains no cycle of length less than v, the number of vertices in the graph. This method always produces an undirected cycle, and it can be traversed in either direction.

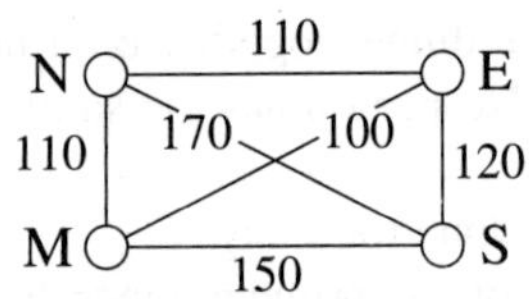

Fig. 2.18. Example of Traveling Salesman Problem

Example. Suppose the costs of travel between St. Louis, Evansville, Nashville and Memphis are as shown in dollars in Figure 2.18. The nearest neighbor algorithm, applied starting from Evansville, starts by selecting the edge EM, because it has the least cost of the three edges incident with E. The next edge must have M as an endpoint, and ME is not allowed (one cannot return to E, it has already been used), so the cheapest of the remaining edges is chosen, namely MN. The cheapest edge originating at N is NE, with cost \$110, but inclusion of this edge would lead back to E, a vertex that has already been visited, so NE is not allowed, and similarly NM is not available. It follows that NS must be chosen. So the algorithm finds route EMNSE, with cost \$520.

A different result is achieved if one starts at Nashville. Then the first edge selected is NE, with cost \$110. The next choice is EM, then MS, then SN, and the resulting cycle NEMSN costs \$530.

To apply the sorted edges algorithm, first sort the edges in order of increasing cost: EM(\$100), EN(\$110), ES(\$120), MN(\$130), MS(\$150), NS(\$170). Edge EM is included, and so is EN. The next choice would be ES, but this is not allowed because its inclusion would give degree 3 to E. MN would complete a cycle of length 3 (too short), so the only other choices are MS and NS, forming route EMSNE (or ENSME) at a cost of \$530.

In this example, the best route is ENMSE, with cost $510, and it does not arise from the nearest neighbor algorithm, no matter which starting vertex is used.

Exercises 2.6

.6.1 How many different Hamilton cycles are there in K_v?

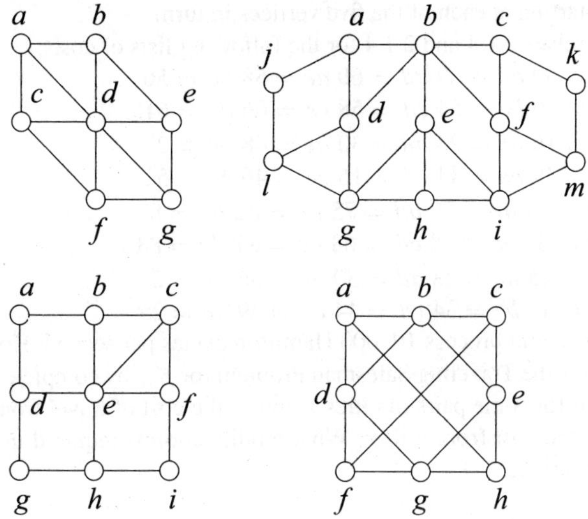

Fig. 2.19. Graphs for Exercise 2.6.2

.6.2 For the graphs shown in Figure 2.19, find all Hamilton cycles.

.6.3 Solve the Traveling Salesman Problem for the following graphs by finding all their Hamilton cycles. The costs of the edges are listed.

(i) 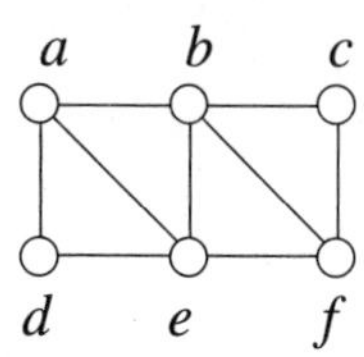

$ab = 14$ $ad = 12$ $ae = 30$
$bc = 19$ $be = 12$ $bf = 29$
$cf = 20$ $de = 16$ $ef = 26.$

(ii) 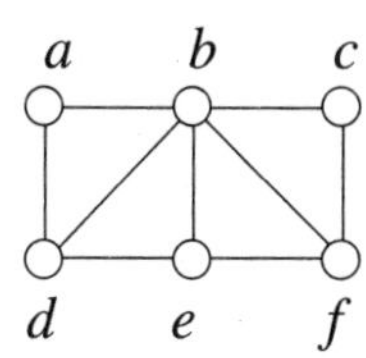

$ab = 19$ $ad = 16$ $bc = 20$
$bd = 18$ $be = 22$ $bf = 27$
$cf = 16$ $de = 18$ $ef = 22.$

.6.4 We consider the complete graph K_5 with vertices a, b, c, d, e. In each part we list the set of costs associated with the edges. Find the costs of the routes generated by the nearest neighbor algorithm starting at a and by the sorted edges algorithms.

(i) $ab = 24$ $ac = 21$ $ad = 30$ $ae = 28$ $bc = 19$
$bd = 32$ $be = 29$ $cd = 18$ $ce = 20$ $de = 26.$

(ii) $ab = 28$ $ac = 79$ $ad = 70$ $ae = 48$ $bc = 57$
$bd = 94$ $be = 66$ $cd = 72$ $ce = 83$ $de = 81$.

(iii) $ab = 24$ $ac = 23$ $ad = 20$ $ae = 21$ $bc = 27$
$bd = 22$ $be = 30$ $cd = 26$ $ce = 27$ $de = 28$.

(iv) $ab = 24$ $ac = 22$ $ad = 30$ $ae = 29$ $bc = 17$
$bd = 19$ $be = 30$ $cd = 18$ $ce = 21$ $de = 25$.

A2.6.5 For the costs in Exercise 2.1.4, find the routes generated by the nearest neighbor algorithm starting at each of the five vertices in turn.

2.6.6 Repeat Exercises 2.1.4 and 2.1.4 for the following lists of costs.

(i) $ab = 59$ $ac = 69$ $ad = 60$ $ae = 58$ $bc = 56$
$bd = 69$ $be = 54$ $cd = 58$ $ce = 66$ $de = 61$.

(ii) $ab = 16$ $ac = 24$ $ad = 30$ $ae = 48$ $bc = 27$
$bd = 29$ $be = 44$ $cd = 16$ $ce = 46$ $de = 51$.

(iii) $ab = 91$ $ac = 79$ $ad = 75$ $ae = 82$ $bc = 87$
$bd = 64$ $be = 78$ $cd = 68$ $ce = 81$ $de = 88$.

(iv) $ab = 45$ $ac = 28$ $ad = 50$ $ae = 36$ $bc = 21$
$bd = 42$ $be = 34$ $cd = 44$ $ce = 39$ $de = 25$.

2.6.7 Suppose one could process 10, 000 Hamilton cycles per second. How long would it take to solve the Traveling Salesman Problem for K_{10} by complete enumeration?

A2.6.8 Suppose that for some pairs of cities x and y, the cost of travel from x to y is not the same as the cost from y to x. What modification is required in the Traveling Salesman Problem?

3

Connectivity

3.1 Cutpoints and Bridges

Among connected graphs, some are connected so slightly that removal of a single vertex or edge will disconnect them. Such vertices and edges are quite important. A vertex x is called a *cutpoint* in G if $G - x$ contains more components than G does; in particular if G is connected, then a cutpoint is a vertex x such that $G - x$ is disconnected. Similarly, a *bridge* (or *cutedge*) is an edge whose deletion increases the number of components. As an example, consider the graphs in Figure 3.1; the left-hand graph has cutpoints u and y and cutedge uy, while the right-hand graph has cutpoint x and no cutedge.

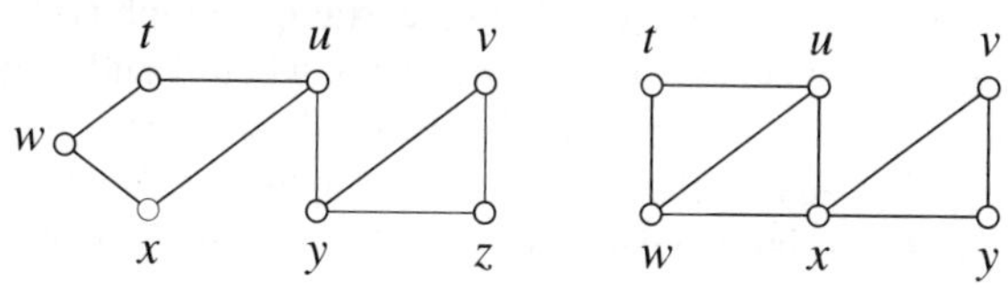

Fig. 3.1. Graphs whose cutpoints and cutedges are listed

A minimal collection of edges whose deletion disconnects G is called a *cutset* in G. A cutset partitions the vertex-set $V(G)$ into two nonempty components, say A and B, such that the edges joining vertices in A to vertices in B are precisely the edges of the cutset. This is the set of edges $[A, B]$, as defined in Section 1.2, and we refer to "the cutset $[A, B]$." (The two sets A and B are not uniquely defined — for example, if there is an isolated vertex in G, it could be allocated to either set — but the *cutset* will be well defined.) We omit set brackets when no confusion arises: for example, if $A = \{x, y\}$, we might write $[A, B]$ or $[xy, B]$, but not $[\{x, y\}, B]$.

If A is the set of all vertices of G other than x, so that $A = V(G)\backslash\{x\}$, then the cutset $[A, x]$, consisting of all edges incident with the vertex x, is called a *trivial* cutset.

Example. Consider the graph

$G =$ 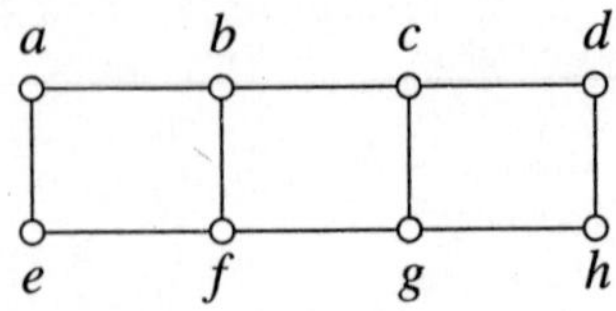

Three of the cutsets in this graph are

$$C = [ae, bcdfgh],$$
$$D = [abef, cdgh],$$
$$E = [b, acdefgh].$$

As graphs, the cutsets and their complements are

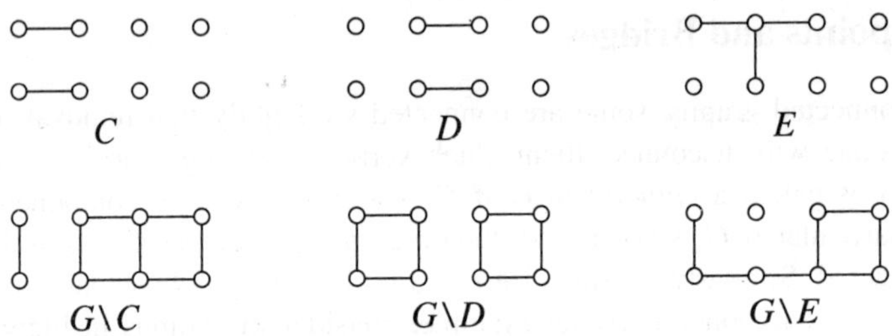

(with vertex labels omitted). E is a trivial cutset.

The notation $[A, B]$ can be used to represent the set of edges with one endpoint in A and the other in B, even when A and B are not complements. In that case $[A, B]$ does not necessarily disconnect the graph. In the example graph G, $[bf, cg]$ is the cutset D, but $[ab, cg]$ is not a cutset. On the other hand, when A and B are complements, $[A, B]$ might not be a cutset: its removal will always disconnect G, but the set of edges might not be minimal. Again referring to our example, $[bf, bcdfgh]$ consists of the four edges ab, bc, ef, fg, but the deletion of ab and ef is sufficient to disconnect G.

Lemma 3.1. *An edge xy in a connected graph G is a bridge if and only if it belongs to no cycle in the graph.*

Proof. Recall that a graph is connected if and only if any two vertices have a walk joining them. Write H for the graph resulting when xy is deleted from C.

(i) Suppose xy is a bridge. G is connected but H is not. So there must exist vertices w and z that are joined by a walk in G but not in H. Then xy must lie on every walk from w to z in G. Consider a walk from w to z that includes xy. If xy lies on a cycle in G, then delete the edge xy from the walk and replace it by the rest of the cycle. The result is a walk from w to z in H, which is a contradiction. So xy belongs to no cycle in G.

(ii) Suppose xy is not a bridge. Then H is connected. Select a walk from x to y in H. By Theorem 2.1, this walk contains an x-y path; since xy is not an edge of H, it is not an edge of the path, and the union of edge xy with the path is a cycle that contains xy. □

The deletion of a cutpoint from a connected graph may yield a graph with any number of components. As an example, consider the star $K_{1,n}$; deletion of its central vertex yields n components. However, the situation with bridges is different.

Lemma 3.2. *The deletion of a bridge from a connected graph yields a graph with exactly two components.*

Proof. Suppose G is a connected graph with bridge xy; write H for G with xy deleted. Define $H(x)$ to be the set of all vertices connected to x in H, and $H(y)$ to be the set of all vertices connected to y. If w is any vertex that does not lie in $H(x)$, then the walk from x to w in G must have contained the edge xy, and w must still be connected to y. So $H(x) \cup H(y)$ equals the vertex-set of G. The induced subgraph of H spanned by $H(x)$ is connected: if w and z are any two members of $H(x)$, there is a walk from w to z via x in H. Similarly $\langle H(y) \rangle$ is connected. So H has at most two components. Since xy is a bridge, H is not connected, so it has exactly two components. □

Exercises 3.1

.1.1 Find all cutpoints in the following graphs.

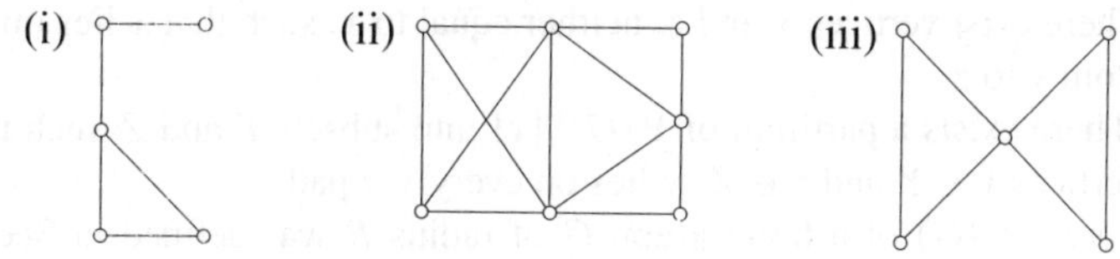

.1.2 Find all cutpoints in the following graphs.

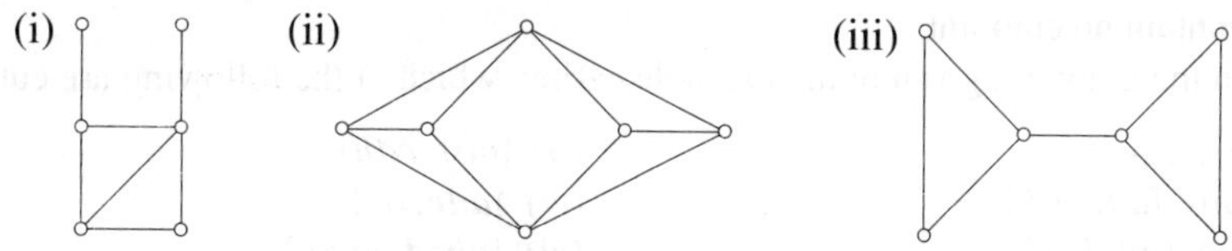

.1.3 What is the maximum number of bridges in a graph on v vertices?

.1.4 Prove that a graph in which every vertex has even degree can have no bridge.

.1.5 Prove that if x is a cutpoint of G, then x is not a cutpoint of $\overline{G}$.

.1.6 List all cutsets in the graphs in Figure 3.2.

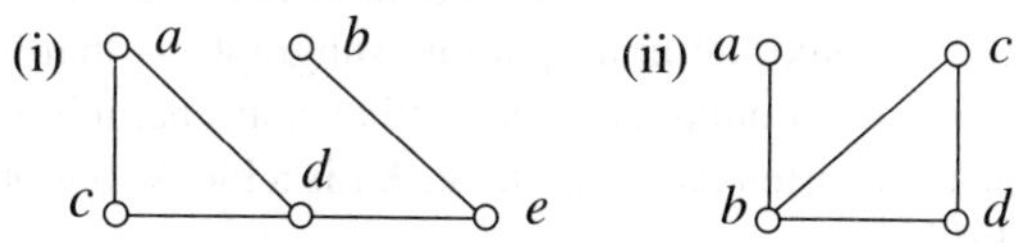

Fig. 3.2. Find all cutsets in these graphs

1.7 Repeat the preceding exercise for the graphs in Figure 3.3.

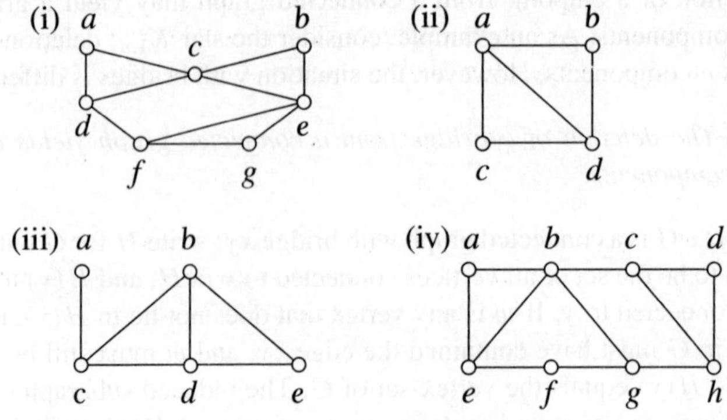

Fig. 3.3. Find all cutsets in these graphs

3.1.8 Let x be a vertex of a connected graph G. Prove that the following statements are equivalent:
(i) x is a cutpoint of G.
(ii) There exist vertices y and z, neither equal to x, such that x lies on every path from y to z.
(iii) There exists a partition of $V(G)\backslash\{x\}$ into subsets Y and Z such that for any vertices $y \in Y$ and $z \in Z$, x lies on every y-z path.

A3.1.9 The *center* $C(G)$ of a finite graph G of radius R was defined in Section 2.2 to consist of all those vertices x that have eccentricity $\varepsilon(x) = R$; the graph G is called *self-centered* if $C(G) = G$. Show that a connected self-centered graph can contain no cutpoint.

3.1.10 In the graph G, given in the example earlier, which of the following are cutsets?

(i) $[ae, bcf]$
(ii) $[abf, cdh]$
(iii) $[ade, bf]$
(iv) $[ade, bc]$
(v) $[ad, bce]$
(vi) $[abcd, efgh]$

3.2 Blocks

A graph is called *nonseparable* if it is connected, nontrivial and contains no cutpoints. A *block* in a graph G is a maximal nonseparable subgraph — that is, a nonseparable subgraph that is not properly contained in any other nonseparable subgraph of G. A nonseparable graph is itself often called a block. K_2 is a block, but obviously no other block can contain a bridge.

Any graph can be considered as a collection of blocks hooked together by its cutpoints. The other vertices are often called internal to their blocks, or simply *internal vertices*.

Example. Partition the following graph into blocks.

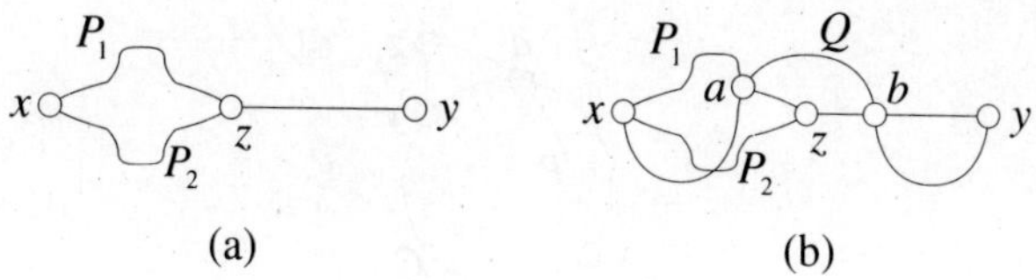

Fig. 3.4. Proof that (i) ⇒ (ii) in Theorem 3.3

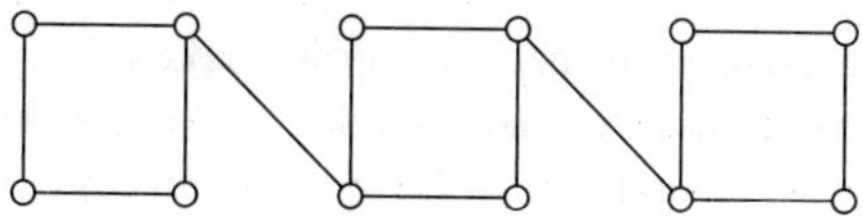

The partition is

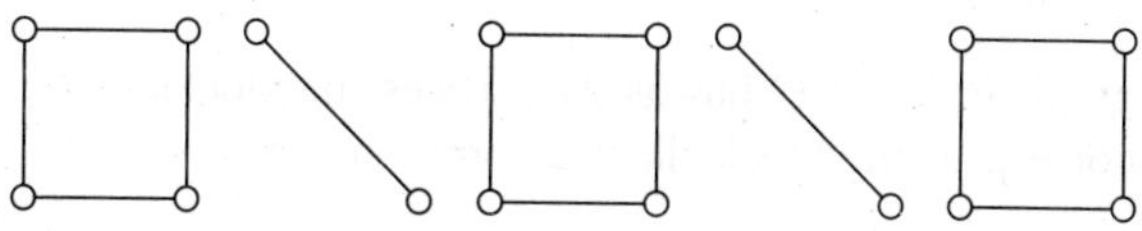

Theorem 3.3. *Suppose G is a connected graph with at least three vertices. Then the following are equivalent:*

(i) *G is a block.*
(ii) *Any two vertices of G lie on a common cycle.*
(iii) *Any vertex and edge of G lie on a common cycle.*
(iv) *Any two edges of G lie on a common cycle.*

Proof. We prove (i) ⇒ (ii), (ii) ⇒ (iii) and (iv) ⇒ (i). (The proof that (iii) ⇒ (iv) is similar to the proof that (ii) ⇒ (iii).)

(i) ⇒ (ii) Assume G is a block. Suppose x and y are distinct vertices of G, and write X for the set of all vertices other than x that lie on a cycle passing through x. Since G has at least three vertices and no cutpoint, it contains no bridge. So every vertex adjacent to x is in X, and X is not empty.

Assume y is not in X; we shall derive a contradiction. Select a vertex z in X such that the distance $d(y, z)$ is minimal; let P_0 be a shortest y-z path, and write P_1 and P_2 for the two disjoint x-z paths that make up a cycle containing x and z. (See Figure 3.4(a).) Since z is not a cutpoint, there will be an x-y path not containing z (see Exercise 3.1.8); say Q is such a path. Let b be the vertex nearest to x in Q that is also in P_0, and a the last vertex in the x-b section of Q that lies in $P_1 \cup P_2$; without loss of generality we can assume a is in P_1. This is illustrated in Figure 3.4(b).

We now construct two x-b paths R and S. To form R, follow P_1 from x to a and Q from a to b. S consists of P_2 followed by P_0 from z to b. Then $R \cup S$ is a cycle containing x and b, whence b is in X. The only vertex in $P_0 \cup X$ is z, so $b = z$, and z is in Q — a contradiction.

(ii) ⇒ (iii) Select a vertex x and an edge yz of G. Let C be a cycle containing x and y. If z is also on C, the required cycle is constructed from the edge yz together with a y-z path that is part of C. Otherwise, select a z-x path that does not contain y (this must be possible since y is not a cutpoint). Let a be the point of $P \cap C$ that

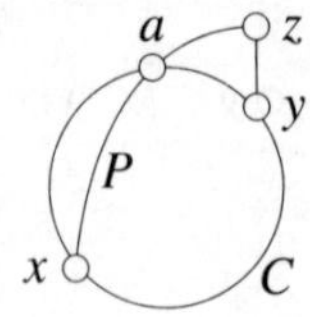

Fig. 3.5. Proof that (ii) ⇒ (iii) in Theorem 3.3

is nearest to z. Then a cycle is formed as follows: take edge yz, followed by the z-a section of P, and the a-y path of C that includes x. (See Figure 3.5.)

(iv) ⇒ (i) Suppose x is a cutpoint in G, and ρ is an edge containing x. From (iv), ρ lies in a cycle, so x is on a cycle. But this contradicts Lemma 3.1. Therefore G contains no cutpoint, so it certainly contains no bridge. □

The block graph $B(G)$ of G has as its vertices the blocks of G; two vertices are adjacent if the corresponding blocks have a common vertex.

Theorem 3.4. [64] *A graph H is the block graph of some graph if and only if every block of H is complete.*

Proof. Let $H = B(G)$, and assume there is a block H_i of H that is not complete. Then there are two vertices in H_i that are nonadjacent and lie on a shortest common cycle C of length at least 4. But the union of the blocks of G corresponding to the points of H_i that lie on C is then connected and has no cutpoint, so it is itself contained in a block, contradicting the maximality property of a block of a graph.

On the other hand, let H be a graph in which every block is complete. Form $B(H)$, and then form a graph G by adding to each vertex H_i of $B(H)$ a number of pendant edges equal to the number of vertices of the block H_i that are not cutpoints of H. Then it is easy to see that $B(G)$ is isomorphic to H. □

Exercises 3.2

A3.2.1 Partition the following graphs into blocks.

(i)

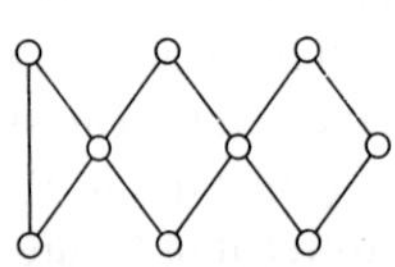

(ii)

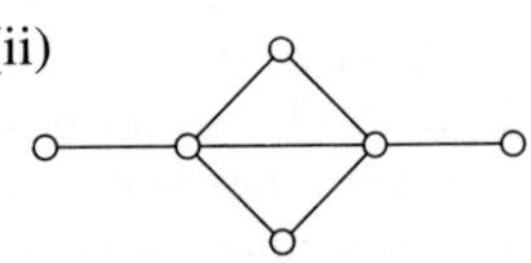

3.2.2 Partition the following graphs into blocks.

(i)

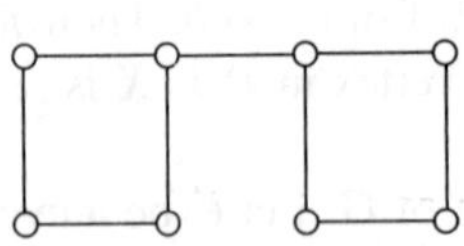

(ii)

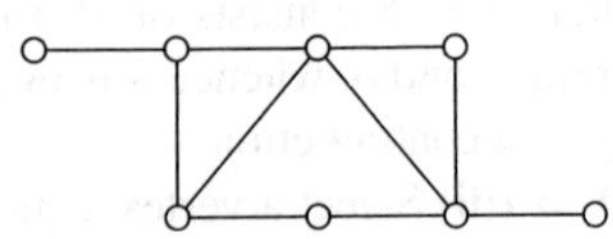

A3.2.3 Prove that a connected graph with at least two edges is a block if and only if any two adjacent edges lie on a cycle.

3.2.4 The square G^2 of the graph G was defined in Exercise 2.1.11. If G is a nontrivial connected graph, prove that G^2 is a block.

3.2.5 Write $b(G)$ for the number of blocks of G, and $b_G(x)$ for the number of blocks of G that contain the vertex x. If G is connected, prove that

$$b(G) - 1 = \sum [b_G(x) - 1]$$

(the sum is taken over all vertices x of G). [63]

3.2.6 G is a nontrivial connected graph. $C(G)$ is the number of cutpoints of G; $c(B)$ is the number of cutpoints of G that are vertices of the block B.
 (i) Prove that $c(B) = C(B)$ if and only if $B = G$.
 (ii) Prove that

$$C(G) - 1 = \sum [c(B) - 1]$$

where the sum is over all blocks B of G. [51]

3.2.7 A graph G is a *critical block* if G is a block but $G - \rho$ is not a block for any edge ρ of G. A *chord* of G is an edge joining two vertices that lie on a cycle but are not adjacent in the cycle. If G is a critical block with $v(G) \geq 4$, prove:
 (i) G has no chords.
 (ii) G contains no subgraphs isomorphic to K_3.
 (iii) $v(G) \leq e(G) \leq 2v(G) - 4$.
 (iv) If G is not a cycle, and if all vertices of degree 2 are deleted from G, the resulting graph is disconnected. [93]

3.3 Connectivity

Generalizing the idea of a cutpoint, we define the *connectivity* $\kappa(G)$ of a graph G to be the smallest number of vertices whose removal from G results in either a disconnected graph or a single vertex. (The latter special case is included to avoid problems when discussing complete graphs.) If $\kappa(G) \geq k$, then G is called k-connected. The *edge-connectivity* $\kappa'(G)$ is defined to be the minimum number of edges whose removal disconnects G (no special case is needed). In other words, the edge-connectivity of G equals the size of the smallest cutset in G. From the definition, it is clear that the connectivity and edge-connectivity of a graph is at least as great as that of any of its subgraphs.

The following theorem is due to Whitney [133]. Recall that $\delta(G)$ denotes the minimum degree of vertices of G.

Theorem 3.5. *For any graph G,*

$$\kappa(G) \leq \kappa'(G) \leq \delta(G).$$

Proof. It is clear that $\kappa'(G) \leq \delta(G)$, because one can disconnect G by removing all edges incident with any one given vertex.

Suppose $T = [X, Y]$ is a cutset of minimal size in G, where $X \cup Y = V(G)$ and $X \cap Y = \emptyset$. Then $\kappa'(G) = |T|$.

If every vertex in X is adjacent in G to every vertex in Y, then the number of edges in G is at least $\kappa'(G) \geq |X| \cdot |Y| \geq v - 1 = \delta(K_v)$, where $v = |V(G)|$. But trivially $\kappa'(G) \leq \delta(G) \leq \delta(K_v)$, so $G = K_v$, and in this case the theorem is easily seen to be true.

So let us assume that there exist vertices $x \in X$ and $y \in Y$ that are not adjacent. Define

$$S = \{p : p \in Y, px \in T\} \cup \{q : q \in X, q \neq x, qy \in T\}.$$

Then $G - S$ is a subgraph of $G - T$. Both x and y are vertices of $G - S$, and they are in different components of $G - T$, so they are in different components of $G - S$, and $G - S$ is not connected. Therefore $\kappa(G) \leq |S|$. But $|S| \leq |T|$, since each vertex of S is incident with at least one edge of T, and each edge of T is incident with exactly one vertex in S. Therefore $\kappa(G) \leq |S| \leq |T| = \delta(G)$. □

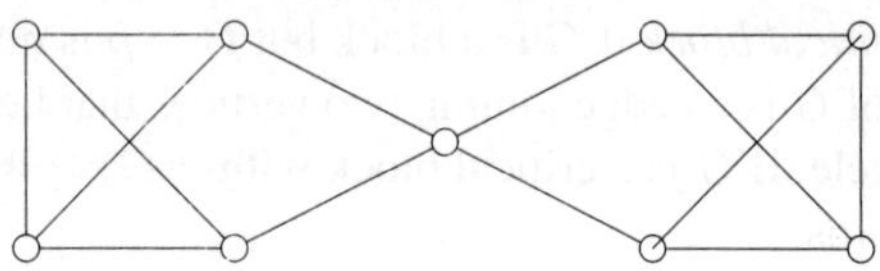

Fig. 3.6. A graph G with $\kappa(G) = 1, \kappa'(G) = 2, \delta(G) = 3$

It is easy to see that all combinations of strictness are possible in Theorem 3.5: both of the inequalities can be strict, or one of them, or neither. For example, Figure 3.6 shows a graph G with $\kappa(G) < \kappa'(G) < \delta(G)$. Graph H in Figure 3.7 has $\kappa(H) = \kappa'(H) = \delta(H)$, while K has $\kappa(K) < \kappa'(K) = \delta(K)$. Chartrand and Harary [24] proved that if ℓ, m and n are any integers such that $0 < \ell \leq m \leq n$, then there is a graph with $\kappa(G) = \ell, \kappa'(G) = m$ and $\delta(G) = n$. However, if the minimum degree is restricted in terms of the number of vertices in the graph, the amount of freedom in assigning connectivities is considerably less; see Exercises 3.3.5 and 3.3.6. Apart from the trivial case of K_1, a graph is 1-connected if and only if it is connected. All 2-connected graphs are blocks, and K_2 is the only block that is not 2-connected. So we have the following corollary to Theorem 3.3.

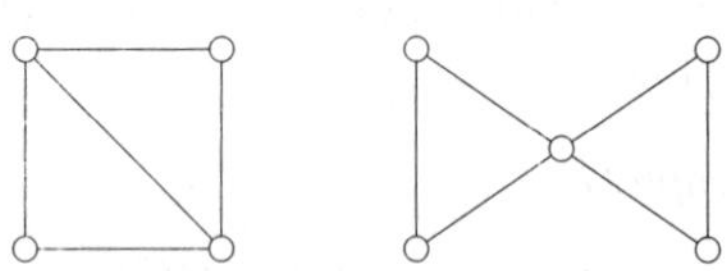

H K

Fig. 3.7. $\kappa(H) = \kappa'(H) = \delta(H) = 2$; $\kappa(K) = 1, \kappa'(K) = \delta(K) = 2$

Theorem 3.6. *G is 2-connected if and only if every two vertices of G lie on a cycle.*

Corollary 3.7. *All Hamiltonian graphs are 2-connected.*

Dirac [36] proved that if G is n-connected, then any n vertices lie on a cycle, but the converse is obviously false for $n > 2$. In fact, the characterization of n-connected graphs in general is a difficult problem. Tutte found a characterization of 3-connected graphs; for details see [116].

Exercises 3.3

3.3.1 Find the connectivity and edge-connectivity of the following graphs.

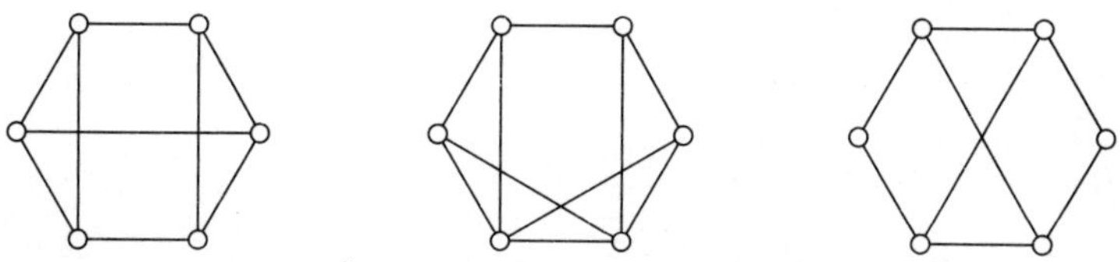

3.3.2 Find the connectivity and edge-connectivity of the following graphs.

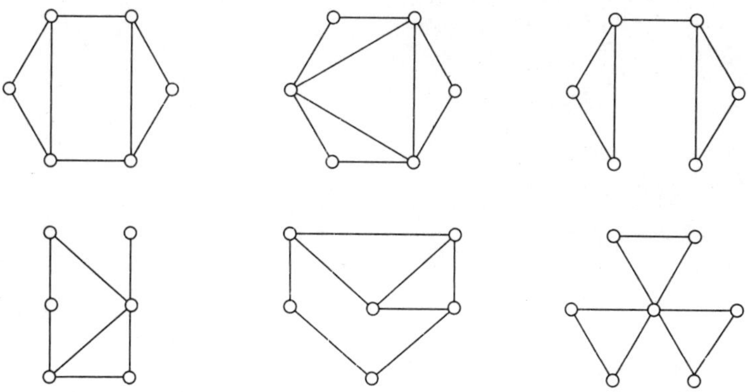

3.3.3 Find examples, different from those in the text, of graphs with:
(i) $\kappa(G) = \kappa'(G) = \delta(G)$;
(ii) $\kappa(G) < \kappa'(G) = \delta(G)$;
(iii) $\kappa(G) = \kappa'(G) < \delta(G)$.

3.3.4 Find a graph G with $\kappa(G) = 2$, $\kappa'(G) = 3$, $\delta(G) = 3$.

3.3.5 Prove that if $\delta(G) \geq v(G) - 2$, then $\kappa(G) = \delta(G)$. Find a graph with $\delta(G) = v(G) - 3$ and $\kappa(G) < \delta(G)$.

3.3.6 Prove that if $\delta(G) \geq \frac{1}{2}v(G)$, then $\kappa'(G) = \delta(G)$. Find a graph with $\delta(G) = \lfloor \frac{1}{2}v(G) - 1 \rfloor$ and $\kappa'(G) < \delta(G)$.

3.3.7 As an extension of Exercise 3.3.6, prove that for every $v \geq 3$ there exists a graph on v vertices with $\delta(G) = \lfloor \frac{1}{2}v - 1 \rfloor$ and $\kappa'(G) < \delta(G)$.

3.3.8 Prove that if $\delta(G) \geq \frac{1}{2}(v(G) + k - 2)$, then $\kappa(G) \geq k$.

4

Trees

4.1 Characterizations of Trees

A *tree* is a connected graph that contains no cycle. Figure 4.1 contains three examples of trees. It is also clear that every path is a tree, and the star $K_{1,n}$ is a tree for every n.

A tree is a minimal connected graph in the following sense: if any vertex of degree at least 2, or any edge, is deleted, then the resulting graph is not connected. In fact it is easy to prove the following stronger theorem; the proof is left as an exercise.

Theorem 4.1. *A connected graph is a tree if and only if every edge is a bridge.*

Trees are also characterized among connected graphs by their number of edges.

Theorem 4.2. *A finite connected graph G with v vertices is a tree if and only if it has exactly v − 1 edges.*

Proof. (i) Suppose that G is a tree with v vertices. We proceed by induction on v. The theorem is true for $v = 1$, since the only graph with one vertex is K_1, which is a tree. Suppose it is true for $w < v$, and suppose G is a tree with v vertices. Select an edge (G must have an edge, or it will be the unconnected graph $\bar{K}_v$) and delete it. The result is a union of two disjoint components, each of which is a tree with less than v vertices; say the first component has v_1 vertices and the second has v_2, where $v_1 + v_2 = v$. By the induction hypothesis, these graphs have $v_1 - 1$ and $v_2 - 1$ edges respectively.

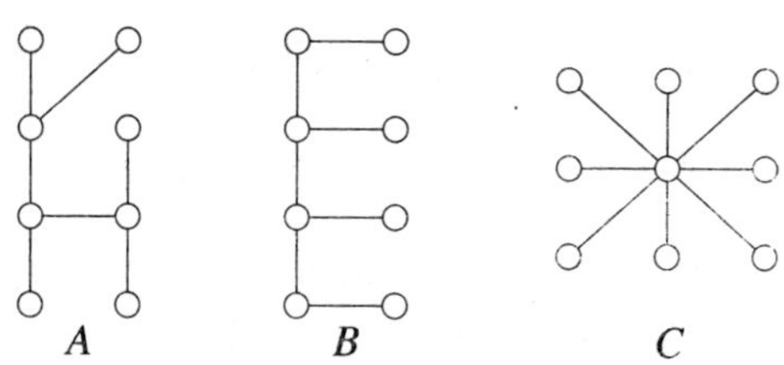

Fig. 4.1. Three trees

Adding one edge for the one that was deleted, we find that the number of edges in G is

$$(v_1 - 1) + (v_2 - 1) + 1 = v - 1.$$

(ii) Conversely, suppose G is not a tree. Select an edge that is *not* a bridge, and delete it. If the resulting graph is not a tree, repeat the process. Eventually there will be only bridges left, and the graph is a tree. From what we have just said it must have $v - 1$ edges, and the original graph had more than $v - 1$ edges. □

The word "leaf" is used to refer to a vertex of degree 1 in a tree, together with the edge incident with it.

Corollary 4.3. *Every tree other than K_1 has at least two leaves.*

Proof. Suppose the tree has v vertices. It then has $v - 1$ edges. So, by Theorem 1.1, the sum of all degrees of the vertices is $2(v - 1)$. There can be no vertex of degree 0, since the tree is connected; if $v - 1$ of the vertices have degree at least 2, then the sum of the degrees is at least $1 + 2(v - 1)$, which is impossible. □

The corollary does not hold if we allow our graphs to have infinite vertex-sets. One elementary example consists of the infinitely many vertices $0, 1, 2, \ldots, n, \ldots$ and the edges $01, 12, 23, \ldots, (n, n+1), \ldots$. The only vertex with degree 1 is vertex 0; every other vertex in the "tree" has degree 2.

The following interesting theorem uses Corollary 4.3.

Theorem 4.4. *Suppose T is a tree with k edges and G is a graph with minimum degree $\delta(G) \geq k$. Then G has a subgraph isomorphic to T.*

Proof. The proof uses induction on k. If $k = 0$, then $T = K_1$, which is a subgraph of every graph. Suppose $k > 0$, and suppose the theorem is true for all nonnegative integers less than k. Select a vertex x of degree 1 in T (the existence of such a vertex is guaranteed by Corollary 4.3). Say wx is the edge of T containing x.

The graph $T - x$ is a tree with $k - 1$ edges, so it is isomorphic to some subgraph H of G (since $\delta(G) \geq k > k - 1$). Suppose y is the vertex of G corresponding to w. Since y has degree at least k in G, and H contains only $K - 1$ edges, there must be at least one edge adjacent to y, say yz, which is not an edge of H. Then $H + yz$ is isomorphic to T. □

Exercises 4.1

4.1.1 Show that there are exactly six nonisomorphic trees on six vertices.

4.1.2 Prove Theorem 4.1.

A4.1.3 Prove that a finite graph on v vertices that contains no cycle is connected if and only if it has $v - 1$ edges.

4.1.4 Prove that a connected graph is a tree if and only if it has the following property: *if x and y are distinct vertices, then there is a unique path in G from x to y.*

A4.1.5 A perfect square was defined in Exercise 2.1.11. Prove that no tree other than K_1 or K_2 is a perfect square.

A4.1.6 Give an example of an infinite "tree" that contains no vertex of degree 1.

4.1.7 Let T be a tree on v vertices, $v \geq 5$, with precisely four vertices of degree 1 each and precisely one vertex of degree 4. Find the degrees of the remaining vertices of T, and show that T can be written as the union of two edge-disjoint simple walks.

4.1.8 Let the vertices of a tree T be labeled with the integers $1, 2, \ldots, v$. As usual, $D(i, j)$ denotes the distance between vertices i and j. Let M_T be the $v \times v$ matrix with (i, j) entry $x^{D(i,j)}$. Show that the determinant of M_T equals $(1 - x^2)^{n-1}$.

4.1.9 A tree T with v vertices has a vertex of degree k. Prove that the longest path in T has at most $v - k + 1$ edges.

4.1.10 Prove that a graph is a tree if and only if every vertex of degree greater than 1 is a cutpoint.

4.1.11 The *center* $C(G)$ of a finite graph G of radius R was defined in Section 2.2 to consist of all those vertices x that have eccentricity $\varepsilon(x) = R$.

H(i) Prove that the center of a tree consists of either one vertex or two adjacent vertices. [73]

(ii) Give examples of trees with centers of size 1 and size 2.

4.1.12 Recall that a graph G is called *self-centered* if $C(G) = G$. Which trees are self-centered?

4.1.13 Let A be the incidence matrix of a tree on t vertices. Consider the t rows of A as vectors over GF[2], by interpreting 0 and 1 as the elements of the two-element field GF[2]. Show that any $t - 1$ rows of A are linearly independent over GF[2].

4.2 Spanning Trees

Recall that a subgraph of a graph G *spans* G if it contains every vertex of G. A *spanning tree* is a spanning subgraph that is a tree when considered as a graph in its own right.

Theorem 4.5. *Every connected graph G has a spanning tree.*

Proof. If G is a tree, then the whole of G is itself the spanning tree. Otherwise G contains a cycle. Let a be an edge in the cycle. Then a is not a bridge in G, so the graph G' obtained by deleting a from G is still connected. We have not deleted any vertex, so G' is a spanning subgraph. If G' contains a cycle, we delete an edge from that cycle. The new graph we obtain is again a connected spanning subgraph of G. This process may be continued until the remaining graph contains no cycle — that is, it is a tree. So, when the process stops, we have found a spanning tree. But the process must stop since G is finite and there are only finitely many edges that could be deleted. □

It is easy to see that Theorem 4.5 generalizes to graphs with loops and multiple edges.

It is clear from the above proof that a given multigraph may have many different spanning trees. In certain applications it is useful to know the exact number. We shall write $\tau(G)$ for the *number of spanning trees* of a graph G.

One can sometimes calculate $\tau(G)$ quite quickly. If G is a tree, then $\tau(G) = 1$. If G is a cycle of length n, then n spanning trees can be constructed, each by deleting one edge, so $\tau(G) = n$. We can consider a general multigraph G: the existence of loops does not change $\tau(G)$, as no loop can contribute to a tree; if one edge is multiple, of multiplicity k, then each spanning tree includes at most one of the k edges, and replacing one edge joining the two vertices by another gives another spanning tree.

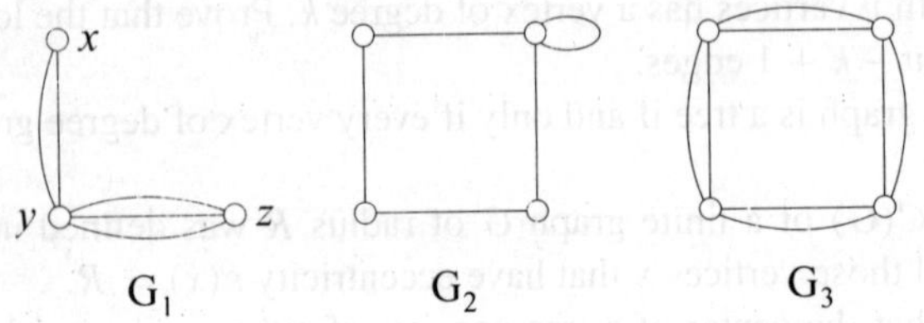

Fig. 4.2. Multigraphs whose trees are to be counted

Example. Figure 4.2 shows multigraphs G_1, G_2, G_3. We find the number of spanning trees for each.

A spanning tree in G_1 must contain one of the edges xy and one of the edges yz. The number of choices is $2 \times 3 = 6$. So $\tau(G_1) = 6$. Since loops do not affect the function τ, $\tau(G_2) = \tau(C_4) = 4$. To calculate $\tau(G_3)$, one first observes that three pairs of vertices must be joined. This can be done in four ways. In each case there are eight trees. So $\tau(G_3) = 4 \times 8 = 32$.

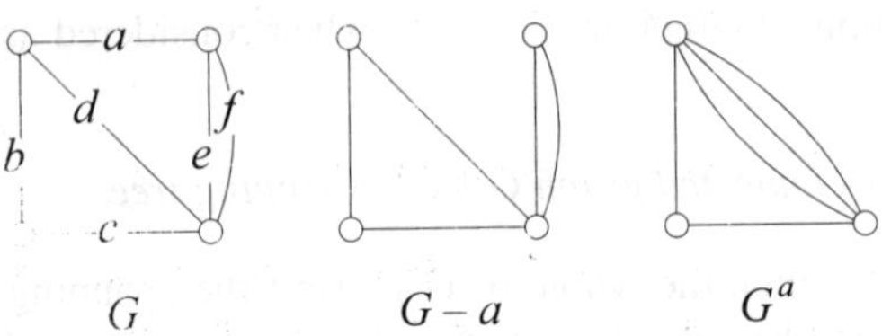

Fig. 4.3. The multigraphs used in counting trees

Calculation of $\tau(G)$ by counting becomes very tedious when G is large. In order to derive a formula for $\tau(G)$, we first introduce a new multigraph G_a. If a is any edge of a multigraph G, G_a is formed by identifying the endpoints of a: if $a = xy$, then G_a is formed by deleting both x and y, inserting a new vertex, and replacing every edge zx and every edge zy by an edge from z to the new vertex. For convenience we assume that every edge from x to y is deleted; alternatively, we could introduce these edges as loops in G_a, but this is unnecessary, as G_a will be used only in counting trees, and loops are immaterial in that context. We also use the multigraph $G - a$, formed from G by deleting a. Examples of G_a and $G - a$ are shown in Figure 4.3.

Suppose a is an edge of G. Then the spanning trees of G either contain a or they do not. A spanning tree that does not contain a is a subgraph of $G - a$, and is still a spanning subgraph, so it is a spanning tree of $G - a$; conversely, the spanning trees of $G - a$ are spanning trees of G and do not contain a. So the two sets, the spanning trees of $G - a$ and the spanning trees of G that do not contain a, are in one-to-one correspondence. So the sets are equal in size, and there are $\tau(G - a)$ spanning trees of G that do not contain a. Similarly, the number of spanning trees of G that do contain a is $\tau(G_a)$; Exercise 4.2.1 asks for a proof of this fact, but we look first at the special case shown in Figure 4.3. The spanning trees of G that contain a must also contain *either* b and c only, *or* one of b and c together with one of d, e, and f. Exactly the same is true of the spanning trees of G_a: they contain b and c only, *or* one of b and c together with one of d, e, and f.

Thus, summing the number of spanning trees of G that do or do not contain edge a, we obtain:

Theorem 4.6.

$$\tau(G) = \tau(G - a) + \tau(G_a).$$

An n-fold path is formed from a path by replacing each edge with a multiple edge of multiplicity n. An n-fold cycle is defined similarly. These multigraphs recur frequently in applications of Theorem 4.6, so it is helpful to know their numbers of spanning trees.

Theorem 4.7. *The number of spanning trees in an n-fold path is*

$$\tau(nP_v) = n^{v-1}.$$

The number of spanning trees in an n-fold cycle is

$$\tau(nC_v) = vn^{v-1}.$$

Proof. For the multiple path, one has n choices of edge for each edge of the underlying path, giving n^{v-1} paths in all. For the multiple cycle, each spanning tree is a path; there are v choices for the pair of adjacent vertices that will not be adjacent in the spanning tree, and for each choice there are again n^{v-1} paths. □

Example. Calculate $\tau(G)$, where G is the graph of Figure 4.3.

The method of decomposing the relevant graphs is indicated in Figure 4.4 (next page).

It is clear that $\tau(G_4) = 3$ (since G_4 is a cycle), that $\tau(G_5) = 1$ (since G_5 is a tree), that $\tau(G_6) = 2$, that $\tau(G_7) = 3$ and that $\tau(G_8) = 4$. So:

$$\begin{aligned}
\tau(G_2) &= \tau(G_7) + \tau(G_8) = 3 + 4 = 7;\\
\tau(G_3) &= \tau(G_5) + \tau(G_6) = 1 + 2 = 3;\\
\tau(G_1) &= \tau(G_3) + \tau(G_4) = 3 + 3 = 6;\\
\tau(G) &= \tau(G_1) + \tau(G_2) = 6 + 7 = 13.
\end{aligned}$$

Suppose G is a graph with v vertices, T is any spanning tree in G, and a is any edge of G that is not in T. Then $T + a$ has v vertices, so it must contain a cycle. Moreover, a must be an edge in that cycle. Select an edge b of the cycle, other than a. Then $T + a - b$ will be acyclic, and it is still connected, so it is a tree.

In particular, suppose R is a spanning tree of G that has k edges in common with t, and suppose a is an edge of R (but not of T). The cycle in $T + a$ must contain an edge that is not in R, because otherwise R would contain a cycle. If such an edge is chosen as b, then the tree $T - a + b$ will have $k + 1$ edges in common with R. Call this tree T_1. One can then construct another tree T_2 that shares $k + 2$ edges with R, and so on. Eventually the number of shared edges will be $v - 1$, so the tree must be R. We have proved:

Theorem 4.8. *If T and R are spanning trees of the v-vertex graph G, then there exists a sequence of spanning trees,*

$$T = T_0, T_1, \ldots, T_n = R,$$

where T_i and T_{i+1} have $v - 2$ common edges for every i.

Exercises 4.2

4.2.1 Prove that there is a one-to-one correspondence between the trees of G containing edge a and the trees of G_a.

4.2.2 A multigraph G consists of a multigraph H, together with one new vertex x and an edge from x to one of the vertices of H. Show that $\tau(G) = \tau(H)$.

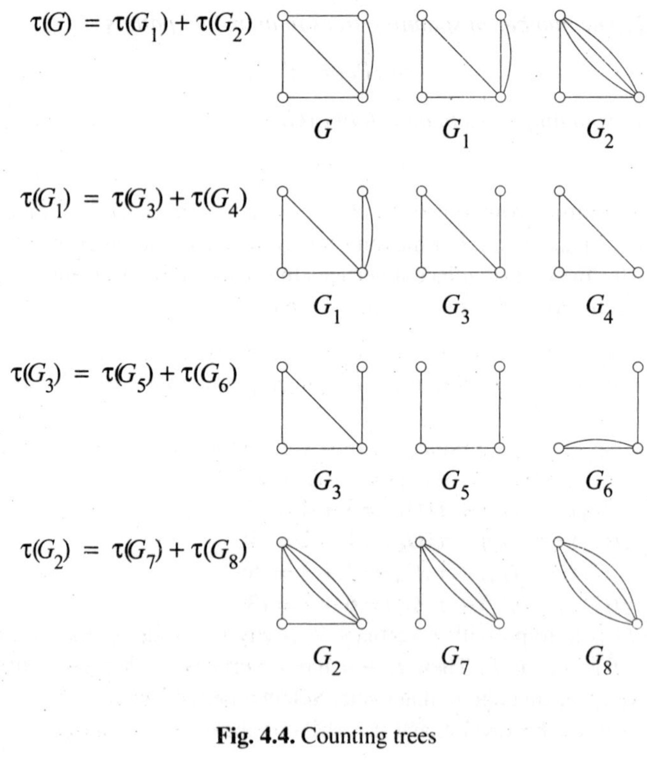

Fig. 4.4. Counting trees

4.2.3 Show that K_v contains a pair of edge-disjoint spanning trees if and only if $v \geq 4$.

4.2.4 Find the number of spanning trees in each of the graphs and multigraphs shown in Figure 4.5.

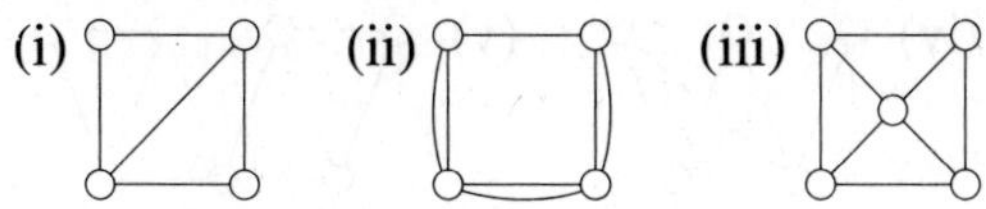

Fig. 4.5. Count the spanning trees

4.2.5 Repeat the preceding exercise for the graphs and multigraphs shown in Figure 4.6.

4.2.6 Recall that a graph is called cubic if every vertex has degree 3.

(i) Prove that if a cubic graph on n vertices contains two edge-disjoint spanning trees, then $n \leq 8$.

(ii) Is there a cubic graph on four vertices containing two edge-disjoint spanning trees? Is there one which does not contain two edge-disjoint spanning trees?

(iii) Repeat part (ii) for $n = 6$ and for $n = 8$.

4.2.7 Let G be the graph with four vertices 1, 2, 3, 4 and two edges (1, 2), (3, 4). Construct multigraphs N_1, N_2, N_3, N_4, with the following properties: each N_i consists of four edges, and four, five or six vertices; each N_i contains G as a subgraph; $\tau(N_i) = i - 1$, for $i = 1, 2, 3, 4$.

4.2.8 Prove that every bridge in a connected graph lies on every spanning tree of the graph.

4.2.9 Find $\tau(K_4)$ and $\tau(K_5)$.

4.2.10 Suppose a graph G is formed by taking two disjoint connected graphs G_1 and G_2 and identifying a vertex in G_1 with a vertex in G_2. Show that $\tau(G) = \tau(G_1)\tau(G_2)$.

4.2.11 Suppose G is formed by taking two disjoint connected graphs G_1 and G_2 and inserting an edge connecting some vertex of G_1 with some vertex of G_2. Use Theorem 4.5 and the result of the preceding Exercise to find an expression for $\tau(G)$.

4.2.12 Let T_1 and T_2 be spanning trees of a connected graph G; show that if a is any edge of T_1, then there exists an edge b of T_2 such that $(T_1 - \{a\}) \cup \{b\}$ (the graph obtained from T_1 on replacing a by b) is also a spanning tree. Show also that T_1 can be "transformed" into T_2 by replacing the edges of T_1 one at a time by edges of T_2 in such a way that at each stage we obtain a spanning tree.

4.2.13 (i) Show that in any connected graph, any cycle must have at least one edge in common with the complement of any spanning tree.

(ii) Show that in any connected graph, any cutset must have at least one edge in common with any spanning tree.

4.2.14 Let H be a subgraph of a connected graph G. Show that H is a subgraph of some spanning tree T of G if and only if H contains no cycle.

4.2.15 If G is any connected graph or multigraph with v vertices, the *tree graph* of G has as its vertices the spanning trees of G; two vertices are adjacent if and only if the

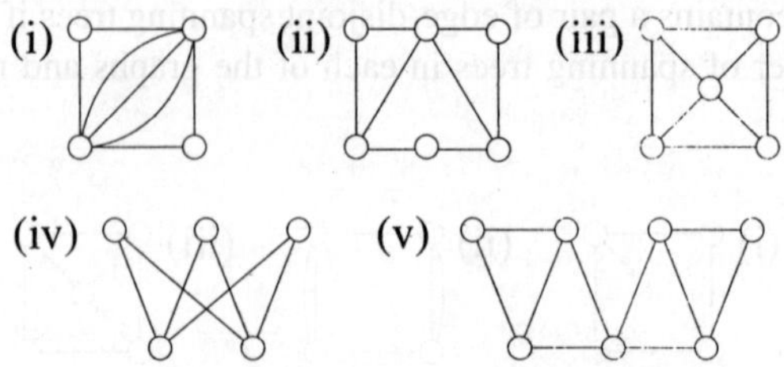

Fig. 4.6. Count the spanning trees

trees have $v-2$ edges in common. Prove that the tree graph of a graph is always connected.

4.3 Minimal Spanning Trees

Consider applications of the kind discussed in Section 2.3, where each edge of a graph has a weight associated with it. It is sometimes desirable to find a spanning tree such that the weight of the tree — the total of the weights of its edges — is minimum. Such a tree is called a *minimal spanning tree*.

It is clear that a finite graph can contain only finitely many spanning trees, so it is possible in theory to list all spanning trees and their weights, and to find a minimal spanning tree by choosing one with minimum weight. This process could take a very long time however, since $\tau(G)$ can be very large. So efficient algorithms that find a minimal spanning tree are useful. We present here an example due to Prim [96].

We assume that G is a graph with vertex-set V and edge set E, and suppose there is associated with G a map $w: E \rightarrow R$ called the *weight* of the edge; when xy is an edge of G we write $w(x, y)$ for the image of xy under w. We could quite easily modify the algorithm to allow for multiple edges, but the notation is slightly simpler in the graph case. The algorithm consists of finding a sequence of vertices $x_0, x_1, x_2, \ldots$, of G and a sequence of sets $S_0, S_1, S_2, \ldots$, where

$$S_i = \{x_0, x_1, \ldots, x_{i-1}\}.$$

We choose x_0 at random from V. When $n > 0$, we find x_n inductively using S_n as follows.

1. Given i, $0 \le i \le n-1$, choose y_i to be a member of $V \setminus S_n$ such that $w(x_i, y_i)$ is minimum, if possible. In other words:
   a) if there is no member of $V \setminus S_n$ adjacent to x_i, then there is no y_i;
   b) if $V \setminus S_n$ contains a vertex adjacent to x_i, then y_i is one of those vertices adjacent to x_i, and if $x_i \sim y$, then $w(x_i, y_i) \le w(x_i, y)$.
2. Provided that at least one y_i has been found in Step (1), then define x_n to be a y_i such that $w(x_i, y_i)$ is minimal; in other words, x_n is the y_i that satisfies

$$w(x_i, y_i) \le w(x_j, y_j) \text{ for all } j.$$

3. Put $S_{n+1} = S_n \cup \{x_n\}$.

This process stops only when there is no new vertex y_i. If there is no new vertex y_i, it must be true that no member of $V \setminus S_n$ is adjacent to a vertex of S_n. It is impossible to partition the vertices of a connected graph into two nonempty sets such that no edge joins one set to the other, and S_n is never empty, so the process stops only when $S_n = V$.

When we reach this stage, so that $S_n = V$, we construct a graph T as follows:
(i) T has vertex-set V;
(ii) if x_k arose as y_i, then x_i is adjacent to x_k in T;
(iii) no edges of T exist other than those that may be found using (ii).
It is not hard to verify that T is a tree and that it is minimal; see Exercise 4.3.3.

Observe that xX_n may not be defined uniquely at Step (2) of the algorithm, and indeed y_i may not be uniquely defined. This is to be expected: after all, there may be more than one minimal spanning tree.

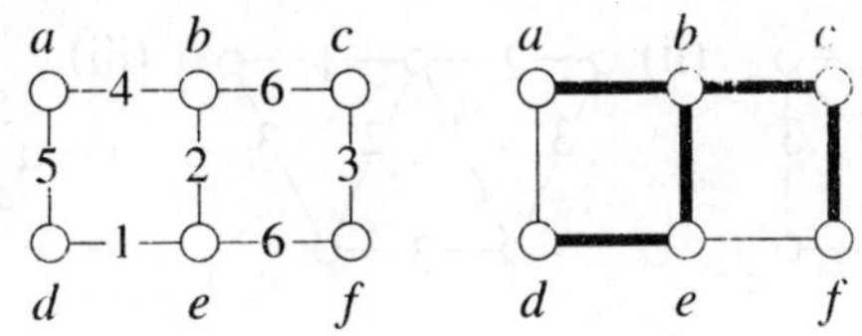

Fig. 4.7. An example of Prim's algorithm

Example. Consider the graph G shown in Figure 4.7. Weights are shown next to the edges.

Select $x_0 = a$. Then $S_1 = \{a\}$. Now $y_0 = b$, and this is the only choice for x_1. So $S_2 = \{a, b\}$. The tree will contain edge ab.

Working from S_2, we get $y_0 = d$ and $y_1 = e$. Since be $(= x_1 y_1)$ has smaller weight than ad $(= x_0 y_0)$, we select $x_2 = e$. Then $S_3 = \{a, b, e\}$ and edge be goes into the tree. Similarly, from S_3, we get $x_3 = d$ and $S_4 = \{a, b, d, e\}$, and the new edge is de.

Now there is a choice. Working from S_4, $y_1 = c$ and $y_2 = f$. In both cases the weight of the edge to be considered is 6. So either may be used. Let us choose c, and use edge bc.

The final vertex is f, and the edge is cf. So the tree has edges ab, bc, be, cf, de and weight 16.

The algorithm might be described as follows. First, choose a vertex x_0. Trivially the minimum weight tree with vertex-set $\{x_0\}$ — the *only* tree with vertex-set $\{x_0\}$ — is the K_1 with vertex x_0. Call this the *champion*. Then find the smallest weight tree with two vertices, one of which is x_0; in other words, find the minimum weight tree that can be formed by adding just one edge to the current champion. This tree is the new champion. Continue in this way: each time a champion is found, look for the cheapest tree that can be formed by adding one edge to it. One can consider each new tree to be an approximation to the final minimal spanning tree, with successive approximations having more and more edges.

Prim's algorithm was a refinement of an earlier algorithm due to Kruskal [81]. In that algorithm, one starts by listing all edges in order of increasing weight. The first approximation is the K_2 consisting of the edge of least weight. The second approximation is formed by appending the next edge in the ordering. At each stage the next approximation is formed by adding on the smallest edge that has not been used, provided only that it does not form a cycle with the edges already chosen. In this case the successive approximations are not necessarily connected, until the last one. The advantage of Prim's algorithm is that, in large graphs, the initial sorting stage of Kruskal's algorithm can be very time consuming.

Exercises 4.3

4.3.1 Find minimal spanning trees in the following graphs, using both Kruskal's and Prim's methods.

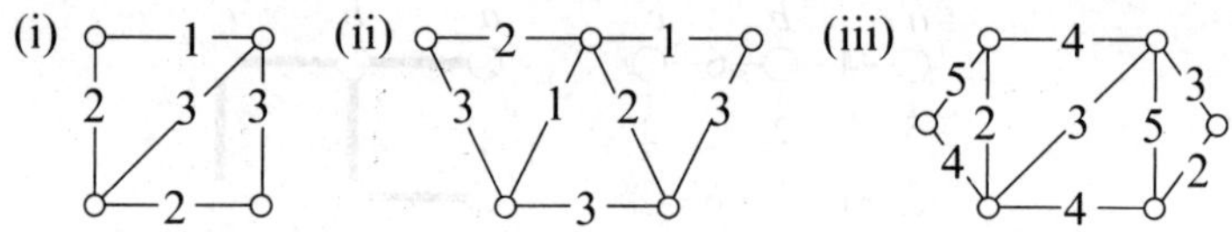

A4.3.2 Find minimal spanning trees in the following graphs, using both Kruskal's and Prim's methods.

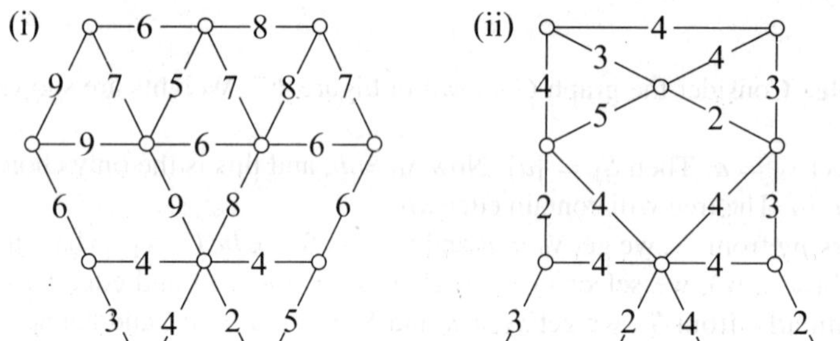

4.3.3 Prove that the graph T constructed in Prim's algorithm *is* in fact a minimal spanning tree.

3.4 Find minimal spanning trees in the following graphs, using both Kruskal's and Prim's methods.

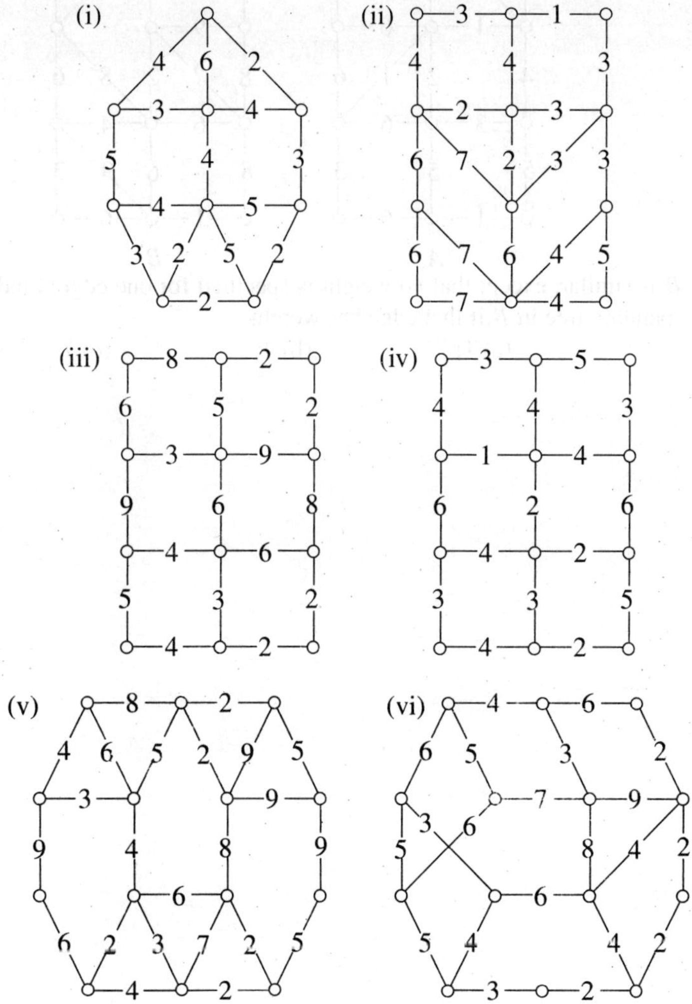

3.5 It is required to find a *maximal* spanning tree in a graph. Suggest a modification of Prim's algorithm for this problem.

3.6 (i) On graph A below, a weight function is shown. Find a minimal spanning tree in A.

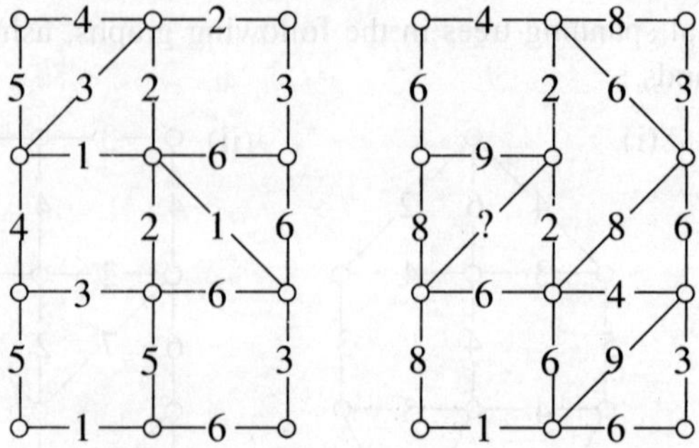

A B

(ii) B is similar, except that no weight is specified for one edge. Find a minimal spanning tree in B if that edge has weight

(a) 1; (b) 4; (c) 7.

5

Linear Spaces Associated with Graphs

5.1 Finite Fields and Vector Spaces

The reader will be familiar with vector spaces over the rational, real and complex fields and with vectors as ordered n-tuples of these numbers. These ideas can be extended to more general fields.

We write $GF[k]$ for the (unique) finite field with k elements. In particular, the two-element field $GF[2]$ is defined by the addition and multiplication tables

+	0	1
0	0	1
1	1	0

×	0	1
0	0	0
1	0	1

A finite-dimensional vector space over a finite field will be called a *finite vector space*. The vector space of dimension n over $GF[k]$ has k^n elements.

As an example, consider the n-dimensional vector space V over $GF[2]$.

To choose a basis for V, one must choose n vectors that form a linearly independent set. First we count the number of ordered linearly independent sets. The first vector x_1 may be chosen in $2^n - 1$ ways, since one may choose any vector except the zero vector. The second vector x_2 may be chosen in $2^n - 2$ ways, since zero is excluded and so is the vector x_1 already chosen. The third vector, x_3, may not equal zero, x_1 or x_2, nor may it equal the sum $x_1 + x_2$. Hence x_3 may be chosen in $2^n - 2^2$ ways. Continuing, x_4 may be chosen in $2^n - 2^3$ ways, and so on; x_n may be chosen in $2^n - 2^{n-1}$ ways. Hence the number of ways to choose an ordered basis is

$$b = (2^n - 1)(2^n - 2)(2^n - 2^2) \cdots (2^n - 2^{n-1}), \tag{5.1}$$

and if the order of the vectors is ignored, there are $b/(n!)$ ways to choose the basis.

These arguments can also be used to count the number of subspaces in a given vector space.

Example. Let V be a vector space of dimension 5 over $GF[2]$. Let us count the number of two-dimensional subspaces of V.

To choose a basis for a two-dimensional subspace it is necessary to choose a set of two linearly independent vectors in V, which can be done in $\frac{31\cdot 30}{2}$ ways. But each of the two-dimensional subspaces has three different bases, so each of the subspaces is found three times in the above choice process. Hence there are $\frac{31\cdot 30}{3\cdot 2} = 155$ two-dimensional subspaces in V.

Exercises 5.1

5.1.1 If p is a prime, then $\mathbb{Z}_p$, the set of integers modulo p, forms a field with respect to addition and multiplication modulo p. Construct the addition and multiplication for fields with:
A(i) three elements;
(ii) five elements.

5.1.2 Show that the integers $\{0, 2, 4, 6, 8\}$ form a field under addition and multiplication modulo 10.

A5.1.3 Let V be a vector space of dimension 4 over $GF[2]$. How many subspaces does V contain of dimensions 0, 1, 2, 3, 4 and 5?

5.1.4 Let V be a vector space of dimension 3 over $GF[2]$. How many subspaces does V contain of dimensions 0, 1, 2, 3 and 4?

5.1.5 Let V be a vector space of dimension 3 over the field $\mathbb{Z}_3 = GF[3]$.
(i) How many subspaces does V contain of dimensions 1 and 2?
(ii) Choose one particular 2-dimensional subspace. How many 1-dimensional subspaces does it contain?
(iii) Choose one particular 1-dimensional subspace. How many 2-dimensional subspaces contain it?

5.2 The Power Set as a Vector Space

If S is a finite set, then its *power set* $\mathcal{P}(S)$ is the set of all subsets of S. Let F be the field $GF[2]$. Then we can consider $\mathcal{P}(S)$ as a vector space over F in the following way: *vector addition* is the operation of *symmetric difference* $X + Y = (X \cup Y)\backslash(X \cap Y)$; *scalar multiplication* is defined by

$$0\cdot X = \emptyset \text{ and } 1\cdot X = X$$

for every $X \in \mathcal{P}(S)$. We verify that this is a vector space, and find a basis for it.

Clearly $\mathcal{P}(S)$ is closed under symmetric difference, which is an associative and commutative binary operation. Since

$$X + \emptyset = (X \cup \emptyset)\backslash(X \cap \emptyset) = X\backslash\emptyset = X = \emptyset + X,$$

$\emptyset$ is an identity under addition. It is easy to see that $X + Y = \emptyset$ if and only if $X = Y$, so each element has a unique negative (additive inverse); in fact each element is its own negative. Hence $\mathcal{P}(S)$ is an abelian group of order t.

Now consider scalar multiplication: for any $X \subseteq S$ and $Y \subseteq S$,

$$0\cdot(X + Y) = \emptyset = \emptyset + \emptyset = 0\cdot X + 0\cdot Y$$

and

$$1\cdot(X+Y) = X+Y = 1\cdot X + 1\cdot Y,$$

so

$$\lambda\cdot(X+Y) = \lambda\cdot X + \lambda\cdot Y$$

for any scalar λ. Also

$$\begin{aligned}(0+0)\cdot X &= 0\cdot X = \emptyset = \emptyset+\emptyset = 0\cdot X + 0\cdot X,\\ (0+1)\cdot X &= 1\cdot X = X = \emptyset + X = 0\cdot X + 1\cdot X,\\ (1+0)\cdot X &= 1\cdot X = X = X+\emptyset = 1\cdot X + 0\cdot X,\\ (1+1)\cdot X &= 0\cdot X = \emptyset = X+X = 1\cdot X + 1\cdot X,\end{aligned}$$

so that

$$(\lambda+\mu)\cdot X = \lambda\cdot X + \mu\cdot X$$

for any scalars λ and μ. Finally for any scalar μ,

$$\begin{aligned}(0\mu)\cdot X &= 0\cdot X = \emptyset = 0\cdot(\mu\cdot X),\\ (1\mu)\cdot X &= \mu\cdot X = \quad 1\cdot(\mu\cdot X),\end{aligned}$$

so

$$(\lambda\mu)\cdot X = \lambda\cdot(\mu\cdot X)$$

for any scalar λ. Since $1\cdot X = X$ by definition, $\mathcal{P}(S)$ is a vector space.

Let B be the collection of all one-element subsets of S. If S is $\{s_1, s_2, \ldots, s_n\}$, the typical linear combination of members of B is

$$\lambda_1\cdot\{s_1\} + \lambda_2\cdot\{x_2\} + \cdots + \lambda_n\cdot\{s_n\}. \tag{5.2}$$

If $\lambda_{i_1} = \lambda_{i_2} = \cdots = \lambda_{i_k} = 1$, and all the other λ's are zero, then expression (5.2) equals

$$\{s_{i_1}\} + \{s_{i_2}\} + \cdots + \{s_{i_k}\} = \{s_{i_1}, s_{i_2}, \ldots, s_{i_k}\}.$$

This reduces to $\emptyset$, the zero vector of $\mathcal{P}(S)$, if and only if $\lambda_i = 0$ for each $i = 1, \ldots, n$ in (5). Hence $\{s_1\}, \{s_2\}, \ldots, \{s_n\}$ are linearly independent. Since $T \subseteq S$ has the form

$$T = \{s_{i_1}, s_{i_2}, \ldots, s_{i_k}\},$$

which can be written as

$$T = \{s_{i_1}\} + \{s_{i_2}\} + \ldots + \{s_{i_k}\},$$

the set B is also a spanning set. Hence B is a basis for $\mathcal{P}(S)$.

Exercises 5.2

5.2.1 If S has n elements, then $\mathcal{P}(S)$ has 2^n elements. Is it always true that a vector space over $GF[2]$ has 2^n elements, for some n?

5.2.2 Does the set of $(n-1)$-element subsets of an n-set S form a basis for $\mathcal{P}(S)$?

5.2.3 Verify that the operation of symmetric difference is associative. (In other words, prove that if X, Y and Z are any three sets, then $(X+Y)+Z = X+(Y+Z)$.)

5.2.4 Let $\mathcal{P}(S) = V$ be a vector space, where S is a finite set. Suppose that $W \subseteq V$, $W \neq \emptyset$, and W is closed under vector addition. Show that W is a subspace of V.

5.3 The Vector Spaces Associated with a Graph

Suppose the graph G has edge-set E. Consider the power set $\mathcal{P}(E)$ of E. We write $\mathcal{E}$ to mean the subset of $\mathcal{P}(E)$ whose members are the singleton sets:

$$\mathcal{E} = \{\{a_i\}|a_i \in E\},$$

and $\mathcal{P}(E)$ is treated as a vector space over $GF[2]$ just as was done for the power set of an arbitrary set in the previous section.

The subgraph H of a graph G is often identified with its edge-set. For example, if G is the graph shown in Figure 5.1, and if H is the subgraph consisting of the triangle contained in G, then one can think of H as $\{ab, bc, ac\}$. However, when $\mathcal{P}(E)$ is treated as a vector space, it is convenient to treat subgraphs as *spanning* subgraphs, containing every vertex in G. In the example, H is treated as the set of edges ab, bc, ac together with the vertices a, b, c and d.

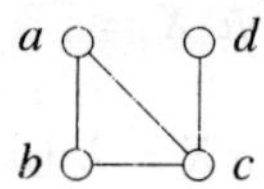

Fig. 5.1. An example graph G

Example. Let G be the graph shown in Figure 5.1. The 16 vectors of $\mathcal{P}(E)$ are G itself, the subgraphs shown in Figure 5.3, the four one-edge subgraphs, and $\emptyset$. (More strictly, the vectors are the edge-sets of these subgraphs.)

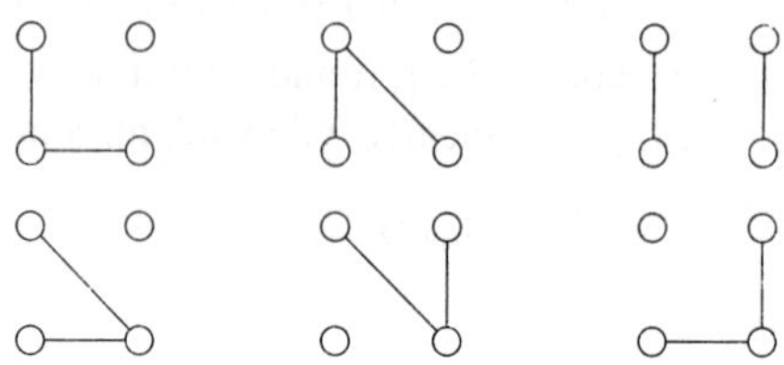

Fig. 5.2. Subgraphs of G with two edges.

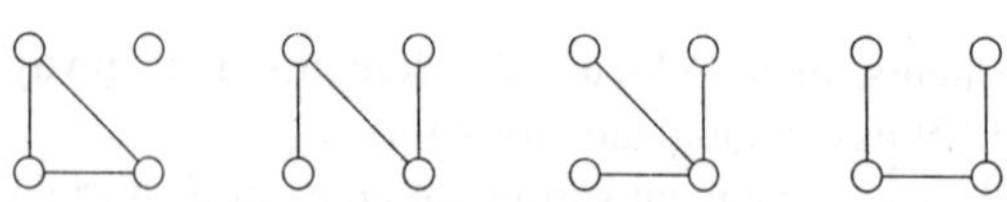

Subgraphs of G with three edges.

Fig. 5.3. Subgraphs of G.

By the definition of symmetric difference,

$$\{ac, bc\} + \{ab, ac, cd\} = \{ab, bc, cd\},$$

since the edge ac belongs to both subsets. This is illustrated in Figure 5.4.

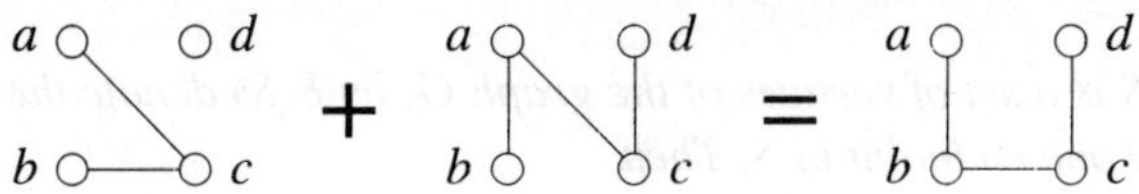

Fig. 5.4. Addition of subgraphs

The other two important vector spaces associated with a graph are subspaces of $\mathcal{P}(E)$, the *cycle subspace* and *cutset subspace* of G. There is a duality relation between the spaces, so that often a theorem relating to one of them implies a corresponding theorem relating to the other. We shall deal with the cycle subspace here and the cutset subspace in the next section.

If E is the edge-set of a graph G, we continue to identify a subset S of E with the subgraph H such that $E(H) = S$. In particular, in the next theorem, cycles are identified with their sets of edges. We shall frequently mention "unions of edge-disjoint subgraphs," by which we mean the sum of subgraphs that have no edge in common. For instance, in Figure 5.3, the graph with edges ab, ac and cd can be considered as the union of the edge-disjoint graphs $\{ab, cd\}$ and $\{ac\}$.

Theorem 5.1. *The set of all cycles and unions of edge-disjoint cycles in G is a subspace of $\mathcal{P}(E)$, known as the cycle subspace $\mathcal{F}(G)$.*
(Observe that the empty set $\emptyset$ belongs to $\mathcal{F}(G)$, since it is the union of no cycles.)

Proof. By Exercise 5.2.4, it is sufficient to show that $\mathcal{F}(G)$ is nonempty (which is true since it contains $\emptyset$) and closed under vector addition.

From Exercise 2.4.4(ii), a connected graph is a cycle or a union of edge-disjoint cycles if and only if all its vertices are of even degree. Suppose x is any vertex of G, and S and T are any two members of $\mathcal{F}(G)$. Let Y and Z be the sets of edges incident with X in S and T respectively. Then the set of edges of $S + T$ incident with x is the symmetric difference $Y + Z$. Now $|Y + Z| = |Y| + |Z| - 2|Y \cap Z|$. Both $|Y|$ and $|Z|$ are even, so $|Y + Z|$ is even also. □

Example. Consider the graph G of the preceding example. Its only cycle is the triangle abc; hence its cycle subspace is

$$\{\emptyset, \{ab, bc, ca\}\},$$

of dimension 1.

Exercises 5.3

5.3.1 Show that the cycle subspace of K_4 has dimension 3.

5.3.2 The graph G is formed from the cycle C_5 by joining two nonadjacent vertices. List the members of the cycle space of G. What is its dimension?

5.4 The Cutset Subspace

Cutsets serve as the dual of cycles in the linear algebra associated with graphs. Theorem 5.3 below is the dual of Theorem 5.1. In order to prove it we use the following lemma:

Lemma 5.2. *If S is a set of vertices of the graph G, let $\delta(S)$ denote the set of edges of G with precisely one endpoint in S. Then*

$$\delta(S+T) = \delta(S) + \delta(T)$$

for any subsets S and T of $V(G)$.

Proof. It is a standard result of set theory that

$$R \cap (S+T) = (R \cap S) + (R \cap T) \tag{5.3}$$

for any sets R, S and T.

Select any edge xy of G; write $R = \{x, y\}$. Then $xy \in \delta(S+T)$ if and only if $|R \cap (S+T)|$ is odd, which by (5.3) will be true if and only if exactly one of $|(R \cap S)|$ and $|(R \cap T)|$ is odd — that is, xy belongs to precisely one of $\delta(S)$ and $\delta(T)$, or equivalently $xy \in \delta(S) + \delta(T)$. So for any edge xy,

$$xy \in \delta(S+T) \Leftrightarrow xy \in \delta(S) + \delta(T),$$

proving the Lemma. □

Theorem 5.3. *Let G be a graph with edge set E. Then the set of all cutsets and unions of edge-disjoint cutsets in G is a subspace of $\mathcal{P}(E)$, known as the cutset subspace $\mathcal{B}(G)$.*

Note that $\emptyset \in \mathcal{B}(G)$, since $\emptyset$ can be considered as the union of no cutsets.

Proof. By analogy with Theorem 5.1, it is sufficient to prove that the sum of two cutsets is a cutset or a union of edge-disjoint cutsets. But this follows easily from the lemma. □

Example. Consider again the graph of Figure 5.1. Among its cutsets are $\{ab, bc\}$, $\{ab, ac\}$, and $\{cd\}$. Checking all partitions of its vertex-set into two disjoint subsets one finds that all its cutsets and unions of edge-disjoint cutsets can be written as linear combinations of these. Hence its cutset subspace is $\{\emptyset, \{ab, bc\}, \{ab, ac\}, \{bc, ac\}, \{cd\}, \{ab, bc, cd\}, \{ab, ac, cd\}, \{ac, bc, cd\}\}$ of dimension 3.

Lemma 5.4. *Every cycle has an even number of edges in common with every cutset.*

Proof. Consider the cutset C that induces the partition of $V = V(G)$ into

$$V = V_1 \cup V_2$$

where the two subsets V_1 and V_2 are those in the two components of G remaining after C has been deleted.

Say L is a cycle in G. If L meets only vertices in V_1 or only vertices in V_2, so that $L \cap C = \emptyset$, then $|L \cap C| = 0$ which is certainly even. Otherwise L must meet vertices in both V_1 and V_2. Follow L from a vertex x in V_1, and suppose y is the first vertex encountered in V_2. Then the path from x to y must cross from V_1 to V_2 using exactly one edge of C. Since the cycle must finish at x, it must cross from V_2 to V_1 and then possibly back and forth from V_1 to V_2 to V_1 and so on. However, each time L leaves V_1 for V_2 and returns to V_1, it uses two edges of C. So $|L \cap C|$ must be even. □

Corollary 5.5. *Every vector in the cycle subspace has an even number of edges in common with every vector in the cutset subspace.*

Corollary 5.6. *Every vector in the cutset subspace has an even number of edges in common with every vector in the cycle subspace.*

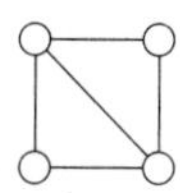

Fig. 5.5. Cycle and cutset spaces are not disjoint

Exercises 5.4

5.4.1 For the graph of Figure 5.1, the cycle subspace and the cutset subspace intersect in $\emptyset$. But this is not necessarily true for all graphs. Show that the graph in Figure 5.5 has a nonzero vector common to the cycle and the cutset subspaces. Name a property of the graph that is necessary for this to happen.

5.4.2 Consider the graphs in Figure 5.6(a) and (b).
- (i) Find the cycle subspace and the cutset subspace for each.
- (ii) For each of the graphs, show that the sum of the dimension of the cycle subspace and the dimension of the cutset subspace equals the dimension of the vector space $\mathcal{P}(E)$ (that is, the number of edges in the graph).

5.4.3 Repeat the above exercise for the graphs in Figures 5.6(c) and 5.5.

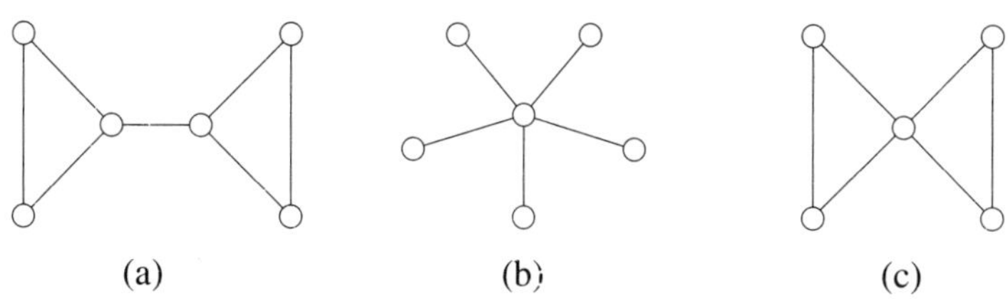

Fig. 5.6. More cycle and cutset subspaces

5.5 Bases and Spanning Trees

Bases for the cycle and cutset subspaces of a graph are closely related to spanning trees in the graph. For this reason we shall now discuss spanning trees. We shall make the convention that if T is a spanning tree of the graph G, then $\overline{T}$ will denote the complement of T in G (not in the corresponding complete graph).

Example. Consider the graph shown in Figure 5.7(a), which has five vertices and eight edges. Any of its spanning trees must contain four edges and the complement of a spanning tree must therefore contain $8 - 4 = 4$ edges. One particular spanning tree is shown in Figure 5.7(b).

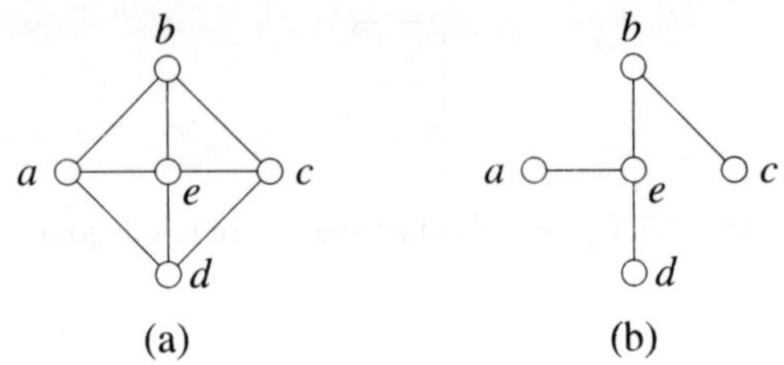

Fig. 5.7. A graph, and a spanning tree

If one edge of its complement is adjoined to a spanning tree, the resulting graph contains exactly one cycle. Figure 5.8 shows the spanning tree of Figure 5.7(b) with each edge of its complement adjoined in turn; in each case the cycle formed is shown below.

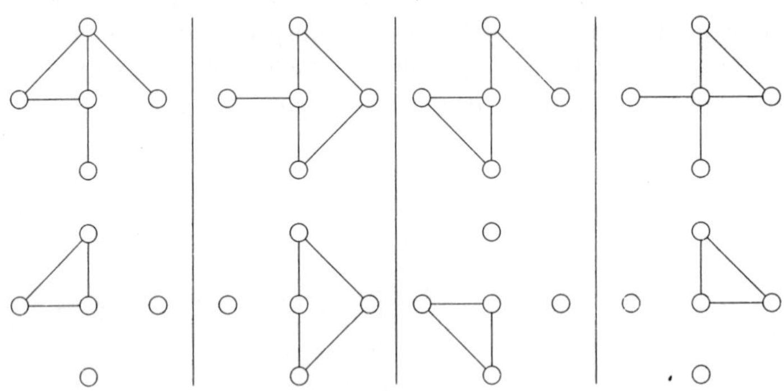

Fig. 5.8. Forming a basis for the cycle subspace

If, instead of adjoining an edge from the complement, one deletes an edge from a spanning tree, the tree becomes exactly two subtrees. The upper half of Figure 5.9

illustrates this: the spanning tree of Figure 5.7(b) is shown with each of its edges deleted in turn. Disconnecting the spanning tree in this way induces a partition of the vertex-set V of G into two disjoint subsets, namely those vertices incident with each of the two sub-trees, and this partition corresponds to a cutset. The lower part of Figure 5.9 shows the corresponding cutsets in each case.

In general, deleting the edge xy from a spanning tree of G partitions the vertex-set $V(G)$ as

$$V = V_1 \cup V_2,$$

where V_1 and V_2 are the vertex-sets of the two subtrees, and $x \in V_1$ and $y \in V_2$. Define C to be the set of edges of G such that

$$C = \{pq | p \in V_1, q \in V_2\}.$$

Then C, the set associated with the edge xy in the given spanning tree T, is a cutset.

If a connected graph G has e edges and v vertices, and if T is any one of its spanning trees, then T contains $v - 1$ edges and $\overline{T}$ contains $e - (v - 1) = e - v + 1$ edges. To each edge of $\overline{T}$ there corresponds a cycle formed by adjoining the edge to T; the set of these $e - v + 1$ cycles is called the *fundamental system of cycles* of G with respect to T, and any cycle in the system is called a *fundamental cycle.* To each edge of T, there corresponds a cutset inducing the same partition of the vertices of G as is induced by deleting the edge from T; the set of these $v - 1$ cutsets is called the *fundamental system of cutsets* of G with respect to T and any cutset in the fundamental system is called a *fundamental cutset.*

We shall determine the bases and characterize the members of the cycle and cutset subspaces of a graph in a series of three theorems. Each theorem has two parts, which are duals; in each case we shall prove the first part and leave the other as an exercise.

Theorem 5.7. *Suppose T is a spanning tree in a connected graph G.*

(i) *Say $L = \{a_1, a_2, \ldots, a_k\}$ is a fundamental cycle, where a_1 is an edge of $\overline{T}$ and $a_2, \ldots, a_k$ are edges of T. Then a_1 is contained in the fundamental cutset corresponding to a_i for each $i = 2, 3, \ldots, k$, and is in no other fundamental cutset.*

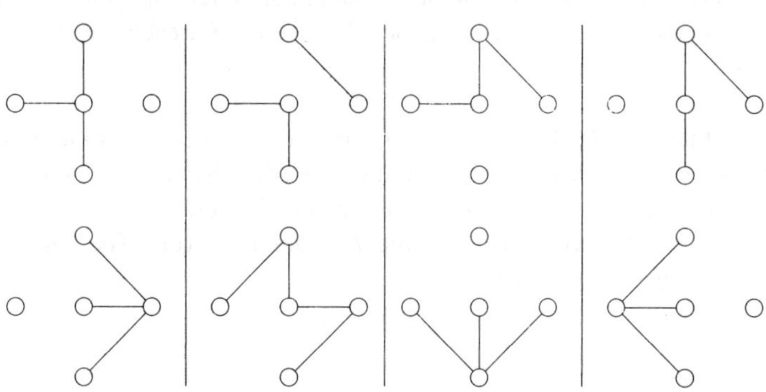

Fig. 5.9. Forming a basis for the cutset space

(ii) *Say* $C = \{a_1, a_2, \ldots, a_k\}$ *is a fundamental cutset, where* a_1 *is an edge of* T *and* $a_2, \ldots a_k$ *are edges of* $\overline{T}$. *Then* a_1 *is contained in the fundamental cycle corresponding to* a_i *for each* $i = 2, 3, \ldots, k$, *and in no other fundamental cycles.*

Proof of (i). Choose i such that $2 \le i \le k$ and let C_i be the fundamental cutset corresponding to the edge a_i. Now a_1 is the only edge of $\overline{T}$ in L and a_i is the only edge of T in C_i. Hence

$$\{a_i\} \subseteq L \cap C_i \subseteq \{a_1, a_i\}.$$

But by Lemma 5.4, $|L \cap C_i|$ is even, and therefore

$$L \cap C_i = \{a_1, a_i\},$$

so that $a_1 \in C_i$.

Now let a_{k+j}, for some $j \ge 1$, be an edge of T, and let C_{k+j} be the corresponding cutset. Since C_{k+j} contains no other edge of T, we have

$$\emptyset \subseteq L \cap C_{k+j} \subseteq \{a_1\}.$$

Again by Lemma 5.4, $|L \cap C_{k+j}|$ is even, and therefore

$$L \cap C_{k+j} = \emptyset,$$

whence $a_1 \notin C_{k+j}$. □

Theorem 5.8. *Let* T *be a spanning tree in the connected graph* G.

(i) *The fundamental system of cycles relative to* T *is a basis for the cycle subspace.*

(ii) *The fundamental system of cutsets relative to* T *is a basis for the cutset subspace.*

Proof of (i). Each cycle in the fundamental system with respect to T contains an edge of $\overline{T}$ not contained in any other fundamental cycle. Hence no fundamental cycle can be written as a linear combination of the other cycles in the fundamental system, so the fundamental system is an independent set of cycles.

To show that it is a spanning set, consider any cycle L in G. Say

$$L = \{a_1, a_2, \ldots, a_j, a_{j+1}, a_{j+2}, \ldots, a_k\},$$

where the edges in L have been labeled so that $a_1, a_2, \ldots, a_j$ are edges of $\overline{T}$ and a_{j+1}, $a_{j+2}, \ldots, a_k$ are edges of T. Now in the fundamental system of cycles with respect to T, we have a cycle, say L_i, containing the edge a_i of $\overline{T}$ for each $i = 1, 2, \ldots, j$. We define L' as

$$L' = L_1 + L_2 + \cdots + L_j$$

and we show that $L = L'$. Consider $L + L'$; it is a sum of cycles, so it belongs to the cycle subspace and is either a cycle or a union of edge-disjoint cycles. But the edges of $\overline{T}$ contained in L are precisely those contained in L'; hence $L + L'$ consists entirely of edges of T, which implies that the tree T contains a cycle. The only member of $\mathcal{F}(G)$ contained in T is $\emptyset$; hence

$$L + L' = \emptyset$$

which means that

$$L = L'$$

and the fundamental system is a spanning set. □

Corollary 5.9. *The dimension of the cycle subspace is* $e - v + 1$.

Corollary 5.10. *The dimension of the cutset subspace is* $v - 1$.

Theorem 5.11. *Let G be a connected graph.*

(i) *A set of edges is a vector in the cycle subspace if and only if it has an even number of edges in common with every vector in the cutset subspace.*

(ii) *A set of edges is a vector in the cutset subspace if and only if it has an even number of edges in common with every vector in the cycle subspace.*

Proof of (i). By Lemma 5.4 and its corollaries, it is sufficient to choose a set L of edges $a_1, \ldots, a_k$ that has an even number of edges in common with every vector in the cutset subspace and to show that L belongs to the cycle subspace. We choose an arbitrary spanning tree T of G and list the edges of L, renumbering if necessary, so that

$$L = \{a_1, a_2, \ldots, a_j, a_{j+1}, a_{j+2}, \ldots, a_k\}$$

where $a_1, a_2, \ldots, a_j$ are edges of $\overline{T}$ and $a_{j+1}, a_{j+2}, \ldots, a_k$ are edges of T. Again let L_i be the fundamental cycle corresponding to the edge a_i of $\overline{T}$ for $i = 1, 2, \ldots, j$, and define L' by

$$L' = L_1 + L_2 + \cdots + L_j.$$

Consider $L + L'$. Since L' belongs to the cycle subspace, it has an even number of edges in common with every vector of the cutset subspace. So does L, by hypothesis, and hence so does $L + L'$.

Now $L + L'$ contains only edges of T. Let a be one such edge of T and let C be the fundamental cutset containing a. Since C contains no other edge of T, we have

$$(L + L') \cap C = \{a\},$$

so that $|(L + L') \cap C|$ is odd, which is impossible. Hence no such edge a of T can exist, which implies that

$$L + L' = \emptyset,$$

so

$$L = L',$$

and L belongs to the cycle subspace. □

Exercises 5.5

.5.1 Prove Theorem 5.7(ii).

.5.2 Prove Theorem 5.8(ii).

.5.3 Prove Theorem 5.11(ii).

.5.4 The Petersen graph, P, was mentioned in Section 2.1 and is shown in Figure 2.4.

(i) Find a spanning tree T in P.

(ii) Describe the fundamental system of cycles with respect to T.

(iii) Describe the fundamental system of cutsets with respect to T.

(iv) Describe the intersection of the cycle and cutset subspaces of P.

5.5.5 Select spanning trees in K_4 and $K_{2,3}$. Find bases for the cycle and cutset subspaces in each case.

5.5.6 Suppose the vertices of K_v are labeled $1, 2, \ldots, v$.

(i) Let T be a spanning tree of K_v. If s_1 is the vertex of degree 1 in T with smallest label, let t_1 denote the vertex adjacent to s_1. Now let s_2 be the vertex of degree 1 in $T - s_1$ with smallest label, and let t_2 denote the vertex adjacent to s_2. Continue in this way, thus defining a sequence $(t_1, t_2, \ldots, t_{v-2})$. Show that different spanning trees give different sequences, and that if vertex x has degree $d_T(x)$ in T, then x appears $(d_T(x)-1)$ times in the sequence associated with T.

(ii) Let $(t_1, t_2, \ldots, t_{v-2})$ be a given $(v-2)$-sequence on $N = \{1, 2, \ldots, v\}$, and consider v vertices labeled 1 to v. Let s_1 be the smallest element of N outside the set $\{t_1, t_2, \ldots, t_{v-2}\}$; join vertex s_1 to vertex t_1. Now let s_2 be the smallest element of $N \backslash \{s_1\}$ outside the set $\{t_2, \ldots, t_{v-2}\}$; join s_2 to t_2. Continue in this way to get $v-2$ edges $s_1t_1, s_2t_2, \ldots, s_{v-2}t_{v-2}$. Adjoin a further edge connecting the two remaining vertices in $N \backslash \{s_1, s_2, \ldots, s_{v-2}\}$. Show that the graph so constructed is a spanning tree of K_v and that different sequences give different spanning trees.

(iii) Hence prove that $\tau(K_v) = v^{v-2}$.

6

Factorizations

6.1 Definitions; One-Factorizations

If G is any graph, then a *factor* or *spanning subgraph* of G is a subgraph with vertex-set $V(G)$. A *factorization* of G is a set of factors of G that are pairwise *edge-disjoint* — no two have a common edge — and whose union is all of G.

Every graph has a factorization, quite trivially: since G is a factor of itself, $\{G\}$ is a factorization of G. However, it is more interesting to consider factorizations in which the factors satisfy certain conditions. In particular a *one-factor* is a factor that is a regular graph of degree 1. In other words, a one-factor is a set of pairwise disjoint edges of G that between them contain every vertex. Similarly, a *two-factor* in a graph G is a union of disjoint cycles that together contain all vertices of G.

A *one-factorization* of G is a decomposition of the edge-set of G into edge-disjoint one-factors.

Example. We find the one-factors of the graph

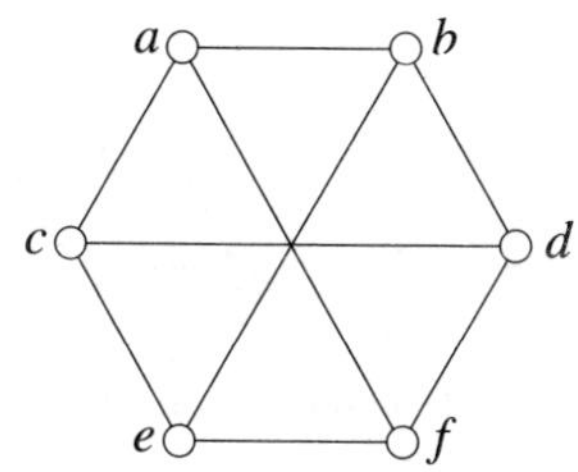

Each factor must contain one edge through a, which must be ab, ac or af. If it is ab, the rest of the factor consists of two edges covering $\{c, d, e, f\}$, and the only possibilities are cd, ef and ce, df. So there are two factors containing ab. In the same way we find two factors containing ac, and two with af. The factors are

$$\begin{array}{lll} ab, cd, ef & ac, be, df & af, bd, de \\ ab, ce, df & ac, bd, cf & af, be, cd. \end{array} \tag{6.1}$$

To find one-factorizations that include ab, cd, ef, we check through this list to find all factors that are edge-disjoint from ab, cd, ef. In this small example it is easy to see that there are exactly two such factors, and together with ab, cd, ef they form a one-factorization that is shown in the first line of (6.1). There is one other one-factorization, shown in the second line.

This graph is one of the two nonisomorphic regular graphs of degree 3 on six vertices. The one-factors and one-factorizations of the other such graph are discussed in Exercise 6.1.3.

Another approach to the study of one-factors is through matchings. A *matching* between sets X and Y is a set of ordered pairs, one member from each of the two sets, such that no element is repeated. Such a matching is a set of disjoint edges of the $K_{m,n}$ with vertex-sets X and Y. The matching is called *perfect* if every member occurs exactly once, so a perfect matching is a one-factor in a complete bipartite graph. One can then define a matching in any graph to be a set of disjoint edges in that graph; in this terminology "perfect matching" is just another phrase for "one-factor."

It is sometimes useful to impose an ordering on the set of one-factors in a one-factorization, or a direction on the edges of the underlying graph. In those cases the one-factorization will be called *ordered* or *oriented* respectively.

Not every graph has a one-factor. In Section 6.3 we shall give a necessary and sufficient condition for the existence of a one-factor in a general graph. For the moment, we note the obvious necessary condition that a graph with a one-factor must have an even number of vertices. However, this is not sufficient; Figure 6.1 shows a 16-vertex graph without a one-factor (see Exercise 6.1.2).

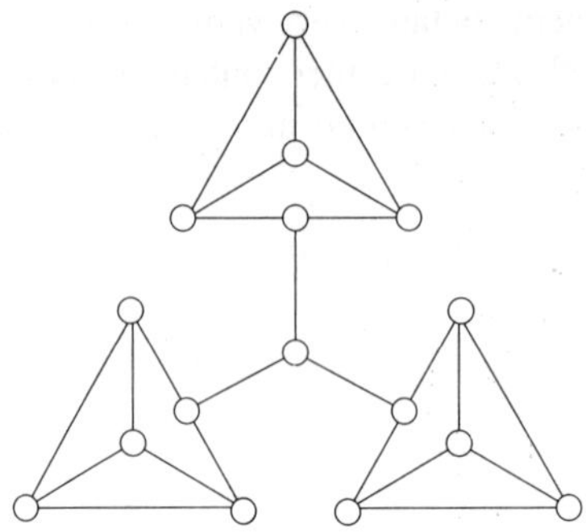

Fig. 6.1. N, the smallest cubic graph without a one-factor

In order to have a one-factorization, a graph not only needs an even number of vertices, but it must also be regular: if G decomposes into d disjoint one-factors, then every vertex of G must lie on precisely d edges. However, the following theorem shows that these conditions are not sufficient.

Theorem 6.1. *A regular graph with a bridge cannot have a one-factorization (except for the trivial case where the graph is itself a one-factor).*

Proof. Consider a regular graph G of degree d, $d > 1$, with a bridge $t = xy$; in $G - e$, label the component that contains x as E and label the component that contains y as F. The fact that t is a bridge implies that E and F are distinct. Suppose G is the edge-disjoint union of d one-factors, $G_1, G_2, \ldots, G_d$; and say G_1 is the factor that contains t. Now t is the only edge that joins a vertex of E to a vertex of F, so every edge of G_2 with one endpoint in E has its other endpoint in E. So G_2 contains an even number of vertices of E. Since G_2 contains every vertex of the original graph, it contains every vertex of E; so E must have an even number of vertices.

On the other hand, consider $G_1 \backslash E$. This contains a number of edges and the isolated vertex y, since t is in G_1. So $G_1 \backslash E$ has an odd number of vertices, and accordingly E has an odd number of vertices. We have a contradiction. □

This theorem can be used, for example, to show that the graph M of Figure 6.2 has no one-factorization, although it is regular and possesses the one-factor $\{ac, be, dg, fi, hj\}$. However, it clearly does not tell the whole story: the Petersen graph P (see Figure 2.4) has no one-factorization, but it also contains no bridge. The Petersen graph *does* contain a one-factor, however. In fact Petersen [94] showed that every bridgeless cubic graph contains a one-factor. We shall give a proof in Section 6.3.

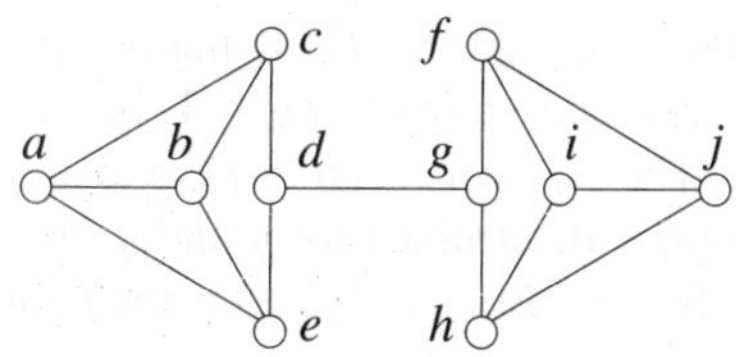

Fig. 6.2. M, the smallest cubic graph without a one-factorization

If the degree increases with the number of vertices, the situation is different. It has been conjectured that a regular graph with $2n$ vertices and degree greater than n will always have a one-factorization, this has only been proved in a very few cases, such as degree $2n - 4$, degree $2n - 5$, and degree at least $12n/7$ (for further details see [103, 25]). On the other hand, one can find regular graphs with degree near to half the number of vertices that do not have one-factorizations.

However, we can prove the existence of one-factorizations in many classes of graphs. Of basic importance are the complete graphs. There are many one-factorizations of K_{2n}. We present one that is usually called $\mathcal{GK}_{2n}$. To understand the construction, look at Figure 6.3. This represents a factor that we shall call F_0. To construct the factor F_1, rotate the diagram through a $(2n - 1)$-th part of a full revolution. Similar rotations provide $F_2, F_3, \ldots, F_{2n-2}$.

Theorem 6.2. *The complete graph K_{2n} has a one-factorization for all n.*

Proof. We label the vertices of K_{2n} as $x_\infty, x_0, x_1, x_2, \ldots, x_{2n-2}$. The spanning subgraph F_i is defined to consist of the edges

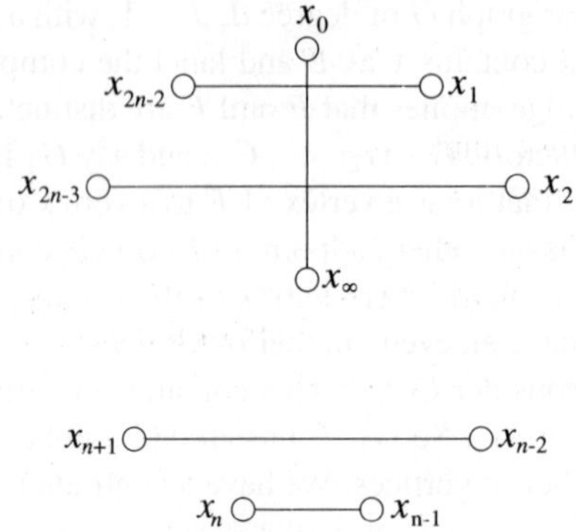

Fig. 6.3. The factor F_0 in $\mathcal{GK}_{2n}$

$$x_\infty x_i,\ x_{i+1}x_{i-1},\ \ldots,\ x_{i+j}x_{i-j},\ \ldots,\ x_{i+n-1}x_{i-n+1} \tag{6.2}$$

where the subscripts other than ∞ are treated as integers modulo $2n-1$. Then F_i is a one-factor: every vertex appears in the list (6.2) exactly once. We prove that $\{F_0, F_1, \ldots, F_{2n-2}\}$ form a one-factorization. First, observe that every edge involving x_∞ arises precisely once: $x_\infty x_i$ is in F_i. If neither p nor q is ∞, then we can write $p+q = 2i$ in the arithmetic modulo $2n-1$, because either $p+q$ is even or $p+q+2n-1$ is even. Then $q = i-(p-i)$, and x_px_q is $x_{i+j}x_{i-j}$ in the case $j = p-i$. Since i is uniquely determined by p and q, this means that x_px_q belongs to precisely one of the F_i. So $\{F_0, F_1, \ldots, F_{2n-2}\}$ is the required one-factorization. $\square$

The complete bipartite graph $K_{n,n}$ is easily shown to have a one-factorization. If $K_{n,n}$ is defined to have vertex-set $\{1, 2, \ldots, 2n\}$ and edge-set $\{(x, y): 1 \le x \le n,\ n+1 \le y \le 2n\}$, then the factors $F_1, F_2, \ldots, F_n$, defined by

$$F_i = \{(x, x+n+i) : 1 \le x \le n\}$$

(where $x+n+i$ is reduced modulo n to lie between $n+1$ and $2n$), form a one-factorization. This will be called the *standard factorization* of $K_{n,n}$.

Another important case is the family of cycles C_n: these have a one-factorization if and only if n is even. This fact will be useful — for example, one common way to find a one-factorization of a cubic graph is to find a spanning subgraph that is a union of disjoint even cycles: a Hamilton cycle will suffice. The complement of this subgraph is a one-factor, so the graph has a one-factorization. Reversing this reasoning, the union of two disjoint one-factors is always a union of disjoint even cycles; if the one-factors are not disjoint, the union consists of some even cycles and some isolated edges (the common edges of the two factors).

The use of unions in the preceding paragraph can be generalized. If G and H both have one-factorizations, then so does $G \oplus H$, the factorization being formed by listing all factors in the factorizations of each of the component graphs. At the other extreme, if G has a one-factorization, then so does nG. However, care must be exercised. If G

and H have some common edges, nothing can be deduced about the factorization of $G \cup H$ from factorizations of G and H.

A number of other types of factorization have been studied. One interesting problem is to decompose graphs into Hamilton cycles. Since each Hamilton cycle can be decomposed into two one-factors, such a Hamiltonian factorization gives rise to a special type of one-factorization. We give only the most basic result.

Theorem 6.3. *If v is odd, then K_v can be factored into $\frac{v-1}{2}$ Hamilton cycles. If v is even, then K_v can be factored into $\frac{v}{2} - 1$ Hamilton cycles and a one-factor.*

Proof. First, suppose v is odd: say $v = 2n + 1$. If K_v has vertices $0, 1, 2, \ldots, 2n$, then a suitable factorization is $Z_1, Z_2, \ldots, Z_n$, where

$$Z_i = (0, i, i+1, i-1, i+2, i-2, \ldots, i+j, i-j, \ldots, i+n, 0).$$

(If necessary, reduce integers modulo $2n$ to the range $\{1, 2, \ldots, 2n\}$.)

In the case where v is even, let us write $v = 2n + 2$. We construct an example for the K_v with vertices $\infty, 0, 1, 2, \ldots, 2n$. The factors are the cycles $Z_1, Z_2, \ldots, Z_n$, where

$$Z_i = (\infty, i, i-1, i+1, i-2, i+2, \ldots, i+n-1, \infty),$$

and the one-factor

$$(\infty, 0), (1, 2n), (2, 2n-1), \ldots, (n, n+1). \qquad \square$$

Exercises 6.1

6.1.1 Verify that every connected graph on four vertices, other than $K_{1,3}$, contains a one-factor.

6.1.2 Prove that the graph N of Figure 6.1 contains no one-factor.

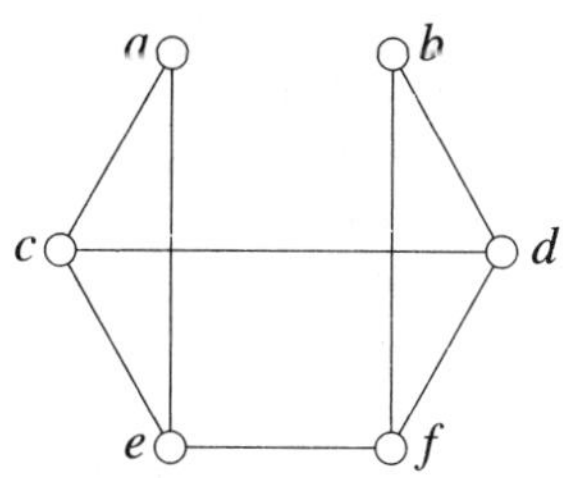

Fig. 6.4. Graph for Exercise 6.1.3

6.1.3 Find all one-factors and one-factorizations in the graph shown in Figure 6.4. Verify that it contains a one-factor that does not belong to any one-factorization.

6.1.4 Verify that the Petersen graph has no one-factorization.

6.1.5 Find all one-factors in the following graphs.

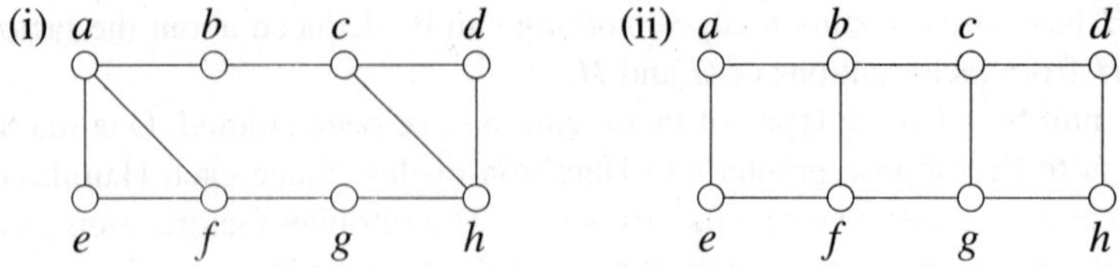

6.1.6 Repeat the preceding exercise for the following graphs.

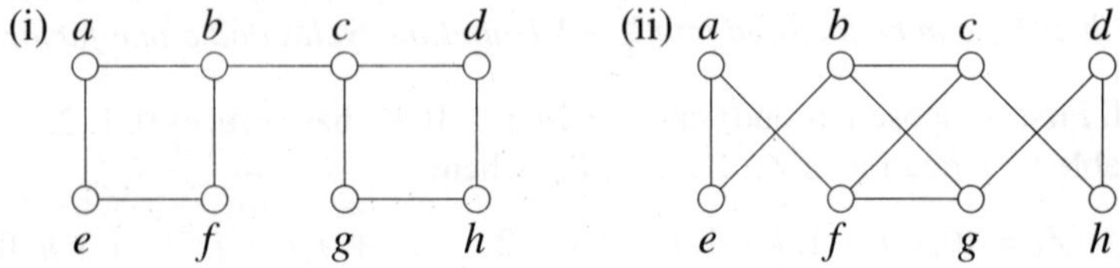

6.1.7 The n-cube Q_n is defined as follows. Q_1 consists of two vertices and one edge; Q_2 is the cycle C_4; in general Q_n is formed by taking two copies of Q_{n-1} and joining each vertex in one copy to the corresponding vertex in the other copy.

(i) How many vertices does Q_n have?

(ii) Prove that Q_n is regular. What is the degree?

(iii) Prove that Q_n has a one-factorization.

(iv) Prove that Q_n has a Hamilton cycle, for $n > 1$.

H6.1.8 Prove that a tree contains at most one one-factor.

6.1.9 What is the number of distinct one-factors in K_{2n}?

HA6.1.10 Suppose G is a connected graph with an even number of vertices, and no *induced* subgraph of G is a star $K_{1,3}$. Prove that G has a one-factor. [112, 113]

6.1.11 In each case either prove the statement or find a counterexample.

(i) If G has a one-factor, then so does its complement $\overline{G}$.

(ii) If G has a one-factorization, then so does its complement $\overline{G}$.

6.1.12 G is a regular graph and every edge of G is contained in at least one one-factor of G. Does G necessarily have a one-factorization? Prove that it does, or provide a counterexample.

6.1.13 We say a two-factor of a graph is of *type* $[a, b, \ldots, c]$ if it consists of one cycle each of lengths $a, b, \ldots, c$. (Repetitions in the list correspond to several cycles of the same length.) For example, a graph on 6 vertices could perhaps have two-factors of types [6] or [3, 3], but no others; for 7 vertices, only [7] and [4, 3]

(i) What are the possible types of two-factors of graphs on 8 vertices?

H(ii) The graph G is formed from K_6 by deleting the edges of a one-factor. Show that G has two-factorizations of the following kinds:

- two factors of type [6];
- one factor of type [6] and one of type [3, 3];

but it has no two-factorization consisting of two factors of type [3, 3].

(iii) Show that K_7 has two-factorizations of the following kinds:

- three factors of type [7];
- three factors of type [4, 3];
- two factors of type [7] and one of type [4, 3];

but it has no two-factorization consisting of two factors of type [4, 3] and one of type [7].

6.2 Tournament Applications of One-Factorizations

Suppose several baseball teams play against each other in a league. The competition can be represented by a graph with the teams as vertices and with an edge xy representing a game between teams x and y. We shall refer to such a league — where two participants meet in each game — as a *tournament*. (The word "tournament" is also used for the directed graphs derived from this model by directing the edge from winner to loser. We shall consider these graphs in Section 11.2.)

Sometimes multiple edges will be necessary; sometimes two teams do not meet. The particular case where every pair of teams plays exactly once is called a *round robin tournament*, and the underlying graph is complete.

A very common situation is when several matches must be played simultaneously. In the extreme case, when every team must compete at once, the set of games held at one time is called a *round*. Clearly the games that form a round form a one-factor in the underlying graph. If a round robin tournament for $2n$ teams is to be played in the minimum number of sessions, we require a one-factorization of K_{2n}, together with an ordering of the factors (this ordering is sometimes irrelevant). If there are $2n-1$ teams, the relevant structure is a near-one-factorization of K_{2n-1}. In each case the (ordered) factorization is called the *schedule* of the tournament.

In many sports a team owns, or regularly plays in, one specific stadium or arena. We shall refer to this as the team's "home field." When the game is played at a team's home field, we refer to that team as the "home team" and the other as the "away team." Often the home team is at an advantage; and more importantly, the home team may receive a greater share of the admission fees. So it is usual for home and away teams to be designated in each match. We use the term *home-and-away schedule* (or just *schedule*) to refer to a round robin tournament schedule in which one team in each game is labeled the home team and one the away team. Since this could be represented by orienting the edges in the one-factors, a home-and-away schedule is equivalent to an oriented one-factorization. It is very common to conduct a double round robin, in which every team plays every other team twice. If the two matches for each pair of teams are arranged so that the home team in one is the away team in the other, we shall say the schedule and the corresponding oriented one-factorization of $2K_{2n}$ are *balanced*.

For various reasons one often prefers a schedule in which runs of successive away games and runs of successive home games do not occur (although there are exceptions: an east coast baseball team, for example, might want to make a tour of the west, and play several away games in succession). We shall define a *break* in a schedule to be a pair of successive rounds in which a given team is at home, or away, in both rounds. A schedule is *ideal* for a team if it contains no break for that team. Oriented factorizations are called ideal for a vertex if and only if the corresponding schedules are ideal for the corresponding team.

Theorem 6.4. [130] *Any schedule for 2n teams is ideal for at most two teams.*

Proof. For a given team x, define its *ground vector* v^x to have $v^x_j = 1$ if x is home in round j and $vv^x_j = 0$ if x is away in round j. If teams x and y play in round j, then $v^x_j \neq v^x_j$, so different teams have different ground vectors. But the ground vector corresponding to an ideal schedule must consist of alternating zeroes and ones. There are only two such vectors possible, so the schedule can be ideal for at most two teams. □

The following theorem shows that the theoretically best possible case can be attained.

Theorem 6.5. [130] *There is an oriented one-factorization of K_{2n} with exactly $2n-2$ breaks.*

Proof. We orient the one-factorization $\mathcal{P} = \{P_1, P_2, \ldots, P_{2n-1}\}$ based on the set $\{\infty\} \cup \mathbb{Z}_{2n-1}$, defined by

$$P_k = \{(\infty, k)\} \cup \{(k+i, k-i) : 1 \le i \le n-1\}. \tag{6.3}$$

Edge (∞, k) is oriented with ∞ at home when k is even and k at home when k is odd. Edge $(k+i, k-i)$ is oriented with $k-i$ at home when i is even and $k+i$ at home when i is odd.

It is clear that ∞ has no breaks. For team x, where x is in $\mathbb{Z}_{2n-1}$, we can write $x = k + (x-k) = k - (k-x)$. The way that x occurs in the representation (6.3) will be: as x when $k = x$, as $k+(x-k)$ when $1 \le x-k \le n-1$, and as $k-(k-x)$ otherwise. The rounds other than P_x where x is at home are the rounds k where $x-k$ is odd and $1 \le x-k \le n-1$, and the rounds k where $k-x$ is even and $1 \le k-x \le n-1$. It is easy to check that factors P_{2j-1} and P_{2j} form a break for symbols $2j-1$ and $2j$, and these are the only breaks. □

Exercises 6.2

A6.2.1 Two chess clubs, each of n members, wish to play a match over n nights. Each player will play one game per night against a different opponent from the other team. What mathematical structure is used? Give an example for $n = 4$.

6.2.2 $v = 3n$ card players wish to play for several nights. Each night, players sit three at a table, and play together for the full session. No two players are to play together (at the same table) on two nights. The players wish to play for as many nights as possible.

A(i) Describe the problem in terms of graph factorizations.

(ii) Prove that no more than $\lfloor \frac{v-1}{2} \rfloor$ nights of play are possible.

(iii) Show that the maximum can be achieved by nine players.

(iv) Show that twelve players cannot achieve five nights of play.

If $v \equiv 3 \bmod 6$, an optimal solution for v players is called a *Kirkman triple system*, and one exists for all such v. An optimal solution when $v \equiv 0 \bmod 6$ is called a *nearly Kirkman triple system*, and one exists for all v except 6 and 12. See, for example, [80, 123].

6.3 A General Existence Theorem

If W is any subset of the vertex-set $V(G)$ of a graph or multigraph G, we write $G - W$ to denote the graph constructed by deleting from G all vertices in W and all edges touching them. One can discuss the components of $G - W$; they are either *odd* (have an odd number of vertices) or *even*. Let $\Phi_G(W)$ denote the number of odd components of $G - W$.

Theorem 6.6. [115] *G contains a one-factor if and only if*

$$\Phi_G(W) \leq |W| \text{ whenever } W \subset V(G). \tag{6.4}$$

Proof (after [83].) First, suppose G contains a one-factor F. Select a subset W of $V(G)$, and suppose $\Phi_G(W) = k$; label the odd components of $G - W$ as $G_1, G_2, \ldots, G_k$. Since G_i has an odd number of vertices, $G_i \backslash F$ cannot consist of $\frac{1}{2}|G_i|$ edges; there must be at least one vertex, say x, of G_i that is joined by F to a vertex y_i that is not in G_i. Since components are connected, y_i must be in W. So W contains at least the k vertices $y_1, y_2, \ldots, y_k$, and

$$k = \Phi_G(W) \leq |W|.$$

So the condition is necessary.

To prove sufficiency, we assume the existence of a graph G that satisfies (6.4) but has no one-factor; a contradiction will be obtained. If such a G exists, we could continue to add edges until we reached a maximal graph G^* such that no further edge could be added without introducing a one-factor. (Such a maximum exists: it follows from the case $W = \emptyset$ that G has an even number of vertices; if we could add edges indefinitely, eventually an even-order complete graph would be reached.)

Moreover the graph G^* also satisfies (6.4) — adding edges may reduce the number of odd components, but it cannot increase them. So there is no loss of generality in assuming that G is already maximal. We write U for the set of all vertices of degree $|V(G)|-1$ in G. If $U = V(G)$, then G is complete, and has a one-factor. So $U \neq V(G)$. Every member of U is adjacent to every other vertex of G. We first show that every component of $G - U$ is a complete graph. Let G_1 be a component of $G - U$ that is not complete. Not all vertices of G_1 are joined, but G_1 is connected, so there must exist two vertices at distance 2 in G_1, say x and z; let y be a vertex adjacent to both. Since $d(y) \neq |V(G)| - 1$ there must be some vertex t of G that is not adjacent to y, and t is in $V(G - U)$ (since every vertex, y included, is adjacent to every member of U). So $G - U$ contains the configuration shown in Figure 6.5 (a dotted line means "no edge").

Since G is maximal, $G + xz$ has a one-factor — say F_1 — and $G + yt$ also has a one-factor — call it F_2. Clearly xz belongs to F_1 and yt belongs to F_2; and also xz is not in F_2.

Consider the graph $F_1 \cup F_2$; let H be the component that contains xz. As xz is not in F_2, H is a cycle of even length made up of alternate edges from F_1 and F_2. Either yt belongs to H or else yt belongs to another even cycle of alternate edges. These two cases are illustrated in Figure 6.6.

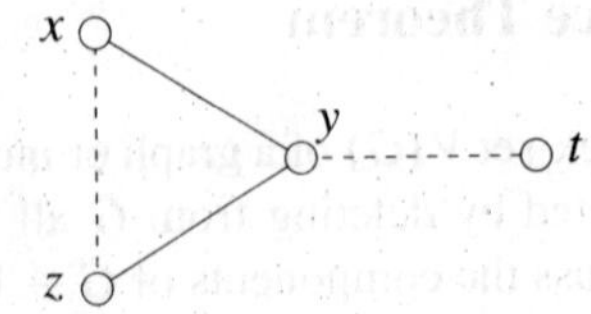

Fig. 6.5. A subgraph arising in the proof of Theorem 6.6

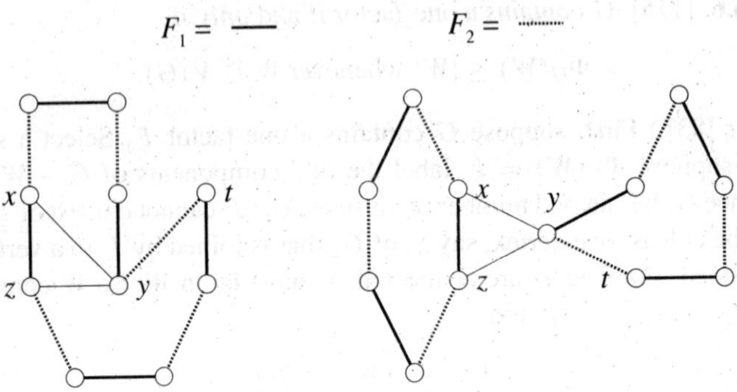

(a) (b)

Fig. 6.6. Two cases needed in Theorem 6.6

In Case (a) we can assume that vertices x, y, t and z appear in that order along the cycle H, as shown in the figure (if not, interchange x and z). Then we can construct a one-factor of G as follows: take the edges of F_1 in the section $y, t, \ldots, z$ of H, the edges of F_2 in the rest of G, and yz. In Case (b) we can construct a one-factor in G by taking the edges of F_1 in H and the edges of F_2 in $G - H$. So in both cases we have a contradiction. So G_1 cannot exist — every component of $G - U$ must be complete.

Now $\Phi_G(U) \leq |U|$, so $G - U$ has at most $|U|$ odd components. We associate with each odd component G_i of $G - U$ a different member u_i of U. We then construct a one-factor of G as follows. From every even component of $G - U$ select a one-factor (possible because the components are complete graphs); for each odd component G_i, select a one-factor of $G_i + u_i$ (again, $G_i + u_i$ is an even-order complete graph). Since G has an even number of vertices, there will be an even number of vertices left over, all in U; as they are all connected, a one-factor can be chosen from them. The totality is a one-factor in G, contradicting the hypothesis. □

Suppose G has v vertices and suppose the deletion of the w-set W of vertices results in a graph with k odd components; then $v - w \equiv k \pmod 2$. If v is even, then $w \equiv k \pmod 2$, and $k > w$ will imply that $k \geq w + 2$. So we have a slight improvement on Theorem 6.6.

Theorem 6.7. [121] *If the graph or multigraph G has an even number of vertices, then G has no one-factor if and only if there is some w-set W of vertices such that $G - W$ has at least $w + 2$ odd components.*

As an application of Theorem 6.6 we prove the following result.

Theorem 6.8. [31] *If n is even, then any regular graph of degree $n - 1$ on $2n$ vertices has a one-factor.*

Proof. If $n = 2$ or $n = 4$, then the result is easily checked by considering all cases. So we assume G is a regular graph of degree $n - 1$ on $2n$ vertices, where n is even and $n > 4$, and W is any set of w vertices of G, and we prove that the graph $G - W$ has at most w odd components.

If $w \geq n$, then $G - W$ has at most w vertices, so it has at most w components. If $w = 0$, then G has an odd component, which is impossible since w has odd degree. So we assume $1 \leq w \leq n - 1$.

The deletion of W cannot reduce the degree of a vertex of G by more than w, so every vertex of $G - W$ has degree at least $n - 1 - w$, and each component has at least $n - w$ vertices. If there are $w + 1$ or more components, then $G - W$ has at least $(w + 1)(n - w)$ vertices, and

$$(w + 1)(n - w) \leq 2n - w,$$

which simplifies to

$$w^2 - wn + n \geq 0.$$

For fixed n, this is a quadratic inequality which will have the solution

$$w \leq w_1 \text{ or } w \geq w_2$$

where w_1 and w_2 are the roots of $w^2 - wn + n = 0$: that is,

$$w_1 = \tfrac{1}{2}(n - \sqrt{n^2 - 4n}), \quad w_2 = \tfrac{1}{2}(n + \sqrt{n^2 - 4n}).$$

Now when $n > 4$,

$$n - 3 < \sqrt{n^2 - 4n} < n - 2,$$

so

$$1 < w_1 < 2, \quad n - 2 < w_2 < n - 1,$$

and the only integer values in the range $1 \leq w \leq n - 1$ that satisfy the inequality are $w = 1$ and $w = n - 1$.

If $w = 1$, then $G - W$ has every vertex of degree at least $n - 2$, so every component has at least $n - 1$ vertices. The only possible case is two components, one with $n - 1$ vertices and one with n. Only one is odd, so $G - W$ has at most w $(= 1)$ odd components.

If $w = n - 1$, then $G - W$ has $n + 1$ vertices. To get $w + 1$ odd components, the only possibility is $n + 1$ components, each of one vertex. So $G - W$ is empty: deletion of W has removed all $n(n - 1)$ edges of G. Since each vertex of W had degree $n - 1$, at most $(n - 1)^2$ edges can have been removed, which is a contradiction. □

Another application is Petersen's theorem that every bridgeless cubic graph contains a one-factor, which we promised in Section 6.1. In fact we prove the more general result, due to Schönberger [108], that there is a factor containing any specified edge. This obviously has Petersen's theorem as a corollary.

If W is any set of vertices of a graph G, define $z_G(W)$ to be the number of edges of G with precisely one endpoint in W.

Lemma 6.9. *If G is a regular graph of degree d and S is any set of vertices of G, then*

$$d|S| = 2e(\langle S\rangle) + z_G(S). \qquad (6.5)$$

Proof. The sum of the degrees of vertices in S, and $2e(\langle S\rangle)$ is the contribution from edges with both endpoints in S; (6.5) follows. □

Corollary 6.10. *If G is a bridgeless cubic graph and S is a set of vertices of G whose order is odd, then $z_G(S) \geq 3$.*

Proof. Apply (6.5). Since d and $|S|$ are both odd, $z_G(S)$ must be odd also. If $z_G(S) = 1$, then the unique edge joining S to $G - S$ would be a bridge, which is impossible in G. So $z_G(S) \geq 3$. □

Theorem 6.11. [108] *If G is a bridgeless cubic graph and e is any edge of G, then G has a one-factor that contains e.*

Proof. Suppose $e = xy$. We prove that $H = G - \{x, y\}$ has a one-factor. Then this factor together with xy is the required one-factor.

Suppose H has no one-factor. From Theorem 6.7, $\Phi_H(X) \geq |X| + 2$ for some subset X of $V(H)$. So

$$\Phi_G(W) \geq |W|$$

where W is the subset $X \cup \{x, y\}$ of $V(G)$. $\langle W\rangle$ contains at least one edge, xy, so from (6.5)

$$z_G(W) \leq 3|W| - 2. \qquad (6.6)$$

Now if S is any odd component of $G - W$, then $z_G(S) \geq 3$, and the three edges coming from S must all have their other endpoints in W, so the number of edges into W from outside is at least 3 times the number of such subsets:

$$z_G(W) \geq 3\Phi_G(W) \geq 3|W|, \qquad (6.7)$$

so (6.6) and (6.7) together give a contradiction. □

Exercises 6.3

A6.3.1 Does Theorem 6.8 apply to multigraphs?

6.3.2 Prove that there are exactly three regular graphs of degree 3 on eight vertices (up to isomorphism), and that each has a one-factor.

HA6.3.3 Prove that if a cubic graph has fewer than three bridges, then it has a one-factor. [108]

6.3.4 Suppose G is a regular graph of degree d that has a one-factorization. Prove that $z_G(W) \geq d-1$ for every odd-order subset W of $V(G)$. Is this necessary condition sufficient?

6.3.5 G is n-connected, regular of degree n, and has an even number of vertices. Prove that G has a one-factor.

6.3.6 Suppose G is a graph with $2n$ vertices and has minimum degree $\delta < n$. Prove that if

$$2n > \binom{\delta}{2} + \binom{2n-2\delta-1}{2} + \delta(2n-\delta),$$

then G has a one-factor. [17]

6.3.7 G is a cubic graph without a bridge. Prove that if xy is any edge of G, then G contains a one-factor that does *not* include xy.

6.3.8 Suppose G is a r-connected graph ($r \geq 1$) with an even number of vertices, and no *induced* subgraph of G is a star $K_{1,r+1}$. Prove that G has a one-factor. [113] (Compare with Exercise 6.1.10.)

6.4 Graphs Without One-Factors

Suppose G is a regular graph of degree d and suppose $G - W$ has a component with p vertices, where p is no greater than d. The number of edges within the component is at most $\frac{1}{2}p(p-1)$. This means that the sum of the degrees of these p vertices in $G - W$ is at most $p(p-1)$. But in G each vertex has degree d, so the sum of the degrees of the p vertices is pd, whence the number of edges joining the component to W must be at least $pd - p(p-1)$. For fixed d and for integer p satisfying $1 \leq p \leq d$, this function has minimum value d (achieved at $p = 1$ and $p = d$). So any odd component with d or fewer vertices is joined to W by d or more edges.

We now assume that G is a regular graph of degree d on v vertices, where v is even. If G has no one-factor, then by Theorem 6.7 there is a set W of w vertices whose deletion leaves at least $w+2$ odd components. We call a component of $G - W$ *large* if it has more than d vertices, and *small* otherwise. The numbers of large and small components of $G - W$ are α_W and β_W, or simply α and β, respectively. Clearly

$$\alpha + \beta \geq w + 2. \tag{6.8}$$

There are at least d edges of G joining each small component of $G - W$ to W, and at least one per large component, so there are at least $\alpha + d\beta$ edges attached to the vertices of W; by regularity we have

$$\alpha + d\beta \leq wd. \tag{6.9}$$

Each large component has at least $d+1$ vertices if d is even, and at least $d+2$ if v is odd, so

$$v \geq w + (d+1)\alpha + \beta \text{ if } d \text{ is even}, \tag{6.10}$$

$$v \geq w + (d+2)\alpha + \beta \text{ if } d \text{ is odd}. \tag{6.11}$$

Since α is nonnegative, (6.9) yields $\beta \le w$, so from (6.8) we have $\alpha \ge 2$; but applying this to (6.9) again we get $\beta < w$, so from (6.8) we have $\alpha \ge 3$. So from (6.10) and (6.11) we see that if $w \ge 1$, then

$$v \ge 3d + 4 \text{ if } d \text{ is even,} \tag{6.12}$$

$$v \ge 3d + 7 \text{ if } d \text{ is odd.} \tag{6.13}$$

In the particular case $d = 4$, the bound in (6.12) cannot be attained. Suppose P is an odd component of $G - W$ with p vertices. Then the sum of the vertices in G of members of P is $4p$, which is even. On the other hand, if there are r edges joining W to P in G and s edges internal to W, the sum of the degrees is $r + 2s$. So r is even. So there are at least two edges from each large component to W, and (6.9) can be strengthened to

$$2\alpha + 4\beta \le 4t,$$

whence $2\beta \le 4k - (2\alpha + 2\beta)$; substituting from (6.8) we get

$$2\beta \le 2k - 4$$

and $\alpha \ge 4$. So (6.10) yields

$$v \ge 2 + 5{\cdot}4 = 22.$$

Summarizing this discussion, we have:

Theorem 6.12. [121] *If a regular graph G with an even number v of vertices and with degree d has no one-factor and no odd component, then*

$$\begin{array}{lll} v \ge 3d + 7 & \textit{if} & d \textit{ is odd, } d \ge 3; \\ v \ge 3d + 4 & \textit{if} & d \textit{ is even, } d \ge 6; \\ v \ge 22 & \textit{if} & d = 4. \end{array} \tag{6.14}$$

The condition of "no odd component" is equivalent to the assumption that $w \ge 1$. The cases $d = 1$ and $d = 2$ are omitted, but in fact every graph with these degrees that satisfies the conditions has a one-factorization.

It follows from the next result that Theorem 6.12 is best possible.

Theorem 6.13. *If v is even and is at least as large as the bound of Theorem 6.12, then there is a regular graph of the relevant degree on v vertices that has no one-factor.*

Proof. We use two families of graphs. The graph $G_1(h, k, s)$ has $2s + 1$ vertices, and is formed from K_{2s+1} as follows. First factor K_{2s+1} into Hamilton cycles, as in Theorem 6.3. Then take the union of $h - 1$ of those cycles. Finally, take another of the cycles, delete k disjoint edges from it, and adjoin the remaining edges to the union. This construction is possible whenever $0 < h \le s$ and $0 \le k \le s$. If a graph has $2s + 1$ vertices, of which $2k$ have degree $2h - 1$ and the rest have degree $2h$, let us call it a 1-(h, k, s) graph; for our purposes, the essential property of $G_1(h, k, s)$ is that it is a 1-(h, k, s) graph.

The graph $G_2(h, k, s)$ has $2s + 1$ vertices. We construct it by taking a Hamilton cycle decomposition of K_{2s+1}, and taking the union of h of the cycles. Then another cycle is chosen from the factorization; from it are deleted paths with $2k+1$ vertices and

$s - k$ edges that contain each of the remaining $2s - 2k$ vertices once each. We define a 2-(h, k, s) graph to be a graph on $2s + 1$ vertices with $2k - 1$ vertices of degree $2h$ and the rest of degree $2h + 1$. Then $G_2(h, k, s)$ is a 2-(h, k, s) graph whenever $0 < h < s$ and $0 < k \leq s$.

Finally, we define the composition $[G, H, J]$ of three graphs G, H and J, each of which have some vertices of degree d and all other vertices of degree $d - 1$, to be the graph formed by taking the disjoint union of G, H and J and adding to it a vertex x and edges joining all the vertices of degree $d - 1$ in G, H and J to x. It is clear that if d is even, say $d = 2h$, and $h \geq 3$, then

$$[G_1(h, 1, h), G_1(h, 1, h), G_1(h, h - 2, h + t)]$$

is a connected regular graph of degree d on $3d + 4 + 2t$ vertices that has no one-factor, since the deletion of x results in three odd components. Similarly, if $d = 2h + 1$, then

$$[G_2(h, 1, h + 1), G_2(h, 1, h + 1), G_2(h, h, h + t + 1)]$$

is a connected regular graph of degree d on $3d + 7 + 2t$ vertices that has no one-factor.

The first construction does not give a connected graph when $h = 2$, so another construction is needed for the case $d = 4$. Take three copies of $G_1(2, 1, 2)$, each of which has five vertices, two of degree 3 and three of degree 4, and one copy of $G_1(2, 1, 2 + t)$, which has two vertices of degree 3 and $3 + 2t$ of degree 4. Add two new vertices, x and y; join one vertex of degree 3 from each of the four graphs to x and the other to y. The result has $22 + 2t$ vertices, degree 4 and no one-factor. □

Exercises 6.4

6.4.1 Prove that $G_1(h, k, h)$ is uniquely defined up to isomorphism and that it is the only 1-(h, k, h) graph. Is the corresponding result true of 2-$(h, k, h + 1)$-graphs?

A6.4.2 What is the smallest value s such that there is a 1-$(h, 1, s)$ graph not of the type $G_1(h, 1, s)$?

A6.4.3 Consider the graph G constructed as follows. Take two copies of K_5. Select one edge from each; join the four endpoints of those edges to a new vertex. Then delete the two edges, and delete one further edge from one K_5. Prove that G is an 1-(2, 1, 5) graph but is not of type $G_1(2, 1, 5)$.

7

Graph Colorings

7.1 Vertex Colorings

In this chapter we discuss partitions of the vertex-set, or edge-set, of a graph into subsets such that no two elements of the same subset are adjacent. Such partitions are called *colorings* for historical reasons (which will become clearer in Section 8.3).

Suppose $C = \{c_1, c_2, \ldots\}$ is a set of undefined objects called *colors*. A *C-coloring* (or C-vertex coloring) ξ of a graph G is a map

$$\xi: V(G) \to C.$$

The sets $V_i = \{x: \xi(x) = c_i\}$ are called *color classes*. Alternatively, a coloring could be defined as a partition of $V(G)$ into color classes.

A *proper coloring* of G is a coloring in which no two adjacent vertices belong to the same color class. In other words,

$$x \sim y \Rightarrow \xi(x) \neq \xi(y).$$

A proper coloring is called an *n-coloring* if C has n elements. If G has an n-coloring, then G is called *n-colorable*.

The *chromatic number* $\chi(G)$ of a graph G is the smallest integer n such that G has an n-coloring. A coloring of G in $\chi(G)$ colors is called *minimal*. We use the phrase "G is *n-chromatic*" to mean that $\chi(G) = n$ (but note that a minority of authors use n-chromatic as a synonym for n-colorable).

Figure 7.1 shows four ways of assigning colors to the Petersen graph. Assignments A and B are proper, a 4-coloring and a 3-coloring respectively. C is not a coloring because one vertex is not assigned a color. D is a coloring, but it is not proper because two adjacent vertices receive color 2.

Of course, G is n-colorable for every n such that $n \geq \chi(G)$. If the vertices of G are sorted into n color classes, it is clear that no edge of G lies entirely within one color class, so G is an n-partite graph with $n = \chi(G)$. Obviously $\chi(G)$ is the smallest integer n for which G is n-partite.

Some easy small cases of chromatic number are

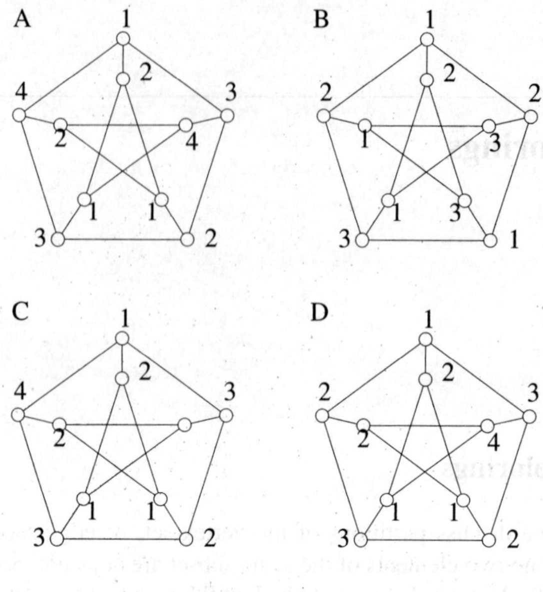

Fig. 7.1. Four color assignments

$$\chi(K_v) = v,$$
$$\chi(P_v) = 2.$$

A cycle of length v has chromatic number 2 if v is even and 3 if v is odd. The star $K_{1,n}$ has chromatic number 2; this is an example of the next theorem.

Clearly $\chi(G) = 1$ if G has no edges and $\chi(G) = 2$ if G has at least one edge.

Theorem 7.1. *$\chi(G) = 2$ if and only if G is not null and G contains no cycles of odd length.*

Proof. $\chi(G) = 2$ is equivalent to G being bipartite. So the theorem follows from the characterization of bipartite graphs in Theorem 2.2. □

Suppose G is a graph with a subgraph H, and suppose G has been n-colored. If all the vertices and edges that are not in H were deleted, there would remain a copy of H that is colored in at most n colors. So $\chi(H) \le \chi(G)$ whenever H is a subgraph of G.

Theorem 7.2. *If $\Delta = \Delta(G)$ is the maximum degree in G, then*

$$\chi(G) \le \Delta + 1.$$

Proof. Suppose G has v vertices $\{x_1, x_2, \ldots, x_v\}$. Then color the vertices with colors $c_0, c_1, \ldots, c_\Delta$ by the following inductive process. Color x_1 with c_0. If $x_1, x_2, \ldots, x_{i-1}$ have been colored, then write C_i for the set of all colors that have been assigned to vertices x_j, where $x_j \sim x_i$ and $j < i$. Then $C \backslash C_i$ is nonempty, since $|C_i| \le d(x_i) \le \Delta$. Color x_i with the element of $C \backslash C_i$ that has smallest subscript. □

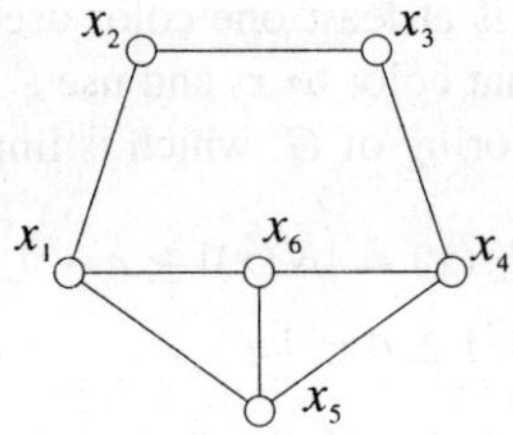

Fig. 7.2. A graph with chromatic number 3

The algorithm used in proving Theorem 7.2 is an example of a greedy algorithm, so this coloring is called a *greedy coloring*, or *the greedy coloring with respect to the given vertex ordering*.

Example. Consider the graph in Figure 7.2. Applying Theorem 7.2 we obtain a 4-coloring with

$$V_0 = \{x_1, x_3\},\ V_1 = \{x_2, x_4\},\ V_2 = \{x_5\},\ V_3 = \{x_6\}.$$

Theorem 7.2 does not necessarily provide a minimal coloring. In fact the graph of Figure 7.2 can be colored in three colors:

$$V_0 = \{x_1, x_4\},\ V_1 = \{x_2, x_5\},\ V_2 = \{x_3, x_6\}.$$

Changing the order in which the vertices are processed can change the coloring, and even change the number of colors needed; see Exercises 7.1.7 and 7.1.8.

We have already noted that if H is a subgraph of G, then $\chi(H) \le \chi(G)$. In some cases equality may hold. If $\chi(H) < \chi(G)$ for every proper subgraph H of G, then G is called *critical* with respect to chromatic number, or *vertex-critical*. When there is no risk of confusion (for example, in the rest of this section), "critical" will mean "vertex-critical." A critical graph G for which $\chi(G) = n$ is called *n-critical*.

Example. The only 1-critical and 2-critical graphs are K_1 and K_2 respectively. The 3-critical graphs are precisely the odd cycles.

Suppose $\chi(G) = n$. If G is not n-critical, then it must contain a proper subgraph G_1 such that $\chi(G_1) = n$. Either G_1 is n-critical or it has a proper subgraph G_2 with $\chi(G_2) = n$. Proceeding in this way, we obtain a sequence

$$G > G_1 > G_2 > \cdots$$

where every term has chromatic number n. This sequence must terminate, since G is finite. So there must be a graph G_k that is n-critical. So every graph of chromatic number n has an n-critical subgraph.

Theorem 7.3. *If G is an n-critical graph, then $\delta(G) \ge n - 1$.*

Proof. Suppose G is an n-critical graph and x is a vertex of G. Then $G - x$ is a proper subgraph of G, so $\chi(G - x) \le n - 1$. Select an $(n - 1)$-coloring ξ of $G - x$. Write $N(x)$ for the neighborhood of x in G, the set of vertices adjacent to x in G.

If $|N(x)| < n - 1$, then there is at least one color used in ξ that is not allocated to any member of $N(x)$. Apply that color to x, and use ξ to color every other vertex of G. The result is an $(n-1)$-coloring of G, which is impossible. So $|N(x)| \geq n - 1$. Therefore

$$d_G(x) = |N(x)| \geq n - 1$$

for every vertex x of G, and $\delta(G) \geq n - 1$. □

Corollary 7.4. [114] *If G is not critical, then*

$$\chi(G) \leq 1 + \max\{\delta(H) : H < G\}.$$

Proof. Select a $\chi(G)$-critical subgraph G_1 of G. Then Theorem 7.3 implies that

$$\chi(G) - 1 \leq \delta(G_1)$$

and obviously

$$\delta(G_1) \leq \max\{\delta(H) : H < G\},$$

giving the result. □

Exercises 7.1

7.1.1 Which of the following assignments are colorings? Which are proper colorings?

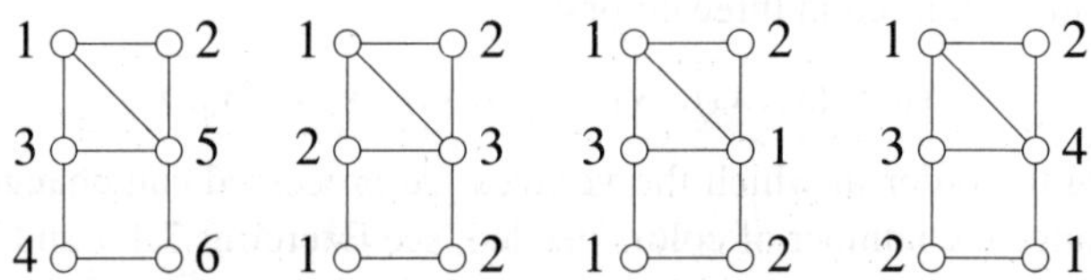

A7.1.2 P is the Petersen graph. What is $\chi(P)$?

7.1.3 For each of the graphs in Figure 7.3, find the chromatic number and find a coloring that uses the minimum number of colors.

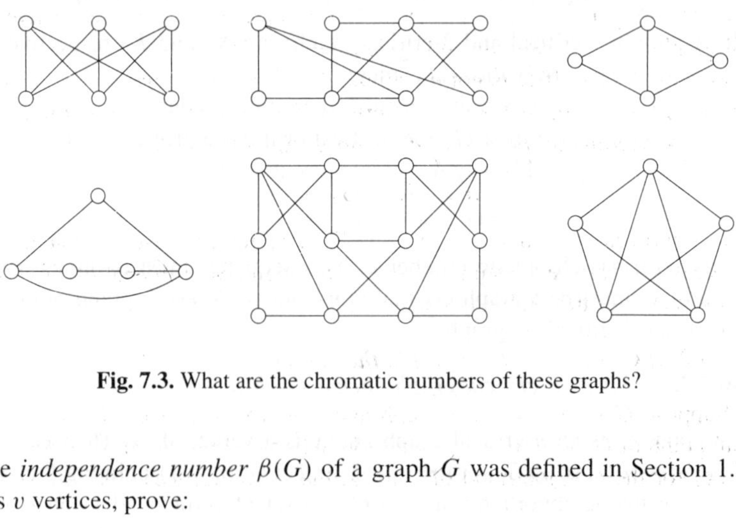

Fig. 7.3. What are the chromatic numbers of these graphs?

A7.1.4 The *independence number* $\beta(G)$ of a graph G was defined in Section 1.2. If G has v vertices, prove:

(i) $v \leq \beta(G) \cdot \chi(G)$;
(ii) $\chi(G) \leq v - \beta(G) + 1$.

.1.5 The wheel W_n was defined in Section 1.2 to consist of an n-cycle together with a further vertex adjacent to all n vertices. What is $\chi(W_n)$?

.1.6 G is the union of two graphs G_1 and G_2 that have one common vertex. Show that $\chi(G) = \max\{\chi(G_1), \chi(G_2)\}$.

.1.7 Find a way of ordering the vertices of the graph of Figure 7.2 so that the resulting greedy coloring uses only three colors.

.1.8 Prove that for any graph G there is a way of ordering the vertices such that the greedy coloring with respect to the ordering uses only $\chi(G)$ colors.

.1.9 A graph G contains exactly one cycle of odd length. Show that $\chi(G) = 3$.

.10 Suppose ξ is a coloring of a graph G. Show that the statement

$$x \sim y \iff \xi(x) \neq \xi(y)$$

holds if and only if ξ is a proper coloring and G is a complete $\chi(G)$-partite graph.

.11 The cartesian product of graphs was defined in Section 1.2.
(i) Draw the graph $K_{1,3} \times P_3$. Find its chromatic number.
(ii) Prove that

$$\chi(G \times H) = \max\{\chi(G), \chi(H)\}.$$

(iii) Prove that G is n-colorable if and only if $G \times K_n$ contains a set of $|V(G)|$ vertices, no two of which are adjacent.

.12 Suppose x is a vertex of G whose degree is less than n. Show that G is n-colorable if and only if $G - x$ is.

.13 A graph G with chromatic number n is called *n-minimal* if $\chi(G - e) < n$ for each edge e of G.
(i) Prove that a connected n-critical graph must be n-minimal.
(ii) Prove that every 2-minimal graph is 2-critical.
(iii) Prove that every 3-minimal graph is 3-critical.
(iv) Prove that not every 4-graph is 4-critical (construct a 4-minimal graph that is not 4-critical).

7.2 Brooks' Theorem

It is clear that the complete graph K_v requires v colors, and its maximum degree is $v - 1$, so equality holds in Theorem 7.2 for those graphs. The odd cycles also have this property (with chromatic number 3 and maximum degree 2). However, Brooks [18] proved that these are the only connected graphs that require more than Δ colors.

If G is any graph and ξ is a proper coloring of G, the *Kempe chains of G* (with respect to ξ) are the graphs induced by all the vertices that receive one of two given colors under ξ. In other words, if c_i and c_j are two colors, then the Kempe chain is the graph G_{ij} with vertex-set $\{x : \xi(x) = c_i \text{ or } c_j\}$ and edge-set all the edges of G that join two such vertices. If H is a connected component of G_{ij}, exchanging colors c_i and c_j on all the vertices of H produces another proper coloring. A proof using this technique is called a *Kempe chain argument*.

Theorem 7.5. [18] *If G is a connected graph other than a complete graph or an odd cycle, then*

$$\chi(G) \leq \Delta(G).$$

Proof. Suppose G is a connected v-vertex graph, other than K_v or an odd cycle, and suppose $\Delta(G) = n$. The only possible case when $n \leq 2$ is for G to be a path or even cycle, in which cases $\chi(G) = \Delta(G) = 2$, so we assume $n \geq 3$.

We now proceed by induction on v. Assume the result is true for all graphs with fewer than v vertices. This is easy to check in the case $v = 4$, because there are only three connected incomplete graphs with $v = 4$ and $\Delta = 3$, and all can be 3-colored (see Exercise 7.2.2).

(i) Suppose G is not regular. Select a vertex x such that $d(x) < n$. Then $G - x$ has fewer than v vertices, so by the induction hypothesis it can be colored in $\Delta(G - x) \leq \Delta(G)$ colors. Certainly it can be colored in at most n colors. n-color $G - x$. Since $d(x) < n$ there will be fewer than n colors on vertices that were adjacent to x in G, so there is a color not used on any of those vertices. Apply that color to x. We have colored G in n colors.

(ii) Now consider a regular graph G of degree n, and suppose G cannot be colored in n colors. Select any vertex x; by induction, $G - x$ has an n-coloring ξ. Moreover, we can assume the neighbors of x receive all n colors (or else we could proceed as in Case (i)). Say the neighbors are $x_1, x_2, \ldots, x_n$, and $\xi(x_i) = c_i$. Select two of these neighbors, x_i and x_j, and consider the Kempe chain G_{ij}. If x_i and x_j were in different components of G_{ij}, one could exchange colors c_i and c_j on all the vertices in the component containing x_i and still have a proper coloring; but in this new coloring there would be no vertex of color c_i adjacent to x, so x could receive c_i. So we need only consider the case in which x_i and x_j belong to the same component of G_{ij}, for every i and j. This means that for every pair of neighbors x_i and x_j of x there is a path P_{ij} from x_i to x_j made up entirely of vertices colored c_i and c_j under ξ.

We next show that $G_{ij} = P_{ij}$. Suppose x_i has degree 2 or greater in Gij. Then x_i has two neighbors of color c_j, and some color c_k is not adjacent to x_i. We could recolor x_i in c_k and x in c_i, and n-color G. So x_i, and similarly x_j, has degree 1 in G_{ij}. Say x_i has neighbor x_{i1} in Gij; either $x_{i1} = x_2$ or x_{i1} has degree greater than 2 or x_{i1} has a unique neighbor x_{i2} other than $X - 1$ in Gij; and so on. If G_{ij} is not a path, there will be a well-defined vertex y of degree 3 or greater that is nearest (in Gij) to x_i. If $\xi(y) = c_i$, then y is adjacent to three c_i vertices, so there must be a color, say c_k, that is not adjacent to y. We can recolor x_i with c_j, $x_i 1$ with c_i, $x_i 2$ with $c_j, \ldots$, y with c_k, and x with $C - i$, giving an n-coloring of G. The case where $\xi(y) = c_j$ can be eliminated similarly.

Each G_{ij} is a path from X_i to x_j. Suppose z were a member of both G_{ij} and G_{ik}, where $k \neq j$. Then $\xi(z) = c_i$, and unless $z = x_i$, z has two neighbors colored c_j and two colored c_k. Again, there is a color not adjacent to z and recoloring is possible as in the preceding paragraph. So the Kempe chains intersect only at their endpoints.

Now suppose two neighbors of x, x_i and x_j, are not adjacent in G. Then they are not adjacent in $G - x$, and the path G_{ij} contains a vertex other than x_j, say y, adjacent to x_i. $\xi(y) = c_j$. Select some color c_k (not c_i or c_j) and interchange the colors of vertices in $G_i k$, so that now x_i receives color c_k. Consider the Kempe chains for this new coloring of $G - x$. Clearly z belongs to the jk-chain, because it is adjacent to x_i, which is an endpoint of the jk-chain. Similarly, y belongs to the ij-chain. This contradicts the preceding paragraph.

So all the neighbors of x are adjacent. Since x could be any vertex, and G is connected, G must be a complete graph. But this contradicts the original assumptions. □

Exercises 7.2

7.2.1 Prove that the only $(n-1)$-regular n-critical graphs are the complete graphs and the odd cycles.

7.2.2 Verify that there are exactly three connected incomplete graphs on 4 vertices with maximum degree 3, and all can be 3-colored.

7.2.3 G is formed from the complete graph K_v by deleting one edge. Prove that $\chi(G) = v - 1$ and describe a way of coloring G in $v - 1$ colors.

7.2.4 Prove that the following statement is equivalent to Brooks' Theorem: if G is an n-critical graph ($n \geq 4$) with v vertices and e edges, and G is not complete, then $2e > v(n-1)$. [17]

7.3 Counting Vertex Colorings

So far we have looked most closely at coloring vertices with the minimum number of colors. Another problem is, given a set of colors, how many ways are there to color a given graph? We count all possible colorings that are not identical, whether or not any two are isomorphic. For example, consider K_2, a single edge with vertices y and z. Write (A, B) to mean the coloring in which y is colored A and z is colored B. K_2 requires two colors, and there are two different colorings, (A, B) and (B, A). If three colors were available, there would be the six colorings (A, B), (A, C), (B, A), (B, C), (C, A) and (C, B). (Not all available colors actually have to be used.)

We write $p_G(x)$ for the number of different colorings of G with x colors. $q_G(x)$ will denote the number of these colorings in which all x colorings are actually used. So $p_{K_2}(2) = q_{K_2}(2) = 2$, $p_{K_2}(3) = 6$ and $q_{K_2}(3) = 0$. Clearly $p_G(x) = 0$ if $x < \chi(G)$. For the null graph on v vertices, $\overline{K}_v$, any color may be assigned to any vertex, so

$$p_{\overline{K}_v}(x) = x^v.$$

Example. To count the proper colorings of K_v, suppose the vertices have been ordered in some way. The first vertex can be assigned any of the x colors. There are then $x - 1$ colors for the second vertex, $x - 2$ for the third, and so on, so

$$p_{K_v}(x) = x(x-1)(x-2)\cdots(x-v+1) = \frac{x!}{(x-v)!}.$$

Example (continued). Consider the cycle C_4. $\chi(C_4) = 2$, so $p_{C_4}(0) = p_{C_4}(1) = 0$. Using two colors, first select any vertex. It can be colored in two ways. The rest of the coloring is then determined. So

$$p_{C_4}(2) = q_{C_4}(2) = 2.$$

With three colors, there are six colorings in which only two of the colors are used — there are three ways to select the color to be omitted, and two ways to apply the remaining colors. If all three colors are actually used, there must be a pair of nonadjacent vertices that receive the same color. The color can be chosen in three ways, and the nonadjacent pair in two ways. The other two colors can be applied to the remaining vertices in two ways. So

$$p_{C_4}(3) = 6 + 3\times2\times2 = 18.$$

The calculation of $p_{C_4}(3)$ in the example could be described as follows. To enumerate colorings in which only two colors appear, multiply the number of ways of choosing two of the three colors ($\binom{3}{2}$, or 3) by the number of colorings using precisely two colors ($q_{C_4}(2)$, or 2). Then add the number of colorings using precisely three colors ($q_{C_4}(3)$). So

$$p_{C_4}(3) = \binom{3}{2} q_{C_4}(2) + q_{C_4}(3).$$

This approach can be generalized. The number of proper colorings of G that use precisely k of x available colors is

$$\binom{x}{k} q_G(k). \tag{7.1}$$

The total number of proper colorings in x colors is found by summing this term over k. It will be zero if $k < \chi(G)$ (no proper colorings exist) or $k > v(G)$ (the number of colors actually used cannot exceed the number of vertices). So

$$p_G(x) = \sum_{k=\chi(G)}^{v(G)} \binom{x}{k} q_G(k). \tag{7.2}$$

Example (continued). There are 24 ways to color C_4 using only four colors. (In general, if G is any graph with v vertices, $q_G(v) = v!$.) So

$$\begin{aligned}
p_{C_4}(4) &= \binom{4}{2} q_{C_4}(2) + \binom{4}{3} q_{C_4}(3) + \binom{4}{4} q_{C_4}(4) \\
&= 6\times2 + 4\times12 + 1\times24 \\
&= 12 + 48 + 24 \\
&= 84.
\end{aligned}$$

In general, we have

$$\begin{aligned}
p_{C_4}(x) &= \binom{x}{2} q_{C_4}(2) + \binom{x}{3} q_{C_4}(3) + \binom{x}{4} q_{C_4}(4) \\
&= 2x(x-1)/2 + 12x(x-1)(x-2)/6 \\
&\quad + 24x(x-1)(x-2)(x-3)/24 \\
&= x(x-1)(x^2 - 3x + 3).
\end{aligned}$$

The function defined in (7.2) is a polynomial in x of degree $v(G)$, called the *chromatic polynomial* of G. It is clear that isomorphic graphs have the same chromatic polynomial. The converse is not true; in fact, all trees of the same order have the same polynomial.

Theorem 7.6. *If T is a tree with v vertices, then*

$$p_T(x) = x(x-1)^{v-1}.$$

Proof. Say T is to be properly colored with x colors. Select any vertex of T, say a_1. There are x ways to color a_1. Now select any vertex a_2 adjacent to a_1. There are $x-1$ colors available for a_2, all colors except the one applied to a_1. Choose one. Now continue in this way. After k vertices have been colored, select a vertex a_{k+1} that has not yet been colored but is adjacent to one of the colored vertices. It cannot be adjacent to two of them, or else T would contain a cycle. If a_{k+1} is adjacent to a_i, it may receive any of the $x-1$ colors other than the one given to a_i. This process may be continued until all vertices are colored. There were x choices at the first stage, and $x-1$ at each of the other $v-1$, so there are $x(x-1)^{v-1}$ colorings. □

There are also examples of graphs other than trees that have the same chromatic polynomial — see Exercises 7.3.3 and 7.3.8(ii).

Multiple edges can obviously be ignored in calculating the chromatic polynomial. If M is a multigraph with underlying graph H, then $p_M(x) = p_H(x)$. The multigraph G_a, formed from the graph G by identifying the endpoints of edge a, was introduced in Section 4.2. $G-a$ is defined as usual by removing the edge a from the edge-set. G_a and $G-a$ occurred together in Theorem 4.6, and they do so again in the following lemma.

Lemma 7.7. *If a is any edge of the graph G, then*

$$p_G(x) = p_{G-a}(x) - p_{G_a}(x). \tag{7.3}$$

Proof. Consider the proper colorings of $G-a$. They are of two types — those in which the two endpoints of a receive the same color, and those in which they receive different colors. Colorings of the first type are in one-to-one correspondence with proper colorings of G_a, while those of the second type are in one-to-one correspondence with proper colorings of G. So

$$p_{G-a}(x) = p_{G_a}(x) + p_G(x),$$

and the lemma follows. □

Example. Deletion of any edge from the cycle C_5 produces a path P_5, and contraction of any edge produces a cycle C_4. So

$$p_{C_5}(x) = p_{P_5}(x) - p_{C_4}(x).$$

Since P_5 is a tree, $p_{P_5}(x) = x(x-1)^4$, and $p_{C_4}(x) = x(x-1)(x^2-3x+3)$ from our previous calculation, so

$$\begin{aligned} p_{C_5}(x) &= x(x-1)^4 - x(x-1)(x^2-3x+3) \\ &= x(x-1)\left((x-1)^3 - (x^2-3x+3)\right) \\ &= x(x-1)(x^3-4x^2+6x-4). \end{aligned}$$

(See also Exercise 7.3.7.)

Theorem 7.8. *Suppose G is a graph with v vertices and e edges. Its chromatic polynomial $p_G(x) = \sum a_k x^k$ has integer coefficients that satisfy:*

(i) *$p_G(x)$ is of degree v;*
(ii) $a_v = 1$;
(iii) $a_0 = 0$;
(iv) $a_{v-1} = -e$;
(v) *the coefficients alternate in sign;*
(vi) *the smallest k such that $a_k \neq 0$ equals the number of components of G.*

Proof. We have already observed that (7.2) implies (i). The only term in (7.2) involving x^v is

$$\binom{x}{v} q_G(v) = \frac{x(x-1)\dots(x-v+1)}{v!} q_G(v),$$

so the coefficient of x^v is $q_G(v)/v! = 1$. One cannot color with no colors, so $a_0 = 0$.

We prove parts (iv) and (v) by induction on e. Suppose they are true for all graphs with less than e edges. In particular, if a is any edge of G, then both $G - a$ and G_a have fewer edges than G. By induction, there exist nonnegative integers $a_1, a_2, \dots, a_{v-2}$ and $b_1, b_2, \dots, b_{v-2}$, such that

$$p_{G-a}(x) = x^v - (e-1)x^{v-1} + a_{v-2}x^{v-2} - a_{v-3}x^{v-3} + \dots + (-1)^{v-1}a_1 x$$

and

$$p_{G_a}(x) = x^{v-1} - b_{v-2}x^{v-2} + b_{v-3}x^{v-3} + \dots + (-1)^{v-2}b_1 x.$$

So, from (7.3),

$$\begin{aligned} p_G(x) &= p_{G-a}(x) - p_{G_a}(x) \\ &= x^v - ex^{v-1} + (a_{v-2} + b_{v-2})\,x^{v-2} - (a_{v-3} + b_{v-3})\,x^{v-3} \\ &\quad + \dots + (-1)^{v-1}(a_1 + b_1)\,x, \end{aligned}$$

which has the required form.

Part (vi) can also be proved by induction, but another proof is indicated in Exercise 7.3.2. □

A number of other properties of the chromatic polynomial are easy to prove. One property that is easy to check is that only an empty graph can be properly colored with one color, so the sum of the coefficients of $p_G(x)$ must be zero whenever G has an edge. Another, not so easy to apply when testing a particular polynomial, is that $p_G(x)$ can never be negative for integer x.

Exercises 7.3

.3.1 Show that if G is a connected graph, then the chromatic polynomial of G satisfies $a_1 \neq 0$.

.3.2 The graph G is the union of two disjoint subgraphs H and K. Prove that

$$p_G(x) = p_H(x) \cdot p_K(x).$$

Hence prove Theorem 7.8(vi).

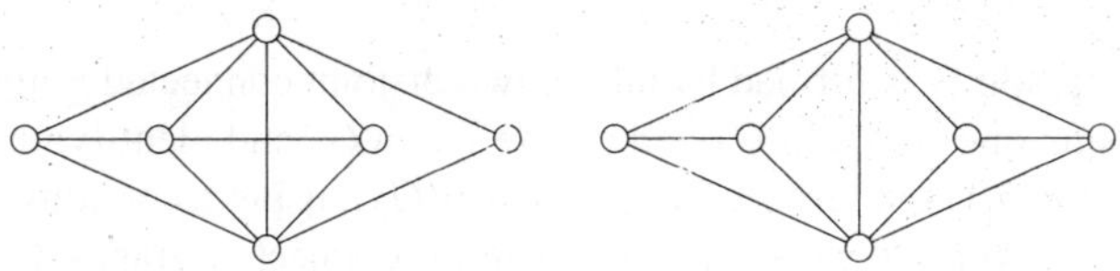

Fig. 7.4. Graphs with the same chromatic polynomial

.3.3 Consider the graphs in Figure 7.4.
  (i) Verify that the graphs are not isomorphic.
  (ii) Show that they have the same chromatic polynomial. (This exercise is significant because it was once conjectured that 2-connected graphs with the same chromatic polynomial might necessarily be isomorphic.)

.3.4 For each of the following polynomials, either prove that it is not a chromatic polynomial or find a graph G for which it is $p_G(x)$:
  (i) $x^5 - 3x^4 + 3x^3 - x^2$.
  (ii) $x^5 - 6x^4 + 5x^3 - 2x^2 + 3x$.
  (iii) $x^5 - 6x^4 + 13x^3 - 12x^2 + 4x$.
  (iv) $x^5 - 4x^4 + 4x^3 - x^2$.
  (v) $x^5 - 2x^4 + x^3$.
  (vi) $x^5 - 5x^4 + 8x^3 - 10x^2 + 6x$.
  (vii) $x^5 - 4x^4 + 5x^3 - 2x^2$.

.3.5 Prove that there is no graph with chromatic polynomial $x^4 - 4x^3 + 3x^2$.

.3.6 Find the chromatic polynomials of the following graphs.

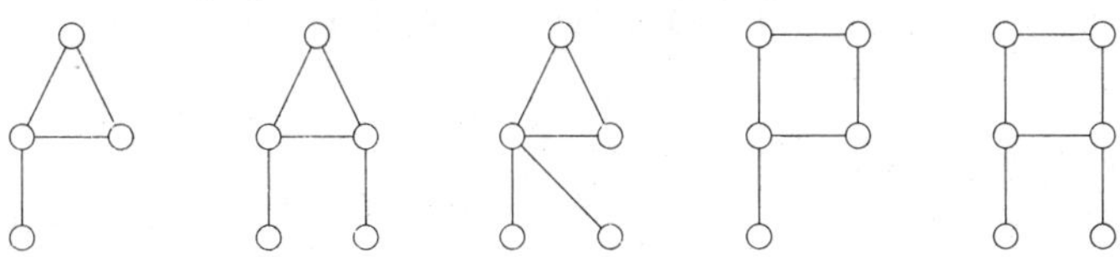

(i) (ii) (iii) (iv) (v)

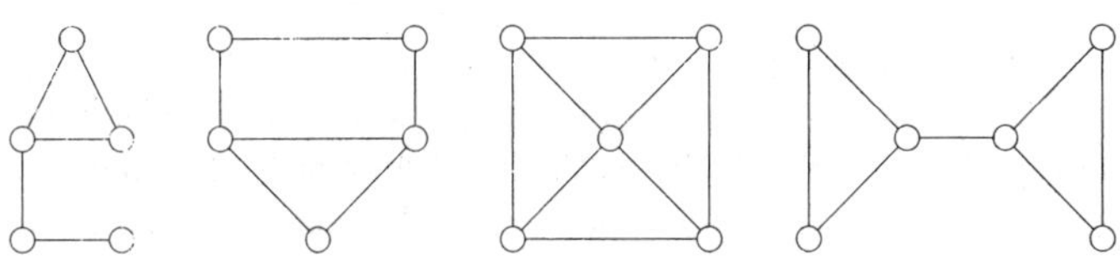

(vi) (vii) (viii) (ix)

7.3.7 In general, deleting an edge of C_v produces a P_v and contracting an edge produces a C_{v-1}. (The multigraph C_2 has the same chromatic polynomial as a K_2, namely $x(x-1)$.) Use these facts to prove that

$$p_{C_v}(x) = (x-1)\left((x-1)^{v-1} + (-1)^v\right)$$

for every $v \geq 2$.

7.3.8 Given a graph G, a new graph H is formed from G by adding a single pendant vertex adjacent to some vertex of G. Find an expression for $p_H(x)$ in terms of $p_G(x)$.

7.3.9 Suppose a graph G is formed by taking two disjoint connected graphs G_1 and G_2 and choosing one vertex in G_1 and one vertex in G_2 and identifying the two. Find a formula for $p(G)$ in terms of $p(G_1)$ and $p(G_2)$, using a counting argument.

7.3.10 Suppose G is formed by taking two disjoint connected graphs G_1 and G_2 and inserting an edge connecting some vertex of G_1 with some vertex of G_2. Use Lemma 7.6 and the results of Exercises 7.3.1 and 7.3.7 to find an expression for $p(G)$ in terms of $p(G_1)$ and $p(G_2)$.

7.4 Edge-Colorings

We now consider partitions of the edge-set of a graph. Again, the terminology of colors is customarily used. A *k-edge-coloring* π of a graph G is a map from $E(G)$ to $\{1, 2, \ldots, k\}$, with the property that if e and f are edges with a common vertex, then $\pi(e) \neq \pi(f)$. G is k-edge-colorable if there is a k-edge-coloring of G; the edge-chromatic number $\chi'(G)$ of G is the smallest number k such that G is k-edge-colorable. $\chi'(G)$ is also called the *chromatic index* of G. Given an edge-coloring π of G, the *spectrum* $S_\pi(x)$ (or simply $S(x)$) is the set of size $d(x)$ defined by

$$S_\pi(x) = \{i : \pi(xy) = i \text{ for some } y \sim x\}.$$

We write $s_\pi(x)$ for $|S_\pi(x)|$ (we simply write $s(x)$ when it is clear that the edge-coloring π is being discussed). The *color classes* under π are the k sets

$$E_i = E_i(\pi) = \{y : y \in E(G), \pi(y) = i\}.$$

The sets E_i partition $E(G)$ (in much the same way as the color classes V_i, defined for vertex coloring, partition $V(G)$).

It is clear that it requires $d(x)$ colors to color the edges at x, so $\chi'(G) \geq \Delta(G)$. However we can say rather more. To do this we introduce a slightly more general type of coloring called a *painting*, and we extend the preceding notations to paintings. A *k-painting* of G is any way of allocating k colors to the edges of G. There might be more than one edge in a given color touching a vertex x. If π is a k-painting but not necessarily an edge-coloring, then $s(x) \leq d(x)$. A painting is an edge-coloring if and only if $s(x) = d(x)$ for all x. The *order* $|\pi|$ of a painting π equals the sum $\sum s(x)$ over all vertices x; π is a *maximal* k-painting if $|\pi|$ equals the maximum for the value of k.

Lemma 7.9. *If G is a connected graph other than an odd cycle, then there is a 2-painting of G in which $s(x) = 2$ whenever $d(x) \geq 2$.*

Proof. First, suppose G is Eulerian. If G is an even cycle, the obvious one-factorization provides an appropriate painting. Otherwise G has at least one vertex x of degree 4 or more. Select an Euler walk with start-finish vertex x whose consecutive edges are $y_1, y_2, \ldots, y_e$. Then the painting

$$E_1 = \{y_1, y_3, \ldots\},\ E_2 = \{y_2, y_4, \ldots\}$$

has the required property, because every vertex (including x) is internal to the walk.

If G is a path of odd length, say $x_1 x_2 x_3 \ldots$, then a suitable painting is

$$E_1 = \{x_1x_2, x_3x_4, \ldots\},\ E_2 = \{x_2x_3, x_4x_5, \ldots\}.$$

Finally, suppose G is not Eulerian but is not an odd path. Construct a new graph H by adding a vertex y to G and adding an edge yz whenever z is a vertex of odd degree. Then H is Eulerian, by Theorem 2.6. Moreover H will not be an odd cycle (if it were, then G would be an odd path). Carry out a painting of H using the Euler walk, and then restrict this painting to G. It is easy to see that if x has degree 2 or greater in G, then the Euler path traverses x at least once using two edges that do not touch y, and those edges will receive different colors. □

Lemma 7.10. *Suppose π is a maximal k-painting of G and suppose vertex x lies on two edges of color 1 and no edges of color 2. Then the component of $E_1 \cup E_2$ that contains x is an odd cycle.*

Proof. Suppose $E_1 \cup E_2$ is not an odd cycle. Then by Lemma 7.9 it has a 2-coloring ρ in which every vertex of degree at least 2 lies on edges of both colors. Replace E_1 and E_2 in π by the two color classes of ρ, and denote the new k-painting by σ. Then

$$s_\sigma(x) = s_\pi(x) + 1,$$

and

$$s_\sigma(y) \geq s_\pi(y)$$

for every $y \neq x$. So $|\sigma| > |\pi|$, which contradicts the maximality of π. □

Theorem 7.11. *If G is a bipartite graph, then*

$$\chi'(G) = \Delta(G).$$

Proof. Suppose G is a graph for which $\chi'(G) > \Delta(G)$. Let π be a maximal Δ-painting of G, and select a vertex x for which $s(x) < d(x)$. (Such a vertex x must exist because no painting with fewer than $\chi'(G)$ colors can be an edge-coloring.) Then x must satisfy the conditions of Lemma 7.10. Consequently G contains an odd cycle, so it is not bipartite. □

Corollary 7.12. *Every regular bipartite graph has a one-factorization.*

The following theorem is due to Vizing [119] and was independently discovered by Gupta [58].

Theorem 7.13. *For any graph G,*

$$\Delta(G) \le \chi'(G) \le \Delta(G) + 1.$$

Proof. The edges incident with a vertex of maximum degree require $\Delta(G)$ colors, so $\Delta(G) \le \chi'(G)$. Let us assume that $\chi'(G) > \Delta + 1$. Select a maximal $(\Delta + 1)$-painting π of G and a vertex x such that $s(x) < d(x)$. (Again, such a vertex x must exist because no painting with fewer than $\chi'(G)$ colors can be an edge-coloring.) There must exist colors c_1 and c_0 such that x is incident with two edges of color c_1 and no edge of color c_0.

Let xy_1 be an edge of color c_1. Since $d(y_1) < \Delta + 1$, there is some color c_2 that is not represented at y_1; define a new painting π_1 that is identical to π except that $\pi_1(xy_1) = c_2$. Clearly there must be an edge x that is of color c_2 in π, for otherwise π_1 would be a $(\Delta + 1)$-painting of greater order than π. Now select a vertex y_2 such that $\pi_1(xy_2) = c_2$, and select a color c_3 missing at y_2 in π_1. Construct a painting π_2 from π_1 by changing the color of xy_2 to c_3. Again x must have been incident with an edge of color c_3 in π_2, and π_2 is maximal. Continuing this process we obtain sequences $\pi_0 = \pi, \pi_1, \ldots$ of maximal paintings, $c_1, c_2, \ldots$ of colors and $y_1, y_2, \ldots$ of vertices such that

$$\pi_{i-1}(xy_i) = c_i, \quad \pi_i(xy_i) = c_{i+1}, \quad \pi_i(z) = \pi_{i-1}(z) \text{ if } z \neq xy_i$$

and no edge incident with y_i has color c_{i+1} in π_{i-1}.

The sequence $c_1, c_2, \ldots$ must eventually contain repetitions, because the set of colors incident with x is the same in each π_i, and the set of colors incident with x is finite. Say c_q is the first such repetition: q is the smallest integer such that, for some p less than q, $c_p = c_q$.

Consider the maximal paintings π_{p-1} and π_{q-1}. By Lemma 7.10, the edges receiving colors c_0 and c_p in π_{p-1} form a cycle in G, say C, that contains x, and the edges receiving colors c_0 and c_q $(= c_p)$ in π_{q-1} form a cycle in G, say C', that also contains x. These cycles are different, because xy_p is in C but not in C'. Now π_{q-1} differs from π_{p-1} only in the colors assigned to edges $xy_p, xy_{p+1}, \ldots, xy_{q-1}$, and only the first of these is in C. So all the other edges of C must belong to C'; in particular, y_p is a vertex of C'. On the other hand, the only edge incident with y_p that receives different colors in π_{p-1} and y_{q-1} is xy_p, so y_p must have degree 1 in C'. So C' is not a cycle — a contradiction. □

Graphs that satisfy $\chi'(G) = \Delta$ are called *class 1*; those with $\chi'(G) = \Delta + 1$ are called *class 2*. Clearly, for regular graphs, being class 1 is equivalent to having a one-factorization. The class 2 graphs include the odd-order complete graphs and the odd cycles.

Exercises 7.4

7.4.1 Prove that $\chi'(K_v) = v$ if v is odd and $\chi'(K_v) = v - 1$ if v is even.

A7.4.2 Up to isomorphism, there are two graphs with five vertices and eight edges. Prove that each is class 1.

.4.3 Prove that every graph with five vertices and seven edges has edge-chromatic number 4.

.4.4 Up to isomorphism, there are two graphs with five vertices and eight edges. Prove that each is class 1.

.4.5 Prove that if P is a color class in an edge-coloring of a graph G, then

$$|P| \leq \left\lfloor \frac{|V(G)|}{2} \right\rfloor.$$

4.6 Suppose G is a graph with km edges, where $k \geq \chi'(G)$. Prove that G can be factored into k factors, each of which is a matching with m edges. [22]

4.7 Prove that all trees are of class 1.

4.8 Prove that every nonempty regular graph with an odd number of vertices is of class 2.

4.9 Find the edge-chromatic numbers of:
(i) the Petersen graph P;
(ii) the graph derived from P by deleting one edge;
(iii) the graph derived from P by deleting one vertex.

.10 A k-edge-colorable graph is called *uniquely k-edge-colorable* if every k-edge-coloring results in the same set of color classes. Prove that every uniquely 3-edge-colorable 3-regular graph is Hamiltonian.

.11 Suppose $\delta(G) \geq 2$. Show that G has a $(\delta(G) - 1)$-painting in which every color is used at least once at every vertex. [58]

7.5 Class 2 Graphs

In general, class 2 graphs appear to be relatively rare, and for that reason we shall investigate some properties of class 2 graphs.

Theorem 7.14. [8] *If G has v vertices and e edges and*

$$e > \left\lfloor \tfrac{v}{2} \right\rfloor \Delta(G), \tag{7.4}$$

then G is class 2.

Proof. Suppose G satisfies (7.4) and is class 1. Select a $\Delta(G)$-coloring of G. Obviously no color class can contain more than $\lfloor \frac{v}{2} \rfloor$ edges, so

$$e = \sum |P_i| \leq \left\lfloor \tfrac{v}{2} \right\rfloor \Delta(G),$$

which is a contradiction. □

The edge e is called *edge-critical* (with respect to edge-coloring) if

$$\chi'(G - e) < \chi'(G).$$

An *edge-critical graph* is defined to be a connected class 2 graph in which every edge is edge-critical. An edge-critical graph of maximum degree Δ is often called Δ-edge-critical. The importance of this index follows from the obvious fact that every class 2 graph contains a Δ-edge-critical subgraph. In fact we can say more:

Theorem 7.15. *If G is a graph of class 2, then G contains a k-edge-critical subgraph for each k such that $2 \le k \le \Delta(G)$.*

Proof. We prove that any class 2 graph G contains an edge-critical graph G_1 with $\Delta(G_1) = \Delta(G)$. We then prove that G_1 contains a class 2 graph H with $\Delta(H) = k$ for each k satisfying $2 \le k < \Delta(G)$. Then H contains a k-edge-critical graph H_1, and H_1 is of course a subgraph of G. If we simply write Δ, we shall mean $\Delta(G)$.

First, suppose G is not edge-critical. Then there will be an edge e_1 such that

$$\chi'(G - e_1) = \chi'(G).$$

Write $G_1 = G - e_1$. Clearly $\Delta(G_1) = \Delta$, and G_1 is class 2. Either G_1 is edge-critical or we can delete another edge, say e_2, and continue. Eventually the process must stop because of finiteness, so eventually a Δ-edge-critical subgraph is constructed.

Now consider G_1. Let k be any integer satisfying $2 \le k < \Delta$. Select any edge uv in G_1, and let π be a Δ-edge-coloring of $G_1 - uv$. (Such a π exists because G_1 is edge-critical.) Clearly $S(u) \cup S(v)$ must equal $\{1, 2, \dots, \Delta\}$ — if any color were missing we could color uv in that color, and Δ-color G_1. But $|S(u)| < \Delta$ because u had degree at most Δ in G and therefore degree at most $\Delta - 1$ in G_1; so there is some color, say i, not in $S(u)$, and similarly there is some color, say j, not in $S(v)$, and $j \neq i$. Without loss of generality we can assume $i = 1$ and $j = 2$. Then write H for the subgraph

$$H = P_1 \cup P_2 \cup \cdots \cup P_k \cup \{uv\}.$$

Clearly $\chi'(H) \ge k + 1$: if H could be colored in k colors, then G_1 could be colored in Δ. But no vertex has degree greater than k in H: if x is not u or v, then it lies on at most one edge in each of $P_1, P_2, \dots, P_k$, while u lies on at most $k - 1$ of those (none in P_1) plus uv, and similarly for v. So $\Delta(H) \le k$. It follows from Theorem 7.13 that $\Delta(H) = k$, $\chi'(H) = k + 1$, and H is the required class 2 subgraph of maximum degree k. □

In the discussion of Δ-edge-critical graphs, the vertices of degree Δ are of special significance. Such vertices are called *major*. A vertex x for which $d(x) < \Delta$ is called *minor*.

Suppose xy is an edge-critical edge in a class 2 graph G, and π is a Δ-coloring of $G - xy$. We define a (π, x, y)-*fan*, or simply *fan*, in G to be a sequence of distinct edges $xy_1, xy_2, \dots, xy_n$ such that the color on edge xy_i does not appear in $S(y_{i-1})$, where, in particular, y_0 is interpreted as y. We write $F(x)$ for the set of all y_i that appear in any (π, x, y)-fan. In the following proofs it is convenient to write $T(v)$ for the complement of $S(v)$, so $T(v)$ is the set of all colors *not* represented at vertex v, and we define P to be the set of all colors not represented on edges from x to $F(x)$.

Lemma 7.16. *The set P is disjoint from $T(y)$ and from each $T(z)$, $z \in F(x)$.*

Proof. Suppose $i \in T(y)$. There must be an edge xz that is colored i, because otherwise we could extend π to a Δ-coloring of G by setting $\pi(xy) = i$. Now xz is a (one-edge) fan, so $i \notin P$. So P and $T(y)$ are disjoint.

Suppose P and $T(z)$ have a common element i. Let us denote a fan containing z by

$$xy_1, xy_2, \ldots, xy_n,$$

where $y_n = z$, and say $\pi(xy_j) = i_j$. If $i \in T(x)$, then one could carry out the recoloring

$$\pi(xy) = i_1, \pi(xy_1) = i_2, \ldots, \pi(xy_{n-1}) = i_n, \pi(xy_n) = i,$$

with other edges unchanged. This yields a Δ-coloring of G. But if $i \notin T(x)$, there is some edge xw such that $\pi(xw) = i$, and i belongs to no fan. However

$$xy_1, xy_2, \ldots, xy_n, xw$$

is clearly a fan — a contradiction. So P is disjoint from each $T(z)$. □

Lemma 7.17. *Suppose $i \in T(x)$ and $j \in T(z)$, where $z \in F(x) \cup \{y\}$. Then x and y belong to the same component of $P_i \cup P_j$.*

Proof. In the case $z = y$ this is easy, because otherwise we could exchange colors i and j in the component that contained x and then color xy with color j. So we assume $z \in F(x)$.

We call a fan $xy_1, xy_2, \ldots, xy_n$ *deficient* if there exist colors i and j such that $i \in T(x)$, $j \in T(y_n)$ and x and y_n belong to different components of $P_i \cup P_j$. Among all deficient fans, consider one for which n has the minimum value. For the relevant i and j, exchange colors i and j in the component containing x. The result is a Δ-coloring of $G - xy$ in which y_n is still a member of $F(x)$ — the minimality of the fan ensures that it is still a fan in the new coloring — and j belongs to $P \cap T(y_n)$, contradicting Lemma 7.16. □

Lemma 7.18. *The sets $T(z)$, where $z \in F(x) \cup \{y\}$, are pairwise disjoint.*

Proof. Suppose $j \in T(v) \cap T(w)$, where v and w are in $F(x) \cup \{y\}$, and select $i \in T(x)$. (Since $|T(x)| = d(x) - 1 < \Delta$, such a color i exists.) Then x, v and w all belong to the same component of $P_i \cup P_j$. But the components of this graph are all paths and cycles. Since x, v and w are all of degree 1 in $P_i \cup P_j$, we must have a path with three endpoints, which is impossible. □

Theorem 7.19. *Suppose xy is an edge-critical edge in a class 2 graph G. Then x is adjacent to at least $\Delta - d(y) + 1$ major vertices other than y.*

Proof. In the preceding notation we have, from the lemmas, that P, $T(y)$ and the $T(z)$, $z \in F(x)$, are pairwise disjoint. Now P has $\Delta - |F(x)|$ elements, $T(y)$ has $\Delta - d(y) + 1$ (since the edge xy received no color), and each $T(z)$ has $\Delta - d(z)$. Since these are disjoint sets of colors,

$$\begin{aligned}\Delta &\geq |P| + |T(y)| + \sum_{z \in F(x)} |T(z)| \\ &= 2\Delta - |F(x)| - d(y) + 1 + \sum_{z \in F(x)} (\Delta - d(z))\end{aligned}$$

whence

$$\Delta - d(y) + 1 \leq \sum_{z \in F(x)} (1 + d(z) - \Delta).$$

Since none of the terms on the right-hand side can be greater than 1, and all are integral, it follows that at least $\Delta - d(y) + 1$ of them must equal 1. So at least $\Delta - d(y) + 1$ of the vertices z in $F(x)$ satisfy $d(z) = \Delta$, and are major. □

Corollary 7.20. *In an edge-critical graph, every vertex is adjacent to at least two major vertices.*

Proof. In the notation of the theorem, either $d(y) < \Delta$ (and $\Delta - d(y) + 1 \geq 2$) or $d(y) = \Delta$ (and x is adjacent to at least one major vertex other than y and also to y). □

If we take x to be a major vertex in the preceding corollary, we see that every edge-critical graph contains at least three major vertices. If G is any class 2 graph, then considering the $\Delta(G)$-edge-critical subgraph we see:

Corollary 7.21. *Every class two graph contains at least three major vertices.*

Corollary 7.22. *An edge-critical graph G has at least $\Delta(G) - \delta(G) + 2$ major vertices.*

Proof. In the theorem, take x to be a major vertex and y to be a vertex of degree $\delta(G)$. □

Theorem 7.19 and the corollaries were essentially proven by Vizing [119] and are collectively known as Vizing's Adjacency Lemma.

Corollary 7.23. *A Δ-edge-critical graph on v vertices has at least $\frac{2v}{\Delta}$ major vertices.*

Proof. Let us count the edges adjacent to the major vertices. There are at least $2v$, by Corollary 7.20, but there are exactly Δ per major vertex. □

Corollary 7.24. *Suppose w is a vertex of a graph G that is adjacent to at most one major vertex, and e an edge containing w. Then*

$$\Delta(G - e) = \Delta(G) \Rightarrow \chi'(G - e) = \chi'(G)$$

and

$$\Delta(G - w) = \Delta(G) \Rightarrow \chi'(G - w) = \chi'(G).$$

Proof. G is either class 1 or class 2. We treat the two cases separately. If G is of class 1, then

$$\Delta(G) = \chi'(G) \geq \chi'(G - e) \geq \Delta(G - e) = \Delta(G)$$

and therefore $\chi'(G - e) = \chi'(G)$. Similarly, $\chi'(G - w) = \chi'(G)$.

If G is of class 2, then by Theorem 7.15, G contains a $\Delta(G)$-edge-critical subgraph, say H. By Corollary 7.23, w cannot be a vertex of $V(H)$. So H is a subgraph of $G - w$, and

$$1 + \Delta(G) = \chi'(H) \leq \chi'(G - w) \leq \chi'(G) = 1 + \Delta(G),$$

so $\chi'(G - w) = \chi'(G)$. Similarly, $\chi'(G - e) = \chi'(G)$. □

We will later need to use the idea of the core of a graph. If G is a graph of maximal degree Δ, the *core* G_Δ is the subgraph induced by the set of all vertices of degree Δ.

Lemma 7.25. *Suppose the core G_Δ of G satisfies the following description.*
(i) $V(G_\Delta) = U \cup V \cup W$, *a disjoint union into three parts, where* $U = \{u_1, u_2, \ldots, u_p\}$ *and* $V = \{v_1, v_2, \ldots, v_p\}$.
(ii) *Whenever* $i \leq j$, *there is an edge* (u_i, v_j).
(iii) *Every other edge joins a member of* V *to a member of* $V \cup W$.
Then G *is class 1.*

The proof is left as an exercise.

Exercises 7.5

7.5.1 Verify that the following graph is 3-edge-critical.

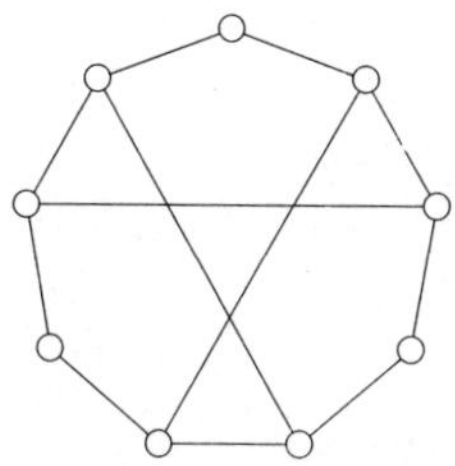

7.5.2 Suppose G is a Δ-edge-critical graph and vw is an edge of G. Prove that $d(v) + d(w) \geq \Delta + 2$.
7.5.3 Show that any edge-critical graph is 2-connected.
7.5.4 Suppose G has maximal degree Δ. Prove that $G \backslash G_\Delta$, the graph obtained by deleting the edges of G_Δ from G, can be Δ-colored.
7.5.5 Prove Lemma 7.25. [26]

8

Planarity

8.1 Representations and Crossings

The two diagrams in Figure 8.1 represent the same graph, a K_4 with vertices a, b, c and d. As diagrams they are quite different: in the version $K_4(1)$, the edges ac and bd cross; in $K_4(2)$ there are no crossings. We shall refer to the two diagrams as different *representations* of the graph in the plane. The *crossing number* of a representation is the number of different pairs of edges that cross; the crossing number $\nu(G)$ of a graph G is the minimum number of crossings in any representation of G. A representation is called *planar* if it contains no crossings, and a *planar graph* is a graph that has a planar representation. In other words, a *planar graph* G is one for which $\nu(G) = 0$. Figure 8.1 shows that $\nu(K_4) = 0$.

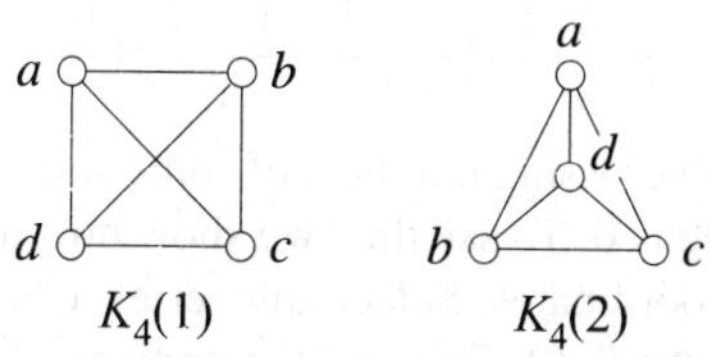

Fig. 8.1. Two representations of K_4

There are many applications of crossing numbers. An early use was in the design of railway yards, where it is inconvenient to have the different lines crossing, and it is better to have longer track rather than extra intersections. An obvious extension of this idea is freeway design. At a complex intersection, fewer crossings means fewer expensive flyover bridges. More recently, small crossing numbers have proven important in the design of VLSI chips; if two parts of a circuit are not to be connected electrically, but they cross, a costly insulation process is necessary.

In 1944, during the Second World War, Turán (see [60]) was forced to work in a brick factory, using hand-pulled carts that ran on tracks to move bricks from kilns to stores. When tracks crossed, several bricks fell from the carts and had to be replaced by hand. The tracks were modeled by a complete bipartite graph with one set of vertices representing kilns and the other representing stores, so to minimize the man-hours lost in replacing bricks, it was necessary to find $\nu(K_{m,n})$ (and find a representation of $K_{m,n}$ that realized the minimum number of crossings). This problem is called "Turán's brick factory problem," and is still unsolved. The best-known bound, which is conjectured to be the best possible, is given in Theorem 8.1.

Theorem 8.1. *The crossing number of $K_{m,n}$ satisfies*

$$\nu(K_{m,n}) \leq \left\lfloor \frac{m}{2} \right\rfloor \left\lfloor \frac{m-1}{2} \right\rfloor \left\lfloor \frac{n}{2} \right\rfloor \left\lfloor \frac{n-1}{2} \right\rfloor. \tag{8.1}$$

For a proof, see Exercise 8.1.8.

Kleitman [77] proved that equality holds in (8.1) when $m \leq 6$. In particular,

$$\nu(K_{6,n}) = 6 \left\lfloor \frac{n}{2} \right\rfloor \left\lfloor \frac{n-1}{2} \right\rfloor. \tag{8.2}$$

The crossing numbers of complete graphs also present difficulties. A very simple (but very bad) upper bound is easily found. If the v vertices of K_v are arranged in a circle, and all the edges are drawn as straight lines, then every set of four vertices contributes exactly one crossing. So

$$\nu(K_v) \leq \binom{v}{4}.$$

More sophisticated constructions (see, for example, [60]) can be used to prove

Theorem 8.2.

$$\nu(K_v) \leq \frac{1}{4} \lfloor \frac{v}{2} \rfloor \lfloor \frac{v-1}{2} \rfloor \lfloor \frac{v-2}{2} \rfloor \lfloor \frac{v-3}{2} \rfloor.$$

It is easy to calculate the crossing numbers of some small graphs. For example, the crossing number of any tree is 0. To see this, we show how to draw a tree in the plane using ordinary Cartesian coordinates. Select any vertex x of the tree, and represent it by the point with coordinates $(0, 0)$. Suppose the vertices adjacent to x are $y_0, y_2, \ldots$. They are represented by $(0, 1), (1, 1), (2, 1), \ldots$. If there are k vertices other than x adjacent to y_i, they are represented by $(i, 2)$, $(i + \frac{1}{k}, 2)$, $(i + \frac{2}{k}, 2), \ldots, (i + \frac{k-1}{k}, 2)$. This process of subdividing continues and provides a representation of the tree with no crossings.

To prove that $\nu(K_5) = 1$, we start by considering smaller complete graphs. The representation of K_3 as a triangle is essentially unique: one can introduce a crossing only by a fanciful, twisting representation of one or more edges. If the K_3 with vertices a, b, c is drawn as a triangle, we obtain a representation of K_4 by introducing a new vertex d and joining it to the other three. If it is inside the triangle, we get the representation shown as $K_4(2)$ in Figure 8.1; if it is outside, it is easy to draw the edges ad, bd, cd so that they do not cross the triangle: for example,

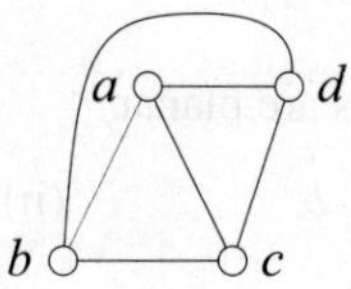

$K_4(3)$

This representation $K_4(3)$ is essentially equivalent to $K_4(2)$, with the vertices relabeled. (Remember, the shape of the edges is unimportant; only the connections matter.). So any planar representation of K_5 can be obtained by introducing another vertex into $K_4(2)$. If a new vertex e is introduced inside the triangle abd, then the representation of ce must cross one of ab, ad or bd. Similarly, the introduction of e inside any triangle causes a crossing involving the edge joining e to the vertex that is not on the triangle. (The "outer area" is considered to be the triangle abc.) So $\nu(K_5) \geq 1$. A representation with one crossing is easy to find, so $\nu(K_5) = 1$.

Another graph with crossing number 1 is $K_{3,3}$; proving this is left as an exercise.

Suppose G is planar. If a new graph were constructed by inserting a new vertex of degree 2 into the middle of an edge (*dividing an edge*), or by deleting a vertex of degree 2 and joining the two vertices adjacent to it (*eliding a vertex*), that new graph will also be planar. Graphs that can be obtained from each other in this way are called *homeomorphic*.

Since K_5 and $K_{3,3}$ are not planar, it follows that a graph having either as a subgraph could not be planar. Moreover, a graph that is homeomorphic to one with a *subgraph* homeomorphic to K_5 or $K_{3,3}$ cannot be planar. In fact, Kuratowski [82] proved that this necessary condition for planarity is also sufficient.

Theorem 8.3. *G is planar if and only if G is homeomorphic to a graph containing no subgraph homeomorphic to K_5 or $K_{3,3}$.*

The proof can be found, for example, in [17] or [66].

Exercises 8.1

8.1.1 Prove that a graph G is planar if and only if every block of G is planar.

8.1.2 Prove that $\nu(K_{1,n}) = \nu(K_{2,n}) = 0$ for any n.

8.1.3 Prove that $\nu(K_{3,3}) = 1$.

8.1.4 What is $\nu(K_{3,4})$?

8.1.5 Prove that the Petersen graph is not planar. What is its crossing number?

8.1.6 Which of the following graphs are planar?

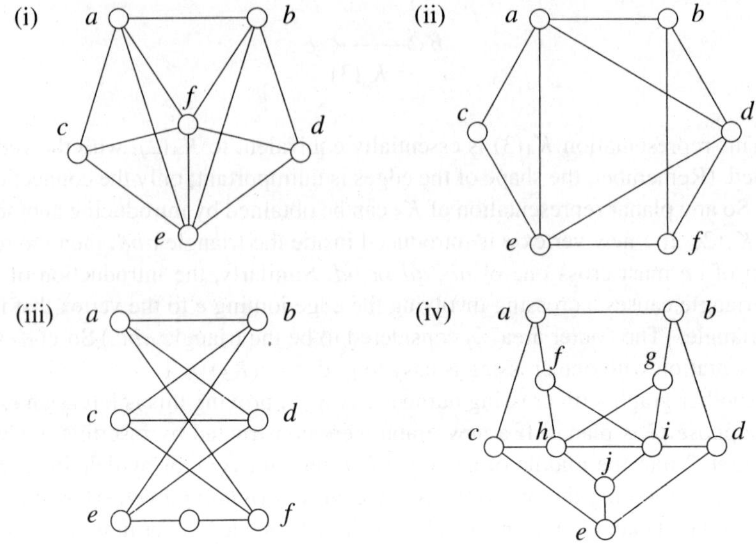

8.1.7 Suppose the graph G has seven or fewer vertices. Prove that either G or its complement $\overline{G}$ is planar.

8.1.8 Consider the complete bipartite graph $K_{m,n}$. In a Cartesian coordinate system, select $\lfloor \frac{m}{2} \rfloor$ points on the positive x-axis, $\lceil \frac{m}{2} \rceil$ points on the negative x-axis, $\lfloor \frac{n}{2} \rfloor$ points on the positive y-axis and $\lceil \frac{n}{2} \rceil$ points on the negative y-axis. The figure formed by joining all pairs with one point on the x-axis and one on the y-axis is a representation of $K_{m,n}$. Use this representation to prove Theorem 8.1.

8.2 Euler's Formula

In each plane representation of a connected graph, the plane is partitioned into regions called *faces*: a face is an area of the plane entirely surrounded by edges of the graph, that contains no edge. It is convenient to define one exterior face, corresponding to the plane outside the representation. For example, the exterior face is $abyz$ in Figure 8.2(a), and $abxyz$ in Figure 8.2(b). The cycle $abxyz$ is not a face in Figure 8.2(a), because it contains the edge bz.

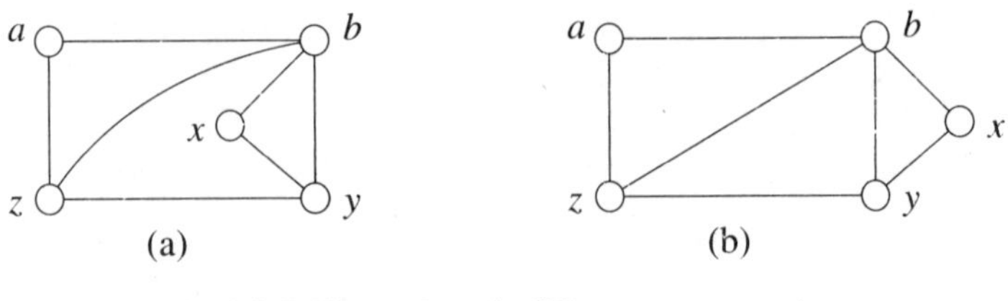

Fig. 8.2. Different faces in different representations

Special care must be taken with bridges. The representations shown in Figure 8.3 have exterior faces $abcx$, $abcxyz$ and abx respectively, and $abcx$ is a face in the third case. A tree has exactly one face.

The edges touching a face are said to *bound* that face, and are collectively called the *boundary*. It is common to think of the boundary as a cycle, but in the case of bridges this need not be true. In particular, any cycle must have at least three edges, but there are three connected graphs (K_1, K_2 and P_3) with faces having boundary smaller than three edges.

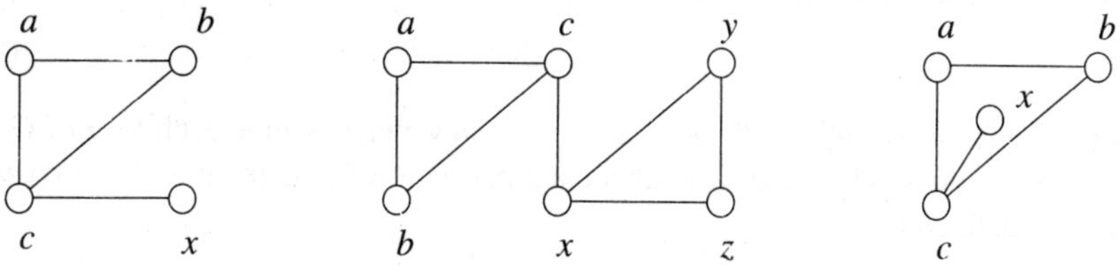

Fig. 8.3. Graphs with bridges

The following theorem was proved by Euler in the eighteenth century.

Theorem 8.4. *Suppose that a plane representation of the connected planar graph G has v vertices, e edges and f faces. Then*

$$v - e + f = 2. \tag{8.3}$$

Proof. By induction on e. It is easy to see that the theorem holds when e is small: if $e = 0$, then the graph must be K_1, which has $v = 1$ and $f = 1$; if $e = 1$ or $e = 2$, we have a path with $e + 1$ vertices and one face. Now assume the theorem is true for all graphs with E or less edges, and suppose $e = E + 1$.

If G is a tree, then $v = e + 1$ and $f = 1$, so

$$v - e + f = e + 1 - e + 1 = 2.$$

Otherwise G contains a cycle. Select an edge that lies in this cycle. That edge will lie separating two faces (one possibly being the exterior face). If it is deleted, one obtains a graph with one fewer edge and one fewer face than the original. It has E edges so, by induction, equation (8.3) is satisfied; in terms of the original graph,

$$v - (e - 1) + (f - 1) = 2,$$

and (8.3) follows. □

Corollary 8.5. *All plane representations of the same connected planar graph have the same number of faces.*

Proof. Suppose a graph has v vertices and e edges; suppose it has two plane representations, with f and f' faces respectively. Then

$$v - e + f = 2 = v - e + f',$$

so $f = f'$. □

Because of Corollary 8.5, one speaks of the number of faces of a connected planar graph, instead of the number of faces in a particular representation of it.

In a planar graph there are various restrictions on the values of v, e and f.

Theorem 8.6. *Suppose a connected planar graph G has v vertices, e edges and f faces, and no component has fewer than three vertices. Then*

$$3f \leq 2e. \tag{8.4}$$

Proof. The result is true for P_2 ($e = 2$, $f = 1$) so we can assume each face of G has at least three edges. The edge-face adjacency matrix A of G is the $e \times f$ matrix with entries a_{ij} defined by

$$a_{ij} = \begin{cases} 1 \text{ if the } i\text{-th edge lies in the boundary of the } j\text{-th face,} \\ 0 \text{ otherwise.} \end{cases}$$

Let σ be the sum of all the entries of A. Each edge bounds at most two faces, so the sum of each row is at most 2. There are e rows, so $\sigma \leq 2e$. As each face has at least three edges in its boundary, $3f \leq \sigma$. The result follows. □

Theorem 8.7. *If a connected planar graph has v vertices and e edges, where $v \geq 3$, then $e \leq 3v - 6$.*

Proof. Suppose the graph has f faces. By Theorem 8.6, $3f \leq 2e$, so $f \leq \frac{2e}{3}$. By Theorem 8.4, $v - e + f = 2$, so $v - e + \frac{2e}{3} \geq 2$. But this implies that $3v - e \geq 6$, and the theorem is proved. □

Example. The complete graph, K_v, is not planar for $v \geq 5$. If $v \geq 5$, then K_v contains K_5 as a subgraph. Hence it is sufficient to show that K_5 is not planar. We already saw this fact in the preceding section, but the following neat proof is now available: Suppose K_5 is planar. We have $v = 5$, $e = 10$, so that $e > 3v - 6$, contradicting Theorem 8.7.

One can refine this example. Suppose G, a graph with v vertices and e edges, can be drawn with c crossings. Suppose you replace each crossing by a new vertex: in other words, if edges xy and zt cross, introduce a new vertex w and replace xy and zt by edges xw, wy, zw and wt. The new graph has $v + c$ vertices and $e + 2c$ edges and is planar, so $e + 2c \leq 3(v + c) - 6$, and $c \geq 6 + e - 3v$. For example, K_6 has 6 vertices and 15 edges, so $c \geq 3$. K_6 can be drawn with 3 crossings, so $\nu(K_6) = 3$.

The following useful consequence of Theorem 8.7 is left as an exercise (see Exercise 8.2.4).

Corollary 8.8. *Every planar graph has at least one vertex of degree smaller than 6.*

Exercises 8.2

8.2.1 Find the values of v, e and f for the following graphs, and verify Euler's formula for them.

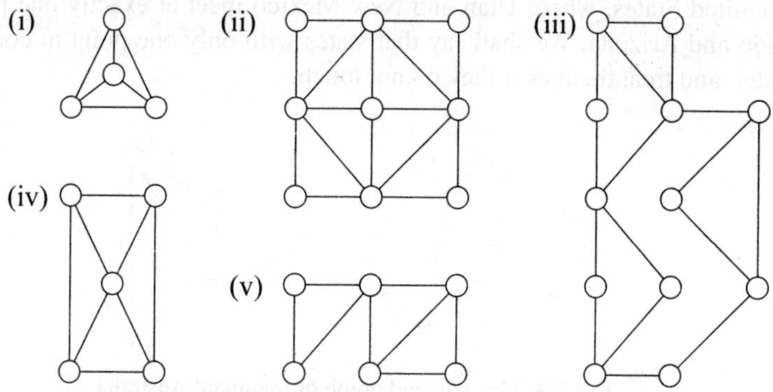

8.2.2 Prove that a connected planar graph with v vertices, $v \geq 3$, has at most $2v - 4$ faces.

8.2.3 Prove that if a connected planar graph has v vertices, $v \geq 3$, and every face in a certain plane representation has four edges, then the graph has $2v - 4$ edges. Hence prove:

(i) If a planar graph with v vertices contains no triangles, then it has at most $2v - 4$ edges;

(ii) The complete bipartite graph $K_{m,n}$ is not planar when $m \geq 3$ and $n \geq 3$.

8.2.4 Suppose G is a graph with v vertices and e edges, and suppose every vertex of G has degree at least 6. Prove that

$$e \geq 3v.$$

Hence prove that every planar graph contains a vertex of degree at most 5.

8.2.5 Prove that in a planar bipartite graph with v vertices, e edges and f faces,

$$2f \leq e.$$

8.2.6 Verify that K_6 can be drawn with 3 crossings.

8.3 Maps, Graphs and Planarity

By a *map* we mean what is usually meant by a map of a continent (showing countries) or a country (showing states or provinces). However we make one restriction. Sometimes one state can consist of two disconnected parts (in the United States, Michigan consists of two separate land masses, unless we consider man-made constructions such as the Mackinaw Bridge). We exclude such cases from consideration.

Given a map, we can construct a graph as follows: the vertices are the countries or states on the map, and the two vertices are joined by an edge precisely when the corresponding countries have a common border. As an example, Figure 8.4 shows a map of the mainland of Australia, divided into states and territories, and its corresponding graph.

Suppose two states have only one point in common. This happens for example, in the United States, where Utah and New Mexico meet at exactly one point, as do Colorado and Arizona. We shall say that states with only one point in common have no border, and treat them as if they do not touch.

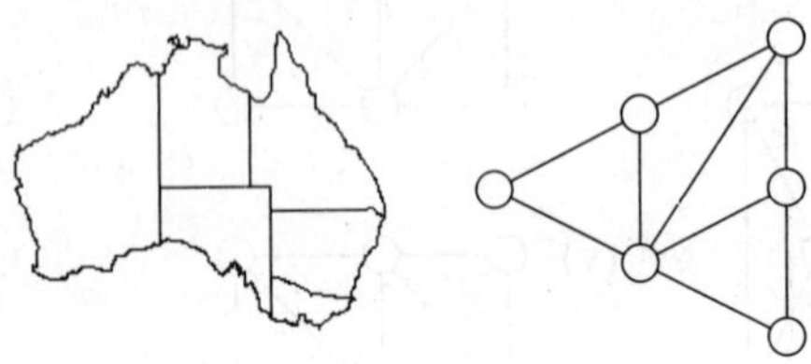

Fig. 8.4. The map and graph of mainland Australia

It is always possible to draw the graph corresponding to a map without crossings. To see this, draw the graph on top of the map by putting a vertex inside each state and joining vertices by edges that pass through common state borders so the graph of any map is planar. Conversely, any planar graph is easily represented by a map. Therefore the theory of maps (with the two stated restrictions) is precisely the theory of planar graphs.

In 1852, William Rowan Hamilton wrote to Augustus de Morgan concerning a problem that had been posed by a student, Frederick Guthrie. (Part of the correspondence is quoted in [14].) Guthrie said: cartographers know that any map (our definition) can be colored using four or less colors; is there a mathematical proof? (Guthrie [59] later pointed out that the question had come from his brother, Francis Guthrie.)

Kempe [75] published a purported proof in 1879. It was thought that the matter was over, but in 1890 Heawood pointed out a fallacy in Kempe's proof. (We explore the problem in Exercise 8.3.3.) Heawood did however repair the proof sufficiently to prove the following weaker result.

Theorem 8.9. [70] *Every planar graph can be colored in five colors.*

Clearly Theorem 8.9 is the five-color analog of Guthrie's four-color map problem.

Proof. We assume the theorem is false, so some planar graphs require six colors. From these, select one that has the minimum number of vertices and has no isolated vertices; call it G. By Corollary 8.8, there is a vertex x in G whose degree is at most 5. $G - x$ has fewer edges than G, so it is 5-colorable. Select a 5-coloring ξ of G. Observe that every color used in ξ must be represented among the neighbors of x: if color c were missing, one could set $\xi(x) = c$ and thus extend ξ to a 5-coloring of G, which is impossible. So $d(x) = 5$. We shall write x_1, x_2, x_3, x_4, x_5 for the five vertices adjacent to x in G, and assume $\xi(x_i) = c_i$. Without loss of generality we shall assume that the vertices x_1, x_2, x_3, x_4, x_5 occur in order around x in some plane representation of G, as shown in Figure 8.5(a).

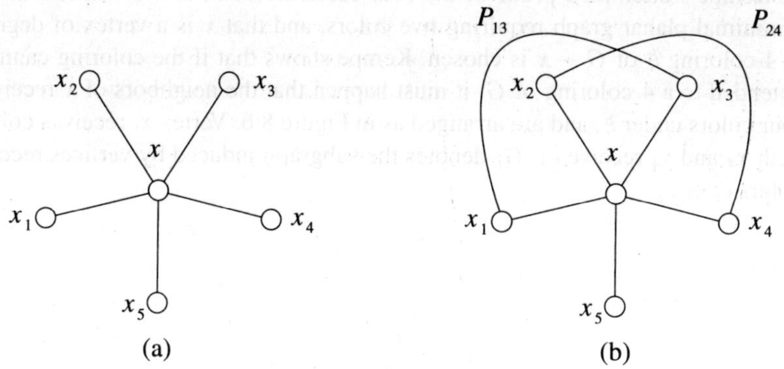

Fig. 8.5. A neighborhood used in the 5-color proof

Consider the induced subgraph G_{ij} of G whose vertex-set consists of all vertices that receive color c_i or color c_j under ξ. If x_i and x_j lie in different components of G_{ij}, then one could exchange colors among all the vertices in one component — say the component containing x_i — and the result would still be a 5-coloring of G. However there would be no vertex of color c_i adjacent to x, so the new coloring could be extended to a 5-coloring of G by allocating c_i to x. This is impossible. Therefore x_i and x_j must lie in the same component of G_{ij}, and there must exist a path P_{ij} from x_i to x_j in G, all of whose vertices receive either c_i or c_j under ξ.

Now consider the paths P_{13} and P_{24}. It is clear from Figure 8.5(b) that these two paths must cross. But the only way two paths in a plane representation can cross is at a common vertex, which is impossible because the vertices in P_{13} do not receive the same colors as those in P_{24}. □

After Heawood's paper appeared, there was renewed interest in the 4-color problem. Because it was easy to state and tantalizingly difficult to prove, it became one of the most celebrated unsolved problems in mathematics, second only to Fermat's Last Theorem. In 1976, Appel and Haken [3] finally proved that any planar graph — and therefore any planar map — can be colored in at most four colors. Their proof involved computer analysis of a large number of cases, so many that human analysis of all the cases is not feasible. We state their theorem as

Theorem 8.10. *Every planar graph can be colored using at most four colors.*

For details of the proof, see [4] or [106].

Exercises 8.3

.3.1 Find the graph of the mainland provinces and territories of Canada. (Treat Labrador as a province.) How many colors does it need?

.3.2 Use the result of Exercise 8.2.4 to prove that any planar graph can be colored in at most six colors (without using any results from this section).

8.3.3 In Kempe's attempted proof of the four color theorem, it is assumed that G is a minimal planar graph requiring five colors, and that x is a vertex of degree 5. A 4-coloring ξ of $G - x$ is chosen. Kempe shows that if the coloring cannot be extended to a 4-coloring of G, it must happen that the neighbors of x receive all four colors under ξ, and are arranged as in Figure 8.6. Vertex x_i receives color c_i; both x_4 and y_4 receive c_4. G_{ij} denotes the subgraph induced by vertices receiving colors c_i or c_j.

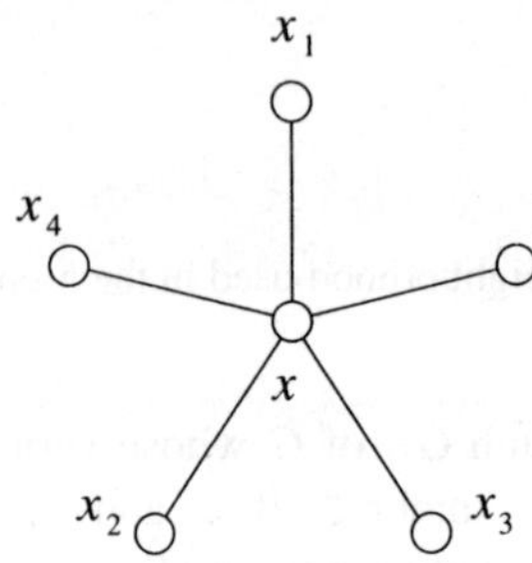

Fig. 8.6. A neighborhood in Kempe's proof

Kempe argues that there must be a path P_{12} from x_1 to x_2 in G_{12}, or else one could interchange colors c_1 and c_2 in the component of G_{12} that contains x_1, and then color x with c_1. Similarly there must be a path P_{13} from x_1 to x_3 in G_{13}.
By planarity, there can be no path from x_2 to y_4 in G_{24}, because such a path would need to cross P_{13}. So one can exchange colors c_2 and c_4 in the component of G_{24} that contains y_4. Similarly, one can exchange colors c_3 and c_4 in the component of G_{324} containing x_4. Now x can receive color c_4.
What is wrong with this argument?

9

Labeling

9.1 Introduction; Graceful Labelings

In general, a *labeling* (or *valuation*) of a graph is a map that carries some set of graph elements to numbers, most often to the positive or nonnegative integers. The most common choices of domain are the set of all vertices and edges (such labelings are called *total* labelings), the vertex-set alone (*vertex*-labelings), or the edge-set alone (*edge*-labelings). Other domains are possible.

We shall call two labelings of the same graph *automorphism-equivalent* if one can be transformed into the other by an automorphism of the graph.

Rosa ([102, 78]) introduced the idea of β-valuations, or as they are now called *graceful labelings*. A graceful labeling of a graph G is a one-to-one mapping γ from the set of all vertices to the integers $S_G = \{0, 1, \ldots, |E(G)|\}$ such that every non-zero member of S_G occurs as the difference between the labels on the endpoints of an edge. That is, if we extend γ to edges by defining $\gamma(xy) = |\gamma(x) - \gamma(y)|$, then every member of $1, 2, \ldots, |E(G)|\}$ arises (exactly) once among the edge-labels. A graph is called graceful if it has a graceful labeling.

Figure 9.1 shows a graceful labeling of K_4. In the right-hand diagram, the corresponding edge-labels are shown.

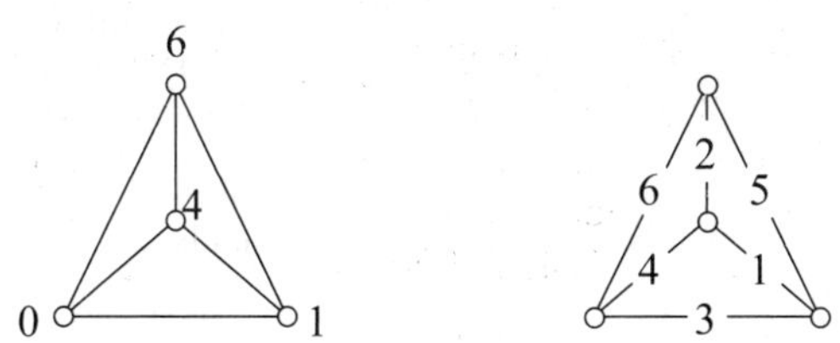

Fig. 9.1. Graceful labeling of K_4

If γ is a graceful labeling on G, define another labeling γ^* by

$$\gamma^*(x) = |E(G)| - \gamma(x)$$

for every vertex x. Then γ^* is also graceful; in fact, γ and γ^* induce the same edge-labeling. γ^* is called the *complementary labeling* or *dual* of γ. We formally define two graceful labelings γ and δ to be *equivalent* if δ is *automorphism-equivalent* to either γ or γ^*.

When testing a small graph with e edges to see whether it is graceful, first notice that there must be adjacent vertices labeled 0 and e in order for edge-label e to occur. Similarly there must be an edge from 0 to $e-1$ or from 1 to e.

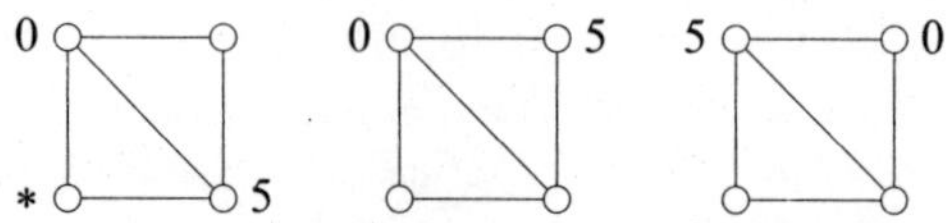

Fig. 9.2. Finding a graceful labeling of $K_4 - e$

Example. We shall find a graceful labeling of $K_4 - e$. This graph has two types of edge. Initially there are three ways of labeling an edge with 0 and 5, as shown in Figure 9.2. However the third is clearly dual to the second.

In order to achieve edge-label 1 in the first case, one of the remaining vertices must be labeled 1 or 4. Assume the vertex marked $*$ is labeled 1 (the 4 case is dual). Then labeling the other vertex 3 results in a graceful labeling.

The reader should check that the partial labeling in the second part of the diagram can also be completed.

Theorem 9.1. *There is no graceful labeling of K_v when $v > 4$.*

Proof. Suppose there is a graceful labeling of K_v, where $v \geq 5$. Let S be the set of vertex-labels used in the labeling. We write e for $\frac{1}{2}v(v-1)$, the number of edges in K_v. It causes no confusion if we use the symbol x for the vertex that receives label x.

The vertex labels must include 0 and e and (to achieve edge-label 1) either 1 or $e-1$; the choices $\{0, 1, e\}$ and $\{0, e-1, e\}$ lead to dual labelings, so we can assume labels $\{0, 1, e\} \subseteq S$.

There can be no vertex labeled 2, for if there were, two edges would receive label 1: $2 \sim 1$ and $1 \sim 0$. Also label $e-1$ is impossible: we would have $e \sim (e-1)$ and $1 \sim 0$. So $(e-2, e)$ is the only possible edge with label 2, and $\{0, 1, e-2, e\} \subseteq S$.

These vertex labels induce edges labeled $1, 2, e-3, e-2, e-1$ and e. Since $e > 6$, edge label $e-4$ is still needed. An argument similar to those above shows that the only suitable new vertex label is 4, and the edge labels are

$$1,\ 2,\ 3,\ 4,\ e-6,\ e-4,\ e-3,\ e-2,\ e-1,\ e.$$

When $e \geq 10$, label $e-5$ is not in this list. To obtain $e-5$ we need vertex-label $3, 5, e-5, e-4$ or $e-1$. But each of these possibilities leads to a duplicated edge-label: $3-0=3$, $5-1=4$, $(e-5)-1=4$, $e-(e-4)=4$, $e-(e-1)=1$. So no labeling is possible. □

Theorem 9.2. *Suppose G is an Eulerian graph with e edges. If G is graceful, then* $e \equiv 0$ *or* $3 \pmod 4$.

Proof. Suppose γ is a graceful labeling of G. Write $z_1, z_2, \ldots, z_e$ for the edges of G; and denote the endpoints of z_i by x_i and y_i, where $\gamma(x_i) > \gamma(y_i)$. Then

$$\sum_{i=1}^{e} \gamma(z_i) = \sum_{i=1}^{e} \gamma(x_i) - \sum_{i=1}^{e} \gamma(y_i)$$

$$= \sum_{i=1}^{e} \gamma(x_i) + \sum_{i=1}^{e} \gamma(y_i) - 2\sum_{i=1}^{e} \gamma(y_i).$$

In the list $x_1, x_2, \ldots, x_e, y_1, y_2, \ldots, y_e$ the number of times each vertex occurs is equal to its degree. Since G is Eulerian, each of these degrees is even. So $\sum_{i=1}^{e} \gamma(x_i) + \sum_{i=1}^{e} \gamma(y_i)$ is even. Therefore $\sum_{i=1}^{e} \gamma(z_i)$ is even.

On the other hand, since γ is graceful, $\sum_{i=1}^{e} \gamma(z_i)$ is the sum of the first e positive integers, $\frac{1}{2}e(e-1)$. This is even only if $e \equiv 0$ or $3 \pmod 4$. □

There has been considerable interest in graceful labelings of trees; Kotzig (quoted in [102]) conjectured that all trees are graceful. However, this conjecture is far from settled. All stars and paths are graceful (see the exercises).

A *caterpillar* is a tree for which, if all *leaves* (vertices of degree 1 and their associated edges) were removed, the result is a path.

Theorem 9.3. [102] *All caterpillars are graceful.*

Proof. Suppose T is a caterpillar with v vertices and H is the path formed from T by deleting all the leaves. Select an endpoint of H (a vertex of degree 1 in the path) and name it x_0; the vertex adjacent to it in H is called x_1, and so on along H. Write X for the set of all vertices of T whose distance from x_0 is even (including x_0 itself), and Y for the set of vertices of odd distance. Every edge connects two vertices, one in X and the other in Y.

Assign label $v-1$ to x_0. Label the neighbors of x_0 with $0, 1, 2, \ldots,$ where the neighbor receiving the greatest label is x_1, the neighbor of x_0 in H. Assign labels $v-2, v-3, \ldots,$ to the neighbors of x_1 other than x_0; the largest label goes to x_2.

Continue as follows: after x_{2i} receives its label, assign increasing integer labels to its neighbors other than x_{2i-1} starting with the smallest unused label, assigning the largest label to x_{2i+1}; then assign labels to the neighbors of x_{2i+1} other than x_{2i} in decreasing order, starting with the largest unused integer smaller than v, ending by labeling x_{2i+2}.

The result will be a labeling in which members of X receive labels $v-1, v-2, \ldots, v-|X|$ and members of Y receive labels $0, 1, \ldots, |Y|$. It is easily checked to be graceful. □

An example is shown in Figure 9.3.

Graceful labelings of trees were first studied in an attempt to prove a conjecture of Ringel [100]. He conjectured that, given any tree T with n edges, it is possible to write the complete graph K_{2n+1} as a union of edge-disjoint copies of T. The connection is shown in the following theorem.

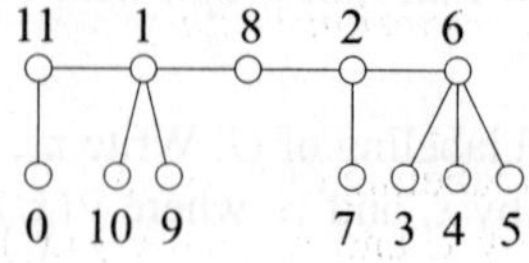

Fig. 9.3. Graceful labeling of a caterpillar

Theorem 9.4. *Suppose T is a graceful tree on $n+1$ vertices. Then K_{2n+1} is a union of $2n+1$ edge-disjoint copies of T.*

Proof. We use a K_{2n+1} whose vertices are the integers $\{0, 1, \dots, 2n\}$ modulo $2n+1$. Suppose γ is a graceful labeling of T. We identify the vertex x of T with the vertex $\gamma(x)$ of the K_{2n+1}. Then T is a subgraph of K_{2n+1} with vertices $\{0, 1, \dots, n\}$.

For each integer $s \in \{0, 1, \dots, 2n\}$ we construct a tree T_s as follows. Edge xy belongs to T_s if and only if $(x-s)(y-s)$ is an edge of T. Thus $T_0 = T$, and if the vertices of K_{2n+1} are written in order equally placed around a circle, T_s is obtained by rotating T_0 through $s/(2n+1)$ of a revolution. Each T_s is isomorphic to T.

Now each edge xy of K_{2n+1} will belong to precisely one of the trees T_s, because there is precisely one edge of T that receives label $\pm(x-y)$ under γ. So we have the required decomposition.

If Kotzig's conjecture is true, that all trees are graceful, this will prove Ringel's conjecture. However, the Kotzig conjecture is stronger. A decomposition of K_{2n+1} into copies of an n-edge tree need not be of a cyclic nature.

Exercises 9.1

9.1.1 Find graceful labelings of P_3 and K_3.

9.1.2 Verify that the partial graceful labeling in Figure 9.2 can be completed.

A9.1.3 Find a graceful labeling of $P_3 \cup K_4$.

H9.1.4 Find a graceful labeling of $P_3 \cup K_3$.

9.1.5 Show that $P_2 \cup K_3$ and $P_3 \cup K_3$ have no graceful labelings.

A9.1.6 Show that no nontrivial forest (that is, a forest containing at least two trees) is graceful.

9.1.7 Show that the star $K_{1,n}$ is graceful for every n.

A9.1.8 Show that the path P_v is graceful for every v.

9.1.9 Suppose G is a graceful graph with e edges. Write X and Y for the sets of vertices with even and odd labels respectively. Show that the set $[X, Y]$ contains precisely $\frac{1}{2}(e+1)$ edges.

9.2 Edge-Magic Total Labeling

A *magic square* of side n is an $n \times n$ array whose entries are an arrangement of the integers $\{1, 2, \dots, n^2\}$, in which all elements in any row, any column, or either the main diagonal or main back-diagonal, add to the same sum. Small examples include

1	15	8	10
12	6	13	3
14	4	11	5
7	9	2	16

1	7	13	19	25
18	24	5	6	12
10	11	23	7	4
22	1	9	15	16
14	20	23	2	8

An excellent reference on magic squares is [2].

Various authors have introduced graph labelings that generalize the idea of a magic square; there has been a lot of duplication of terminology. To avoid confusion we define a labeling to be *edge-magic* if the sum of all labels associated with an edge equals a constant independent of the choice of edge, and *vertex-magic* if the same property holds for vertices. (This terminology could be extended to other substructures: face-magic, for example.) The domain of the labeling is specified by a modifier on the word "labeling." We shall always require that the labeling is a one-to-one map onto the appropriate set of consecutive integers starting from 1.

For example, Kotzig and Rosa [78] defined a magic labeling to be a total labeling in which the labels are the integers from 1 to $|V(G)|+|E(G)|$ and the sum of labels on an edge and its two endpoints is constant. In 1996 Ringel and Llado [101] redefined this type of labeling and called the labelings *edge-magic*. We shall call them *edge-magic total labelings*. On the other hand, Stewart (see, for example, [111]), called such a labeling *supermagic*. Sedláček [110] originally proposed the study of edge-labelings with the magic property on vertices, but did not restrict the values of the labels in any way; they could be any reals.

To discuss these labelings, we define the *weight* of a graph element to be the sum of all labels associated with the element. For example, the weight of vertex x under an edge labeling α is

$$wt(x) = \sum_{y \sim x} \alpha(xy).$$

The weight of x under a total labeling λ is

$$wt(x) = \lambda(x) + \sum_{y \sim x} \lambda(xy),$$

while

$$wt(xy) = \lambda(x) + \lambda(xy) + \lambda(y).$$

If necessary, the labeling can be specified by a subscript, as in $wt_\lambda(x)$.

We formally define an *edge-magic total labeling* or EMTL on a graph G to be a one-to-one map λ from $V(G) \cup E(G)$ onto the integers $1, 2, \ldots, v+e$, where $v = |V(G)|$ and $e = |E(G)|$, with the property that, given any edge (xy),

$$\lambda(x) + \lambda(xy) + \lambda(y) = k$$

for some constant k. In other words, $wt(xy) = k$ for any choice of edge xy. k is called the *magic sum* of G. A graph is called *edge-magic* if it has an edge-magic total labeling.

As an example, Figure 9.4 shows an edge-magic total labeling of $K_4 - e$.

We shall frequently refer to the sum of consecutive integers, so we define

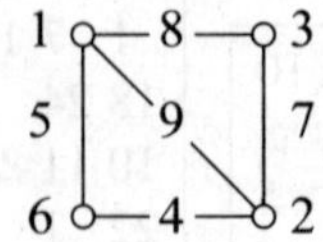

Fig. 9.4. An EMTL of $K_4 - e$ with $k = 12$.

$$\sigma_i^j = (i+1) + (i+2) + \cdots + j = i(j-i) + \tbinom{j+1}{2}. \tag{9.1}$$

Suppose the graph G has v vertices $\{x_1, x_2, \ldots, x_v\}$ and e edges. Vertex x_i has degree d_i and receives label a_i. Among the labels, write S for the set $\{a_i : 1 \le i \le v\}$ of vertex labels, and s for the sum of elements of S. Then S can consist of the v smallest labels, the v largest labels, or somewhere in between, so

$$\sigma_0^v \le s \le \sigma_e^{v+e},$$

$$\binom{v+1}{2} \le s \le ve + \binom{v+1}{2}. \tag{9.2}$$

Clearly, $\sum_{xy \in E}(\lambda(xy) + \lambda(x) + \lambda(y)) = ek$. This sum contains each label once, and each vertex label a_i an additional $d_i - 1$ times. So

$$ke = \sigma_0^{v+e} + \sum (d_i - 1)a_i. \tag{9.3}$$

If e is even, every d_i is odd and $v + e \equiv 2 (\text{mod } 4)$, then (9.3) is impossible. We have

Theorem 9.5. [101] *If G has e even and $v + e \equiv 2 (\text{mod } 4)$, and every vertex of G has odd degree, then G has no EMTL.*

Corollary 9.6. *The complete graph K_n is not edge-magic when $n \equiv 4 (\text{mod } 8)$. The n-spoke wheel W_n is not edge-magic when $n \equiv 3 (\text{mod } 4)$.*

(We shall see in Section 9.9.3 that K_n is never edge-magic for $n > 6$, so the first part of the Corollary really only eliminates K_4.)

Equation (9.3) may be used to provide bounds on k. Suppose G has v_j vertices of degree j, for each i up to Δ, the largest degree represented in G. Then the ke cannot be smaller than the sum obtained by applying the v_Δ smallest labels to the vertices of degree Δ, the next-smallest values to the vertices of degree $\Delta - 1$, and so on; in other words,

$$ke \ge (d_\Delta - 1)\sigma_0^{v_\Delta} + (d_{\Delta-1} - 1)\sigma_{v_\Delta}^{v_\Delta+(v_{\Delta-1})} + \sigma_{v_\Delta+(v_{\Delta-1})+\cdots+v_3}^{v_\Delta+(v_{\Delta-1})+\cdots+v_2} + \binom{v+e+1}{2}.$$

An upper bound is achieved by giving the *largest* labels to the vertices of highest degree, and so on.

In particular, suppose G is regular of degree d. Then (9.3) becomes

$$ke = (d-1)s + \sigma_0^{v+e} = (d-1)s + \tfrac{1}{2}(v+e)(v+e+1) \tag{9.4}$$

or, since $e = \frac{1}{2}dv$,

$$kdv = 2(d-1)s + (v+e)(v+e+1). \tag{9.5}$$

Given a labeling λ, its *dual* labeling λ' is defined by

$$\lambda'(x_i) = (v+e+1) - \lambda(x_i),$$

and for any edge xy,

$$\lambda'(xy) = (v+e+1) - \lambda(xy).$$

It is easy to see that if λ is an edge-magic labeling with magic sum k, then λ' is an edge-magic labeling with magic sum $k' = 3(v+e+1) - k$. The sum of vertex labels in the dual is $s' = v(v+e+1) - s$. Just as in the case of graceful labelings, we define two EMTLs λ and μ to be *equivalent* if λ is *automorphism-equivalent* to either μ or μ'.

Either s or s' will be less than or equal to $\frac{1}{2}v(v+e+1)$. This means that, in order to see whether a given graph has an EMTL, it suffices to check either all cases with $s \le \frac{1}{2}v(v+e+1)$ or all cases with $s \ge \frac{1}{2}v(v+e+1)$ (equivalently, either check all cases with $k \le \frac{3}{2}(v+e+1)$ or all with $k \ge \frac{3}{2}(v+e+1)$).

The cycle C_v is regular of degree 2 and has v edges. In that case, (9.2) becomes

$$v(v+1) \le 2s \le 2v^2 + v(v+1) = v(3v+1),$$

and (9.4) is

$$kv = s + v(2v+1),$$

whence v divides s; in fact $s = (k-2v-1)v$. When v is odd, s has $v+1$ possible values $\frac{1}{2}v(v+1), \frac{1}{2}v(v+3), \ldots, \frac{1}{2}v(v+2i-1), \ldots, \frac{1}{2}v(3v+1)$, with corresponding magic sums $\frac{1}{2}(5v+3), \frac{1}{2}(5v+5), \ldots, \frac{1}{2}(5v+2i+1), \ldots, \frac{1}{2}(7v+3)$. For even v, there are v values $s = \frac{1}{2}v^2 + v, \frac{1}{2}v^2 + 2v, \ldots, \frac{1}{2}v^2 + iv, \ldots, \frac{3}{2}v^2$, with corresponding magic sums $\frac{5}{2}v + 2, \frac{5}{2}v + 3, \ldots, \frac{5}{2}v + i + 1, \ldots, \frac{7}{2}v + 1$.

Kotzig and Rosa [78] proved that all cycles are edge-magic, producing examples with $k = 3v + 1$ for v odd, $k = \frac{5}{2}v + 2$ for $v \equiv 2 (\mathrm{mod}\, 4)$ and $k = 3v$ for $v \equiv 0 (\mathrm{mod}\, 4)$. In [55], labelings are exhibited for the minimum values of k in all cases. We present two proofs here, and leave another as an exercise. In each case the proof consists of exhibiting a labeling.

Theorem 9.7. *If v is odd, then C_v has an edge-magic total labeling with $k = \frac{1}{2}(5v+3)$.*

Proof. Say $v = 2n+1$. Consider the cyclic vertex labeling $(1, n+1, 2n+1, n, \ldots, n+2)$, where each label is derived from the preceding one by adding n (mod $2n+1$). The successive pairs of vertices have sums $n+2, 3n+2, 3n+1, \ldots, n+3$, which are all different. If $k = 5n+4$, the edge labels are $4n+2, 2n+2, 2n+3, \ldots, 4n+1$, as required. We have an edge-magic total labeling with $k = 5n+4 = \frac{1}{2}(5v+3)$ and $s = \frac{1}{2}v(v+1)$ (the smallest possible values).

By duality, we have:

Corollary 9.8. *Every odd cycle has an edge-magic total labeling with $k = \frac{1}{2}(7v+3)$.*

Theorem 9.9. *If v is even, then C_v has an edge-magic total labeling with $k = \frac{1}{2}(5v+4)$.*

Proof. Write $v = 2n$. If n is even,

$$\lambda(u_i) = \begin{cases} (i+1)/2 & \text{for } i = 1, 3, \ldots, n+1 \\ 3n & \text{for } i = 2 \\ (2n+i)/2 & \text{for } i = 4, 6, \ldots, n \\ (i+2)/2 & \text{for } i = n+2, n+4, \ldots, 2n \\ (2n+i-1)/2 & \text{for } i = n+3, n+5, \ldots, 2n-1, \end{cases}$$

while if n is odd,

$$\lambda(u_i) = \begin{cases} (i+1)/2 & \text{for } i = 1, 3, \ldots, n \\ 3n & \text{for } i = 2 \\ (2n+i+2)/2 & \text{for } i = 4, 6, \ldots, n-1 \\ (n+3)/2 & \text{for } i = n+1 \\ (i+3)/2 & \text{for } i = n+2, n+4, \ldots, 2n-1 \\ (2n+i)/2 & \text{for } i = n+3, n+5, \ldots, 2n-2 \\ n+2 & \text{for } i = 2n. \end{cases}$$

□

Corollary 9.10. *Every even cycle has an edge-magic total labeling with $k = \frac{1}{2}(7v+2)$.*

Figure 9.5 shows examples with $v = 7$ and $v = 8$ of the constructions in Theorems 9.7 and 9.9; they have $k = 19$ and 22 respectively. (Only the vertex labels are shown in the figure; the edge labels can be found by subtraction.)

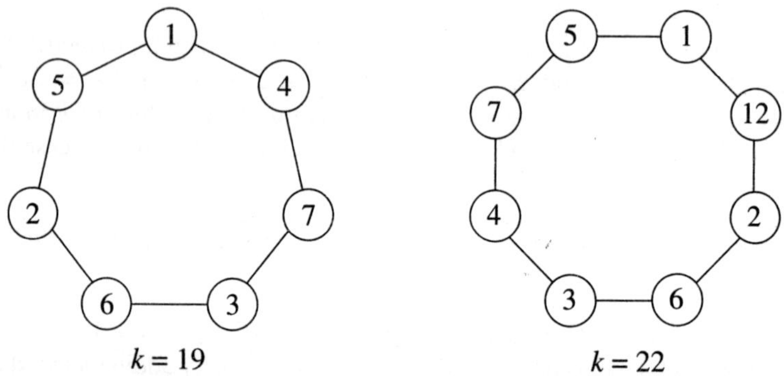

Fig. 9.5. Edge-magic total labelings of C_7 and C_8.

Theorem 9.11. *Every cycle of length divisible by 4 has an edge-magic total labeling with $k = 3v$.*

The path P_n can be viewed as a cycle C_n with an edge deleted. Say λ is an EMTL of C_n with the property that label $2n$ appears on an edge. If that edge is deleted, the result is a P_n with an EMTL.

For every n, there is a labeling of C_n in which $2n$ appears on an edge. Deleting this edge yields a path, on which the labeling is edge-magic. So:

Theorem 9.12. *All paths have EMTLs.*

Theorem 9.13. [78] *The complete bipartite graph $K_{m,n}$ is edge-magic for any m and n.*

Proof. The sets $S_1 = \{n+1, 2n+2, \ldots, m(n+1)\}$, $S_2 = \{1, 2, \ldots, n\}$, define an EMTL with $k = (m+2)(n+1)$. □

In particular, all EMTLs of stars $K_{1,n}$ are easily described.

Lemma 9.14. *In any EMTL of a star, the center receives label 1, $n+1$ or $2n+1$.*

Proof. Suppose the center receives label x. Then

$$kn = \binom{2n+2}{2} + (n-1)x. \tag{9.6}$$

Reducing (9.6) modulo n we find

$$x \equiv (n+1)(2n+1) \equiv 1$$

and the result follows. □

Theorem 9.15. *There are $3 \cdot 2^n$ EMTLs of $K_{1,n}$, up to equivalence.*

Proof. Denote the center of a $K_{1,n}$ by c, the peripheral vertices by $v_1, v_2, \ldots, v_n$ and edge (c, v_i) by e_i. From Lemma 9.14 and (9.6), the possible cases for an EMTL are $\lambda(c) = 1, k = 2n+4$, $\lambda(c) = n+1, k = 3n+3$ and $\lambda(c) = 2n+1, k = 4n+2$. As the labeling is edge-magic, the sums $\lambda(v_i) + \lambda(e_i)$ must all be equal to $M = k - \lambda(c)$ (so $M = 2n+3, 2n+2$ or $2n+1$). Then in each case there is exactly one way to partition the $2n+1$ integers $1, 2, \ldots, 2n+1$ into $n+1$ sets

$$\{\lambda(c)\}, \{a_1, b_1\}, \{a_2, b_2\}, \ldots, \{a_n, b_n\}$$

where every $a_i + b_i = M$. For convenience, choose the labels so that $a_i < b_i$ for every i and $a_1 < a_2 < \cdots < a_n$. Then up to isomorphism, one can assume that $\{\lambda(v_i), \lambda(e_i)\} = \{a_i, b_i\}$. Each of these n equations provides two choices, according as $\lambda(v_i) = a_i$ or b_i, so each of the three values of $\lambda(c)$ gives 2^n EMTLs of $K_{1,n}$. □

It is conjectured ([78], also [101]), that all trees are edge-magic. Unfortunately, this has proven just as intractible as the corresponding conjecture for graceful labelings. Kotzig and Rosa [78] proved that all caterpillars are edge-magic. The proof is left as an exercise; as a hint, we give an example of an EMTL of a caterpillar in Figure 9.6.

Enomoto et al [38] carried out a computer search to show that all trees with fewer than 16 vertices are edge-magic.

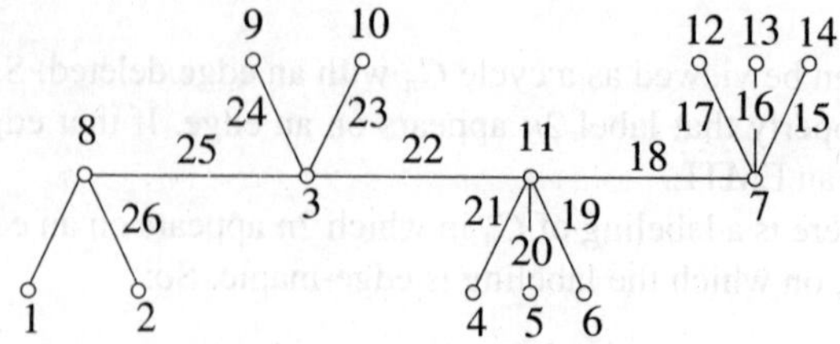

Fig. 9.6. An EMTL of a caterpillar

Vertex-magic total labelings, in which the sum of the labels of all edges adjacent to the vertex x, plus the label of x itself, is constant, have also been studied (see [85]). It is conceivable that the same labeling could be both vertex-magic and edge-magic for a given graph (not necessarily with the same constant). In that case the labeling, and the graph, are called *totally magic*. Totally magic graphs appear to be very rare.

Exercises 9.2

A9.2.1 Find all edge-magic total labelings of K_3.

A9.2.2 Show that every odd cycle has an edge-magic total labeling with $k = 3v + 1$.

9.2.3 Prove that the graph tK_4, consisting of t disjoint copies of K_4, has no edge-magic total labeling when t is odd.

9.2.4 Suppose a regular graph G of degree d is edge-magic. Prove

$$ke = (d-1)s + \sigma_0^{v+e} = (d-1)s + \tfrac{1}{2}(v+e)(v+e+1),$$
$$kdv = 2(d-1)s + (v+e)(v+e+1).$$

A9.2.5 A *triangular book* $B_{3,n}$ consists of n triangles with a common edge. Prove that all triangular books are edge-magic.

9.2.6 An n-sun is a cycle C_n with an edge terminating in a vertex of degree 1 attached to each vertex. Show that all suns are edge-magic. [7]

A9.2.7 What is the range of possible magic sums for an edge-magic total labeling of the *Petersen graph P*? Prove that P is edge-magic.

9.2.8 [7] An (n, t)-kite consists of a cycle of length n with a t-edge path (the *tail*) attached to one vertex. Show that an $(n, 1)$-kite (a kite with tail length 1) is edge-magic. [7]

9.2.9 Prove that all caterpillars are edge-magic.

9.2.10 Find a vertex-magic total labeling of $K_4 - e$.

9.2.11 Prove that C_5 has no totally magic labeling.

9.3 Edge-Magic Labelings of Complete Graphs

The discussion of edge-magic labelings of complete graphs is significantly harder than the "graceful" case, so we devote a section to it.

Suppose the graph G has an edge-magic total labeling λ, and suppose G contains a complete subgraph (or *clique*) H on n vertices. Let us write $x_1, x_2, \ldots, x_n$ for the vertices of H, and denote $\lambda(x_i)$ by a_i. Without loss of generality we can assume the names x_i to have been chosen so that $a_1 < a_2 < \cdots < a_n$.

If k is the magic sum, then $\lambda(x_i x_j) = k - a_i - a_j$, so the sums $a_i + a_j$ must all be distinct. This property is called being *well-spread*; this property will be used in discussing EMTLs of complete graphs. A *well-spread set* $A = \{a_1, a_2, \ldots, a_n)$ is a set of integers in which, if a_i, a_j, a_k, a_ℓ are all different, it never happens that $a_i + a_j = a_k + a_\ell$. If the elements are arranged in order, so that $0 < a_1 < a_2 < \cdots < a_n$, A is called a *well-spread sequence* or *Sidon sequence* of *length n*. Such sequences arise are related to the work of S. Sidon; their study was initiated by Erdős and Turan [40].

In discussing Sidon sequences (or, equivalently, cliques in edge-magic graphs), we write d_{ij} for $|a_j - a_i|$, the absolute difference between the i-th and j-th terms (the labels on the endpoints of the edge $x_i x_j$).

Lemma 9.16. *Suppose A is a Sidon sequence of length n. If $d_{ij} = d_{pq}$, then $\{a_i, a_j\}$ and $\{a_p, a_q\}$ have a common member. No three of the differences d_{ij} are equal.*

Proof. Suppose $d_{ij} = d_{pq}$. We can assume that $i > j$ and $p > q$. Without loss of generality we can also assume $p \geq i$. Then $a_i - a_j = a_p - a_q$, so $a_i + a_q = a_j + a_p$. Therefore $a_i = a_q$ and $i = q$ ($a_j = a_p$ is impossible), and $p > i > j$ — the common element is the middle one in order of magnitude.

Now suppose three pairs have the same difference. By the above reasoning there are two possibilities: the pairs must have a common element, or form a triangle. In the former case, suppose the differences are d_{ij}, d_{ik} and $d_{i\ell}$. From $d_{ij} = d_{ik}$ we must have either $k > i > j$ or $j > i > k$; let us assume the former. Then $d_{ij} = d_{i\ell}$ implies that $i > j > \ell$. So j is greater than both k and ℓ. But $d_{ik} = d_{i\ell}$ must mean that either $k > j > \ell$ or $\ell > j > k$, both of which are impossible. On the other hand, suppose the three pairs form a triangle, say $d_{ij} = d_{ik} = d_{jk}$. We can assume $i > j$. Then $d_{ij} = d_{jk}$ implies $i > j > k$, and $j > k$ and $d_{ik} = d_{jk}$ imply $k > i$, again a contradiction. □

Lemma 9.17. *Suppose A is a Sidon sequence of length n. If $d_{ij} = d_{ik}$, then $d_{ij} < \frac{1}{2} d_{1n}$.*

Proof. Suppose $d_{ij} = d_{ik}$, and assume $j < k$. Then $d_{jk} = a_k - a_j = a_i - a_j + a_k - a_i = d_{ij} = d_{ik} = 2d_{ij}$. But $a_1 \leq a - j$ and $a_k \leq a_n$, so $d_{jk} \leq d_{1n}$, giving the result. □

Theorem 9.18. *In any Sidon sequence of length n, $\binom{n}{2} \leq \lfloor \frac{3}{2} d_{1n} \rfloor$, or equivalently $d_{1n} \geq \lceil \frac{1}{3} n(n-1) \rceil$.*

Proof. There are $\binom{n}{2}$ unordered pairs of elements in the sequence, so there are $\binom{n}{2}$ differences. From Lemmas 9.16 and 9.17, the collection of values of these differences can contain the integers $1, 2, \ldots, \lfloor \frac{1}{2} r \rfloor$ at most twice each, and $\lfloor \frac{1}{2} r \rfloor + 1, \ldots, d_{1n}$ at most once each. The result follows. □

We now use Sidon sequences to show that only five complete graphs are edge-magic.

Theorem 9.19. *The complete graph K_v does not have an edge-magic total labeling if $v > 6$.*

Proof. One can show that K_7 has no EMTL by a complete search (or see Exercise 9.3.4 below). So we assume $v \geq 8$, and suppose there is an edge-magic total labeling of K_v. The vertex labels will form a Sidon sequence of length v, A say. Let us denote the edge labels by $b_1, b_2, \ldots, b_e$, where $b_1 < b_2 < \cdots < b_e$; of course, $e = \binom{v}{2}$. If the magic sum is k, then

$$\begin{aligned} k &= a_1 + a_2 + b_e && (9.7)\\ &= a_1 + a_3 + b_{e-1} && (9.8)\\ &= a_v + a_{v-1} + b_1 && (9.9)\\ &= a_v + a_{v-2} + b_2. && (9.10) \end{aligned}$$

Subtracting (9.7) from (9.8),

$$a_3 - a_2 = b_e - b_{e-1}, \qquad (9.11)$$

while (9.9) and (9.10) yield

$$a_{v-1} - a_{v-2} = b_2 - b_1. \qquad (9.12)$$

Suppose labels 1, 2, $v + e - 1$ and $v + e$ are all edge labels. Then $b_1 = 1, b_2 = 2, b_{e-1} = v+e-1$ and $b_e = v+e$. So, from (9.11) and (9.12), $a_3 - a_2 = a_{v-1} - a_{v-2} = 1$. But 2, 3, $v - 2$ and $v - 1$ are all distinct, so this contradicts Lemma 9.16. So one of $1, 2, v + e - 1, v + e$ is a vertex label. Without loss of generality we can assume either 1 or 2 is a vertex label (otherwise, the dual labeling will have this property). So $a_1 = 1$ or 2.

Equations (9.7) and (9.9) give

$$a_v = b_e - (a_{v-1} - a_2) - (b_1 - a_1). \qquad (9.13)$$

Since $(a_2, a_3, \ldots, a_{v-1})$ is a Sidon sequence of length $v - 2$ (any subsequence of a Sidon sequence is also well-spread), Lemma 9.18 applies to it, and $(a_{v-1} - a_2) \geq \lceil \frac{1}{3}(v-2)(v-3) \rceil$, which is at least 10 because $v \geq 8$. Also $(b_1 - a_1) \geq -1$, $(b - 1$ is at least 1 and a_1 is at most 2), and $b_e \leq v + e$. So, from (9.13),

$$a_v \leq v + e - 9.$$

So the six largest labels are all edge labels:

$$b_{e-5} = v + e - 5, b_{e-4} = v + e - 4, \ldots, b_e = v + e.$$

From (9.7) and (9.8) we get

$$k = a_1 + a_2 + v + e = a_1 + a_3 + v + e - 1,$$

so $a_3 = a_2 + 1$. The next smallest sum of two vertex-labels, after $a_1 + a_2$ and $a_1 + a_3$, may be either $a_2 + a_3$ or $a_1 + a_4$.

If it is $a_2 + a_3$, then

$$k = a_2 + a_3 + v + e - 2$$

and by comparison with (9.8), $a_2 = a_1 + 1$. The next-smallest sum is $a_1 + a_4$, so

$$k = a_1 + a_4 + v + e - 3$$

and $a_4 = a_3 + 2$. Two cases arise. If $a_1 = 1$, then $a_2 = 2, a_3 = 3, a_4 = 5$. Also, a_5 cannot equal 6, because that would imply $a_1 + a_5 = 7 = a_2 + a_4$, contradicting the well-spread property. Every integer up to $v + e$ must occur as a label, so $b_1 = 4$ and $b_2 = 6$. So (9.12) is $a_{v-1} - a_{v-2} = b_2 - b_1 = 2$. But $a_4 - a_3 = 2$, so $d_{v-1,v-2} = d_{34}$, in contradiction of Lemma 9.16. In the other case, $a_1 = 2$, we obtain $a_2 = 3, a_3 = 4, a_4 = 6$, so $b_1 = 1, b_2 = 5$, and $a_{v-1} - a_{v-2} = 4 = a_4 - a_1$, again a contradiction.

If $a_1 + a_4$ is the next-smallest difference, we have

$$k = a_1 + a_4 + v + e - 2,$$

so $a_4 = a_3 + 1$. If $a_1 = 1$ and $a_2 = 3$, it is easy to see that $b_1 = 2, b_2 = 6$, and we get the contradiction $a_{v-1} - a_{v-2} = a_4 - a_1 = 4$. Otherwise $a_2 \geq 4$, so 3 is an edge-label. If $a_1 = 1$, then $b_1 = 2, b_2 = 3$, and $a_{v-1} - a_{v-2} = 1 = a_3 - a_2$. If $a_1 = 2$, then $b_1 = 1, b_2 = 3$, and $a_{v-1} - a_{v-2} = 2 = a_4 - a_2$. In every case, a contradiction is obtained. Therefore we have the result. □

This theorem was first proven in [79]; the above proof follows that in [32].

We know K_4 is not edge-magic; edge-magic total labelings of K_1, K_2 and K_3 are easy to find. One solution for K_5 is to use vertex labels $\{1, 2, 3, 5, 9\}$; one for K_6 is $\{1, 3, 4, 5, 9, 14\}$. (A complete list of solutions is given in [126].)

If $A = (a_1, a_2, \ldots, a_n)$ is any Sidon sequence of length n, we define the *size* of A as $\sigma(A) = a_n - a_1 + 1$. One usually assumes $a_1 = 1$ when constructing a sequence, and then the size equals the largest element. Another useful parameter is

$$\rho(A) = a_n + a_{n-1} - a_2 - a_1 + 1 = \sigma(A) + a_{n-1} - a_2.$$

We denote by $\sigma^*(n)$ and $\rho^*(n)$ the minimum values of $\sigma(A)$ and $\rho(A)$ respectively, taken over all Sidon sequences A of length n.

It is useful to know that there exist Sidon sequences of all positive lengths. The recursive construction $a_1 = 1, a_2 = 2, a_n = a_{n-1} + a_{n-2}$ gives a well-spread sequence. This is the well-known Fibonacci sequence, except that the standard notation for the Fibonacci numbers has $f_1 = f_2 = 1, f_3 = 2, \ldots$. So we have a well-spread sequence with its largest element equal to the $(n + 1)$-th term of the Fibonacci sequence: $a_n = f_{n+1}$. Therefore $\sigma^*(n) \leq f_{n+1}$, and

$$\rho^*(n) \leq f_{n+1} + f_n - 2 = f_{n+2} - 2.$$

It is well known (see texts such as [19]) that

$$f_n = \frac{1}{\sqrt{5}}\left(\frac{1+\sqrt{5}}{2}\right)^n - \frac{1}{\sqrt{5}}\left(\frac{1-\sqrt{5}}{2}\right)^n,$$

so we have

$$\sigma^*(n) \leq \frac{1}{\sqrt{5}}\left(\frac{1+\sqrt{5}}{2}\right)^{n+1} - \frac{1}{\sqrt{5}}\left(\frac{1-\sqrt{5}}{2}\right)^{n+1}$$

and

$\sigma^*(3) = 3$	$f_4 = 3$	$\rho^*(3) = 3$	$f_5 - 2 = 3$
$\sigma^*(4) = 5$	$f_5 = 5$	$\rho^*(4) = 6$	$f_6 - 2 = 6$
$\sigma^*(5) = 8$	$f_6 = 8$	$\rho^*(5) = 11$	$f_7 - 2 = 11$
$\sigma^*(6) = 13$	$f_7 = 13$	$\rho^*(6) = 19$	$f_8 - 2 = 19$
$\sigma^*(7) = 19$	$f_8 = 21$	$\rho^*(7) = 30$	$f_9 - 2 = 32$
$\sigma^*(8) = 25$	$f_9 = 34$	$\rho^*(8) = 43$	$f_{10} - 2 = 53$
$\sigma^*(9) = 35$	$f_{10} = 55$	$\rho^*(9) = 62$	$f_{11} - 2 = 79$
$\sigma^*(10) = 46$	$f_{11} = 81$	$\rho^*(10) = 80$	$f_{12} - 2 = 134$

Table 9.1. Sidon sequence bounds compared to Fibonacci numbers

$$\rho^*(n) \le \frac{1}{\sqrt{5}}\left(\frac{1+\sqrt{5}}{2}\right)^{n+2} - \frac{1}{\sqrt{5}}\left(\frac{1-\sqrt{5}}{2}\right)^{n+2} - 2.$$

However these are not the best-possible values. Table 9.1 shows the results of exhaustive computations.

Suppose G has an edge-magic total labeling λ and G contains a complete subgraph H on n vertices $x_1, x_2, \ldots, x_n$; denote $\lambda(x_i)$ by a_i. Assume $a_1 < a_2 < \cdots < a_n$, so $A = (a_1, a_2, \ldots, a_n)$ is a Sidon sequence of length n. Then

$$\lambda(x_n x_{n-1}) = k - a_n - a_{n-1},$$

and since $\lambda(x_n x_{n-1})$ is a label,

$$k - a_n - a_{n-1} \ge 1. \tag{9.14}$$

Similarly

$$\lambda(x_2 x_1) = k - a_2 - a_1,$$

and since $\lambda(x_2 x_1)$ is a label,

$$k - a_2 - a_1 \le v + e. \tag{9.15}$$

Combining (9.14) and (9.15) we have

$$v + e \ge a_n + a_{n-1} - a_2 - a_1 + 1 = \rho(A) \ge \rho^*(n).$$

Theorem 9.20. [79] *If the edge-magic graph G contains a complete subgraph with n vertices, then the number of vertices and edges in G is at least $\rho^*(n)$.*

Suppose $G = K_n + tK_1$. In other words, G consists of K_n together with t isolated vertices. The smallest t such that G is edge-magic is called the *magic number* $M(n)$. Theorem 9.20 enables us to find a lower bound

$$M(n) \ge \rho^*(n) - n - \binom{n}{2}.$$

(See Exercise 9.9.3.1, and also Exercise 9.9.3.2.)

An edge-magic injection is like an edge-magic total labeling, except that the labels can be any positive integers. We define an $[m]$-edge-magic injection of G to be an edge-magic injection of G in which the largest label is m, and call m the *size* of the injection. The *edge deficiency* $def_e(G)$ of G is the minimum value of $m - v(G) - e(G)$, such that an $[m]$-edge-magic injection of G exists.

Theorem 9.21. *Every graph has an edge-magic injection.*

Proof. Suppose G is a graph with v vertices and e edges. The empty graph is trivially edge-magic, so we assume that G has at least one edge. Let$(a_1, a_2, \ldots, a_v)$ be any Sidon sequence of length v with first element $a_1 = 1$. Define $k = a_{v-1} + 2a_v + 1$.

We now construct a labeling λ as follows. Select any edge of G and label its endpoints with a_{v-1} and a_v, and label the remaining vertices with the other members of the Sidon sequence in any order. If xy is any edge, define $\lambda(xy) = k - \lambda(x) - \lambda(y)$. Every edge weight will be equal to k. The smallest edge label will be $k - a_{v-1} - a_v = a_v + 1$, which is greater than any vertex label. If two edge labels were equal, say $\lambda(xy) = \lambda(zt)$, then $\lambda(x) + \lambda(y) = \lambda(z) + \lambda(t)$, and since the labels of vertices are members of a Sidon sequence this implies that $xy = zt$. The vertex labels are distinct by definition. So λ is an edge-magic injection. □

The proof of Theorem 9.21 gives us an upper bound on the deficiency:

Corollary 9.22. *If G is a graph with v vertices and $(a_1, a_2, \ldots, a_v)$ is any Sidon sequence of length v with $a_1 = 1$, then*

$$def_e(G) \leq a_{v-1} + 2a_v - a_2 - v - e(G).$$

Proof. In the above construction, no label can be greater than $k - 1 - a_2$. □

This upper bound will not usually be very good. For example, consider the graph constructed from C_5 by joining two inadjacent vertices. Using the Sidon sequence $(1, 2, 3, 5, 8)$, a labeling with $k = 22$ is obtained, and the best assignment of the sequence to the vertices gives largest label 17, and deficiency 6. However, the graph is actually edge-magic. See Figure 9.7.

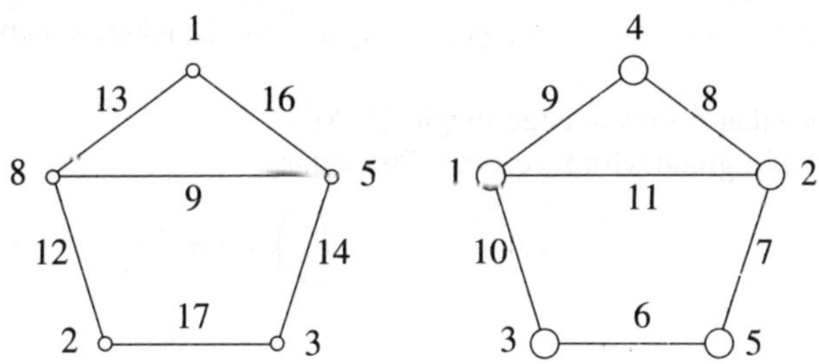

Fig. 9.7. Deficiency 6 on the left; magic on the right

Theorem 9.23. *The edge-magic deficiency of K_v equals the magic number $M(v)$.*

Proof. Consider a edge-magic total labeling λ of $K_v + M(v)K_1$. This graph has $v+M(v)$ vertices and $e(K_v)$ edges, so the largest label is $v+M(v)+e(K_v)$, and clearly this label occurs on a vertex or edge of K_v. The labeling constructed by restricting λ to

K_v is an $[v+M(v)+e(K_v)]$-edge-magic injection of K_v. Obviously any injection of size $v+m+e(K_v)$ gives rise to an edge-magic total labeling of K_v+mK_1 (apply the m unused labels to the extra vertices), so $v+M(v)+e(K_v)$ is the smallest possible size, and $def_e(K_v) = M(v)$. □

Exercises 9.3

A9.3.1 Suppose $G = K_n + tK_1$. Prove that if G is edge-magic, then

$$t \geq \rho^*(n) - n - \binom{n}{2}.$$

That is, $M(n) \geq \rho^*(n) - n - \binom{n}{2}$.

A9.3.2 Find an upper bound for $M(n)$. (It does not have to be a good upper bound. The point is to show that *some* upper bound exists.)

9.3.3 If G is an incomplete graph with v vertices and $(a_1, a_2, \ldots, a_v)$ is any Sidon sequence of length v with $a_1 = 1$, prove that

$$def_e(G) < a_{v-1} + 2a_v - a_2 - v - e(G).$$

H9.3.4 (i) Suppose K_v has an edge-magic total labeling with magic sum k. The number p of vertices that receive even labels satisfies the following conditions:

(i) If $v \equiv 0$ or $3 \pmod 4$ and k is even, then $p = \frac{1}{2}(v - 1 \pm \sqrt{v+1})$.

(ii) If $v \equiv 1$ or $2 \pmod 4$ and k is even, then $p = \frac{1}{2}(v - 1 \pm \sqrt{v-1})$.

(iii) If $v \equiv 0$ or $3 \pmod 4$ and k is odd, then $p = \frac{1}{2}(v + 1 \pm \sqrt{v+1})$.

(iv) If $v \equiv 1$ or $2 \pmod 4$ and k is odd, then $p = \frac{1}{2}(v + 1 \pm \sqrt{v+3})$.

(ii) Prove the following necessary conditions for K_v to have an edge-magic total labeling: if $v \equiv 0$ or $3 \pmod 4$, then $v+1$ is a perfect square; if $v \equiv 1$ or $2 \pmod 4$, then either $v-1$ is a perfect square and the magic sum of the labeling is even, or $v+3$ is a perfect square and the magic sum of the labeling is odd.

(iii) Deduce that K_7 is not edge-magic. [118]

9.3.5 Suppose G is a graph with v vertices. Prove that

$$def_e(G) \leq M(v) + \binom{v}{2} - e(G).$$

10

Ramsey Theory

10.1 The Graphical Case of Ramsey's Theorem

Suppose the edges of a graph G are painted in k colors. We say a subgraph H of G is *monochromatic* if all its edges receive the same color. We say a k-painting of G is *proper* with respect to H if G contains no monochromatic subgraph isomorphic to H in that painting. If no subgraph is specified, "proper" will mean proper with respect to triangles — graphs isomorphic to K_3.

For example, suppose G is a complete graph and its vertices represent people at a party. An edge xy is colored red if x and y are acquaintances, and blue if they are strangers. An old puzzle asks: given any six people at a party, prove that they contain either a set of three mutual acquaintances or a set of three mutual strangers. In graph-theoretic terms, the puzzle asks for a proof that there is no proper 2-painting of K_6.

To observe that the result is *not* true for fewer than six people, consider the complete graph K_5. It is easy to see that K_5 has a proper 2-painting: take all edges of a copy of C_5 in red and all other edges (they will form another copy of C_5) in blue. (See Figure 10.1.)

On the other hand, there is no proper 2-painting of K_6. To see this, select a vertex x in any 2-painting of K_6. There are five edges touching x, so there must be at least three of them that receive the same color, say red. Suppose xa, xb and xc are red edges. Now consider the triangle abc. If ab is red, then xab is a red triangle. Similarly, if ac or bc is red, there will be a red triangle. But if none are red, then all are blue, and abc is a blue triangle.

This proves that any 2-painting of K_v must contain a monochromatic triangle whenever $v \geq 6$: if $v > 6$, simply delete all but six vertices. The resulting 2-painted K_6 must contain a monochromatic triangle, and that triangle will also be a monochromatic triangle in K_v.

The same argument can be used when there are more than two colors, and applies to general graphs, not only to triangles. The general result is the graphical version of Ramsey's theorem. We first prove a particular case.

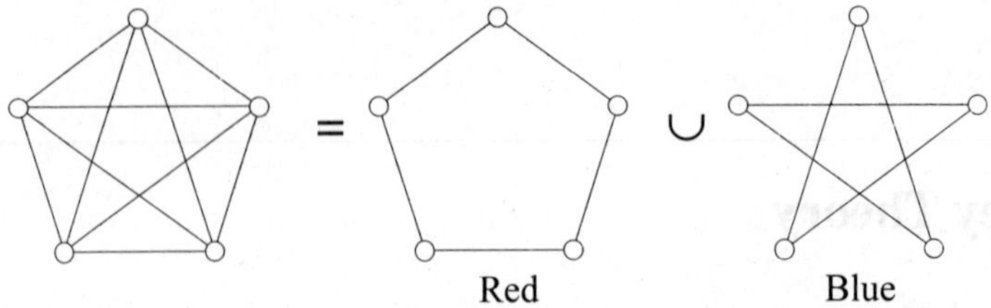

Fig. 10.1. Proper 2-painting of K_5

Lemma 10.1. *There exists a number $R(p, q)$ such that any painting of $K_{R(p,q)}$ in two colors c_1 and c_2 must contain either a K_p with all its edges in color c_1 or a K_q with all its edges in c_2.*

Proof. We proceed by induction on $p+q$. The lemma is clearly true when $p+q = 2$, since the only possible case is $p = q = 1$ and obviously $R(1, 1) = 1$. Suppose it is true whenever $p + q < N$, for some integer N. Consider any two positive integers P and Q that add to N. Then $P + Q - 1 < N$, so both $R(P - 1, Q)$ and $R(P, Q - 1)$ exist.

Consider any painting of the edges of K_v in two colors c_1 and c_2, where $v \geq R(P-1, Q)+R(P, Q-1)$, and select any vertex x of K_v. Then x must either lie on $R(P - 1, Q)$ edges of color c_1 or on $R(P, Q - 1)$ edges of color c_2. In the former case, consider the $K_{R(P-1,Q)}$ whose vertices are the vertices joined to x by edges of color c_1. Either this graph contains a K_{P-1} with all edges of color c_1, in which case this K_{P-1} together with x forms a K_p with all edges in c_1, or it contains a K_Q with all edges in c_2. In the latter case, the K_v again contains one of the required monochromatic complete graphs. So $R(P, Q)$ exists, and in fact $R(P, Q) \leq R(P, Q-1) + R(P-1, Q)$. □

Theorem 10.2. *Suppose $H_1, H_2, \ldots, H_k$ are any k graphs. Then there exists an integer $R(H_1, H_2, \ldots, H_k)$ such that whenever $v \geq R(H_1, H_2, \ldots, H_k)$, any k-painting of K_v must contain a subgraph isomorphic to H_i that is monochromatic in color i, for some i, $1 \leq i \leq k$.*

The numbers $R(H_1, H_2, \ldots, H_k)$ are called *Ramsey numbers*. In particular, if all the H_i are complete graphs, say $H_1 = K_{p_1}, H_2 = K_{p_2}, \ldots$, then the Ramsey number $R(K_{p_1}, K_{p_2}, \ldots, K_{p_k})$ is written $R(p_1, p_2, \ldots, p_k)$. If the p_i are all equal, with common value p, the notation $R_k(p)$ is used.

Proof of Theorem 10.2. It is sufficient to prove the theorem in the case where all the H_i are complete. Then, if v is sufficiently large that a k-painted K_v must contain a monochromatic $K_{v(H_i)}$ in color c_i, for some i, it must certainly contain a monochromatic copy of H_i in color c_i, so

$$R(H_1, H_2, \ldots, H_k) \leq R(v(H_1), v(H_2), \ldots, v(H_k)).$$

We proceed by induction on k to prove that $R(p_1, p_2, \ldots, p_k)$ exists for all parameters. In the case $k = 2$, the result follows from Lemma 10.1. Now suppose it is true

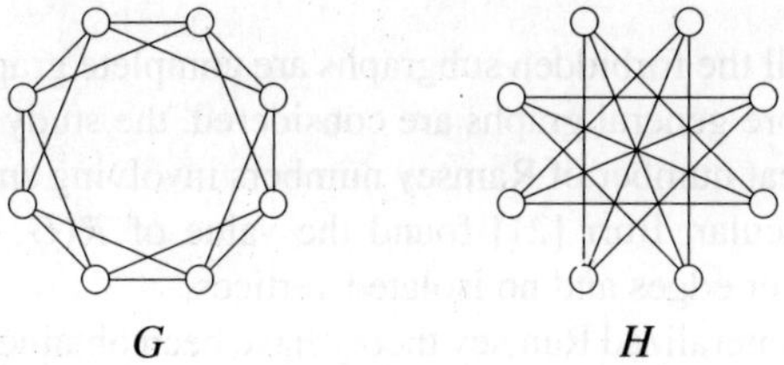

Fig. 10.2. Decomposition of K_8 proving $R(3,4) \geq 9$

for $k < K$, and suppose integers $p_1, p_2, \ldots, p_K$ are given. Then $R(p_1, p_2, \ldots, p_{K-1})$ exists.

Suppose

$$v \geq R(R(p_1, p_2, \ldots, p_{K-1}), p_K).$$

Select any k-painting of K_v. Then recolor by assigning a new color c_0 to all edges that received colors other than c_k. The resulting graph must contain either a monochromatic $K_{R(p_1,p_2,\ldots,p_{K-1})}$ in color c_0 or a monochromatic K_{p_K} in color c_K. In the former case, the corresponding $K_{R(p_1,p_2,\ldots,p_{K-1})}$ in the original painting has edges in the $K-1$ colors $c_1, c_2, \ldots, c_{K-1}$ only, so by induction it contains a monochromatic K_{p_i} in color c_i for some i. □

In discussing individual small Ramsey numbers, it is often useful to consider the graphs whose edges are precisely those that receive a given coloring in a painting of a complete graph. These are called the *monochromatic subgraphs*.

As an example, consider $R(3,4)$. Suppose K_v has been colored in red and blue so that neither a red K_3 nor a blue K_4 exists. Select any vertex x. Define R_x to be the set of vertices connected to x by red edges — that is, R_x is the neighborhood of x in the red monochromatic subgraph, and similarly define B_x in the blue monochromatic subgraph.

If $|R_x| \geq 4$, then either $\langle R_x \rangle$ contains a red edge yz, whence xyz is a red triangle, or else all of its edges are blue, and there is a blue K_4. So $|R_x| \leq 3$ for all x.

Next suppose $|B_x| \geq 6$. Then $\langle B_x \rangle$ is a complete graph on six or more vertices, so it contains a monochromatic triangle. If this triangle is red, it is a red triangle in K_9. If it is blue, then it and x form a blue K_4 in K_9.

It follows that every vertex x has $|R_x| \leq 3$ and $|B_x| \leq 5$, so $v \leq 9$. But $v = 9$ is impossible. If $v = 9$, then $|R_x| = 3$ for every x, and the red monochromatic subgraph has nine vertices each of (odd) degree 3, in contradiction of Corollary 1.3.

On the other hand, K_8 can be colored with no red K_3 or blue K_4. The graph G of Figure 10.2 has no triangle, and can be taken as the red monochromatic subgraph, while its complement $\overline{G}$ is the blue graph. (The construction of this graph will be discussed in Section 10.3, below.) So we have

Theorem 10.3. $R(3,4) = 9$.

The case where all the forbidden subgraphs are complete graphs is called *classical Ramsey theory*; if more general graphs are considered, the study is called *generalized Ramsey theory*. A great number of Ramsey numbers involving small graphs have been investigated; in particular, Burr [21] found the value of $R(G, G)$ whenever G is a graph with six or fewer edges and no isolated vertices.

Many results of generalized Ramsey theory have been obtained by *ad hoc* methods. We illustrate by finding $R(K_3, C_4)$. Clearly $R(K_3, C_4) \leq R(3, 4) = 9$. However, we can do rather better. Suppose K_v has been colored with no red K_3 and no blue C_4. As in the discussion of $R(3, 4)$, we see that no vertex can belong to more than three red edges. Suppose some vertex x was on four blue edges (if $R(K_3, C_4) = 9$, then every vertex must have this property). The graph generated by the other four endpoints of those edges can contain no blue path of length 2 and no red triangle. It is easy to see that the graph is the union of a red C_4 and two independent blue edges, as is shown in Figure 10.3(a) (blue edges are solid, red edges broken). Now suppose another vertex, y, is added. Since xy must be red, y can be joined to at most two other vertices by red edges, and those vertices cannot be adjacent in the red cycle. So y must lie on at least two blue edges of the type shown in Figure 10.3(b). But that graph contains a blue C_4. It follows that if any vertex lies on four blue edges, the graph has at most five vertices. If there is a solution for $v = 7$, then every vertex lies on three red and three blue edges, and both monochromatic subgraphs have an odd number of vertices and are regular of odd degree, which is impossible. So the maximum is $v = 6$. This can be attained: take the red subgraph to be $K_{3,3}$ and the blue one to be $2K_3$. So $R(K_3, C_4) = 7$.

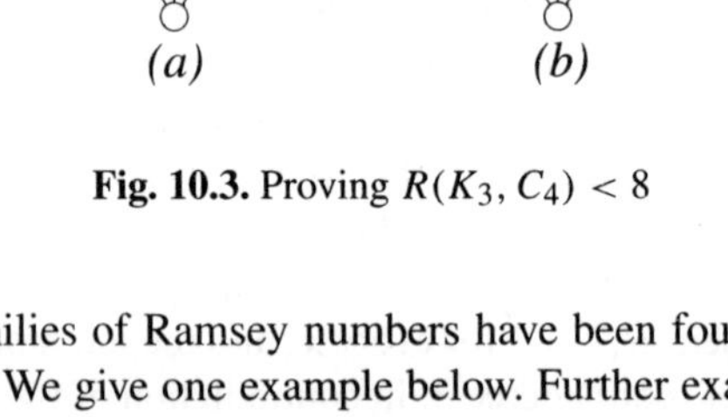

(a) (b)

Fig. 10.3. Proving $R(K_3, C_4) < 8$

A good many families of Ramsey numbers have been found, but many more remain to be discussed. We give one example below. Further examples are given in the exercises, and in surveys of generalized Ramsey theory such as [20], [92] and [65].

Theorem 10.4. [28] . *If T is a tree with m vertices, then*

$$R(T, K_n) = (m - 1)(n - 1) + 1.$$

Proof. To see that $R(T, K_n) > (m-1)(n-1)$, consider a graph consisting of $m-1$ disjoint copies of K_{n-1}, with all edges colored red. Complete this graph to a $K_{(m-1)(n-1)}$ by coloring all remaining edges blue. Since the red subgraph contains no m-vertex component, it contains no copy of T. The blue graph is $(n-1)$-partite, so it can contain no K_n.

Equality is proved using induction on n. The case $n = 1$ is trivial. Suppose $n > 1$ and suppose the theorem is true of $R(T, K_s)$ whenever $s < n$. Suppose there is a coloring of the edges of $K_{(m-1)(n-1)+1}$ in red and blue that contains neither a red T nor a blue K_n, and examine some vertex x. If x lies on more than $(m-1)(n-2)$ blue edges, then the subgraph of G induced by the "blue" neighbors of x contains either a red copy of T or a blue K_{n-1}, by the induction hypothesis. In the former case $K_{(m-1)(n-1)+1}$ contains a red T; in the latter the blue K_{n-1} together with x forms a blue K_n. Therefore x lies on at most $(m-1)(n-2)$ blue edges, so it lies on at least $m-1$ red edges. Since x could be any vertex of the $K_{(m-1)(n-1)+1}$, the red subgraph has minimum degree at least $m-1$. Since T has $m-1$ edges, this red subgraph will contain a subgraph isomorphic to T, by Theorem 4.4. So the $K_{(m-1)(n-1)+1}$ contains a red copy of T, a contradiction. □

Exercises 10.1

0.1.1 Consider $R(P_3, K_3)$.

(i) Show that $R(P_3, K_3) \leq 6$.

(ii) Prove that any graph containing no P_3 consists of some disjoint edges together with some isolated vertices.

(iii) Prove that if the graph described in part (ii) has at least four vertices and contains an isolated vertex, then its complement contains a triangle.

(iv) Find $R(P_3, K_3)$.

0.1.2 Find $R(P_3, K_4)$.

0.1.3 Find $R(P_4, P_4)$, $R(P_4, C_4)$, $R(P_4, K_4)$ and $R(C_4, C_4)$.

0.1.4 Prove that $R(3, 5) \leq 14$.

0.1.5 Prove that $R(4, 4) \leq 18$.

0.1.6 Suppose K_v can be colored in red and blue so that there is no red K_3 or blue $K_{1,s}$.

(i) Prove that the red monochromatic subgraph has maximum degree s.

(ii) Prove that the blue monochromatic subgraph has maximum degree $s-1$.

(iii) Prove that $R(K_3, K_{1,s}) = 2s+1$.

0.1.7 Prove that if m or n is odd, then $R(K_{1,m}, K_{1,n}) = m+n$, and that if both m and n are even, then $R(K_{1,m}, K_{1,n}) = m+n-1$. [65]

0.1.8 Chvátal and Harary [29] conjectured that if G and H are graphs with no isolated vertices, then

$$R(G, H) \geq \min\{R(G, G), R(H, H)\}.$$

Disprove this conjecture by using $G = K_{1,3}$ and $H = P_5$. (This result is due to Galvin; see [65].)

0.1.9 If M is any matrix, its *principal $k \times k$ submatrices* are the submatrices formed by the intersection of rows $i_1, i_2, \ldots, i_k$ with columns $i_1, i_2, \ldots, i_k$ for some selection of k indices. (The principal submatrices are also called the *symmetrically placed* submatrices.) Prove that if k is an integer ≥ 2, and s is sufficiently large, then any $s \times s$ matrix M with entries from $\{0, 1\}$ contains a $k \times k$ principal submatrix with one of the following forms:

- all entries off the main diagonal are 0;
- all entries above the main diagonal are 0, all entries below the main diagonal are 1;
- all entries above the main diagonal are 1, all entries below the main diagonal are 0;
- all entries off the main diagonal are 1.

10.2 Ramsey Multiplicity

We know that any 2-painting of K_6 must contain a monochromatic triangle. However, it is not possible to find a painting with exactly one triangle. (This is easily checked by exhaustion, and follows from Theorem 10.5 below.) More generally, one can ask: what is the minimum number of monochromatic triangles in a k-painting of K_v?

Such questions are called *Ramsey multiplicity problems*. The k-Ramsey multiplicity $N_{k,v}(H)$ of H in K_v is defined to be the minimum number of monochromatic subgraphs isomorphic to H in any k-painting of K_v. Clearly $N_{k,v}(H) = 0$ if and only if $v < R_k(H)$.

The 2-Ramsey multiplicity of K_3 was investigated by Goodman, who proved the following Theorem in [56]. Our proof follows that given by Schwenk [109].

Theorem 10.5.

$$N_{2,n}(K_3) = \binom{n}{3} - \left\lfloor \frac{n}{2} \left\lfloor \left(\frac{n-1}{2}\right)^2 \right\rfloor \right\rfloor.$$

Proof. Suppose K_v is colored in two colors, red and blue. Write R for the number of red triangles, B for the number of blue triangles, and P for the number of *partial* triangles — triangles with at least one edge of each color. There are $\binom{n}{3}$ triangles in K_v, so

$$R + B + P = \binom{v}{3}.$$

Since $N_{2,v}(K_3)$ equals the minimum possible value of $R + B$,

$$N_{2,v}(K_3) = \binom{v}{3} - \max(P).$$

Suppose the vertices of K_v are $x_1, x_2, \ldots, x_v$, and x_i is incident with r_i red edges. Then it is adjacent to $v - 1 - r_i$ blue edges. Therefore the K_v contains $r_i(v - 1 - r_i)$ paths of length 2 in which one edge is red and the other blue. Let us call these mixed paths. The total number of mixed paths in the K_v is

$$\sum_{i=1}^{v} r_i(v - 1 - r_i).$$

The triangle xyz can be considered as the union of the three paths xyz, yzx and zxy. Moreover, the paths corresponding to different triangles will all be different. If the triangle is monochromatic, no path is mixed, but a partial triangle gives rise to two mixed paths. So there are $2P$ mixed paths in the K_v, and

$$P = \frac{1}{2}\sum_{i=1}^{v} r_i(v-1-r_i).$$

If v is odd, the maximum value of $r_i(v-1-r_i)$ is $(v-1)^2/4$, attained when $r_i = (v-1)/2$. If v is even, the maximum of $v(v-2)/4$ is given by $r_i = v/2$ or $(v-2)/2$. In either case, the maximum is

$$\left\lfloor \left(\frac{v-1}{2}\right)^2 \right\rfloor,$$

so

$$\begin{aligned} P &\le \tfrac{1}{2}\textstyle\sum_{L=1}^{v} \left\lfloor \left(\frac{v-1}{2}\right)^2 \right\rfloor \\ &\le \tfrac{v}{2}\left\lfloor \left(\frac{v-1}{2}\right)1^2 \right\rfloor, \end{aligned}$$

and since P is an integer,

$$P \le \left\lfloor \frac{v}{2}\left\lfloor \left(\frac{v-1}{2}\right)^2 \right\rfloor \right\rfloor.$$

So

$$N_{2,v}(K_3) \ge \binom{v}{3} - \left\lfloor \frac{v}{2}\left\lfloor \left(\frac{v-1}{2}\right)^2 \right\rfloor \right\rfloor.$$

It remains to show that equality can be attained.

If v is even, say $v = 2t$, then partition the vertices of K_v into the two sets $\{x_1, x_2, \ldots, x_t\}$ and $\{x_{t+1}, x_{t+2}, \ldots, x_{2t}\}$ of size t, and color an edge red if it has one endpoint in each set, blue if it joins two members of the same set. (The red edges form a copy of $K_{t,t}$.) Each r_i equals $v/2$. If $v = 2t+1$, carry out the same construction for $2t$ vertices, except color edge x_ix_{t+i} blue for $1 \le i \le \lfloor \frac{t}{2} \rfloor$. Then add a final vertex x_{2t+1}. The edges x_ix_{2t+1} and $x_{i+t}x_{2t+1}$ are red when $1 \le i \le \lfloor \frac{t}{2} \rfloor$ and blue otherwise. In both cases it is easy to check that the number of triangles equals the required minimum. □

Substituting into the formula gives $N_{2,v}(K_3) = 0$ when $v \le 5$, $N_{2,6}(K_3) = 2$, $N_{2,7}(K_3) = 4$, and so on.

The 3-Ramsey multiplicity of K_3 has not been fully investigated. We know that $R_3(3) = 17$, so the number $N_{3,17}(K_3)$ is of special interest. It is shown in [107] that $N_{3,17}(K_3) = 5$; the argument involves discussion of many special cases. A sketch of a proof that $N_{3,17}(K_3) \ge 3$ appears in Exercise 10.2.1.

The following theorem, which appears in a simplified form in [120], provides a recursive bound on $N_{k,v}(K_3)$.

Theorem 10.6.

$$N_{k,v+1}(K_3) \le \left\lfloor \frac{v-1}{k} \right\rfloor + \left\lfloor \left(1 + \frac{3}{v}\right) N_{k,v}(K_3) \right\rfloor.$$

Proof. Suppose F is a k-painting of K_v that contains $N_{k,v}(K_3)$ monochromatic triangles. Select a vertex x of F that lies in the minimum number of monochromatic triangles. Since there are v vertices, one can assume that x lies on at most $\lfloor \frac{3}{v} N_{k,v}(K_3) \rfloor$ monochromatic triangles. Since x has degree $v - 1$ in K_v, there will be a color — say R — such that x lies on at most $\lfloor \frac{v-1}{k} \rfloor$ edges of color R.

Construct a k-painting of K_{v+1} from F by adjoining a new vertex y. If z is any vertex other than x, then yz receives the same color as xz, and xy receives color R. Then xy lies in $\lfloor \frac{v-1}{k} \rfloor$ or fewer monochromatic triangles, all in color R. The original K_v contained $N_{k,v}(K_3)$ monochromatic triangles, so this is the number not containing x. Finally, the number of monochromatic triangles with y as a vertex but not x is at most $\lfloor \frac{3}{v} N_{k,v}(K_3) \rfloor$. So the maximum number of monochromatic triangles in the K_{v+1} is

$$\left\lfloor \frac{n-1}{k} \right\rfloor + \left\lfloor \left(1 + \frac{3}{v}\right) N_{k,v}(K_3) \right\rfloor. \qquad \square$$

This theorem provides the upper bounds 2 and 5 for $N_{2,6}(K_3)$ and $N_{3,17}(K_3)$, both of which can be met.

Exercises 10.2

10.2.1 Prove that $N_{2,3}(P_3) = 1$ and $N_{2,4}(P_3) = 4$. Suppose there exists a painting of K_{17} in the three colors red, blue and green that contains two or less monochromatic triangles. If v is any vertex, write $R(v)$, $B(v)$ and $G(v)$ for the sets of vertices joined to v by red, blue and green edges respectively, and write $r(v) = |R(v)|$, and so on.

(i) Select a vertex x that lies in no monochromatic triangle. Prove that one of $\{r(x), b(x), g(x)\}$ equals 6 and the other two each equal 5.

(ii) Without loss of generality, say $r(x) = 6$. Let S be the set of all vertices of K_{17} that lie in monochromatic triangles. Prove that $S \subseteq R(x)$.

(iii) If y is any member of $B(x)$, it is clear that S lies completely within $R(y)$, $B(y)$ or $G(y)$. Prove that, in fact, $S \subseteq R(y)$.

(iv) Prove that there must exist two vertices y_1 and y_2 in $B(x)$ such that $y_1 y_2$ is red.

(v) Use the fact that $S \subseteq R(y_1) \cap R(y_2)$ to prove that K_{17} contains more than two red triangles. So $N_{3,17}(K_3) \geq 3$. [120]

A10.2.2 It follows from Exercise 10.1.7 that $R(K_{1,n}, K_{1,n}) = 2n$ when n is odd and that $R(K_{1,n}, K_{1,n}) = 2n - 1$ when n is even. Prove that

$$N_{2,2n}(K_{1,n}) = 2n - 1, n \text{ odd},$$
$$N_{2,2n-1}(K_{1,n}) = 1, n \text{ even}.$$

10.3 Application of Sum-Free Sets

To introduce this section we derive the construction of the red graph of Figure 10.2. The vertices of the graph are labeled with the elements of the cyclic group $\mathbb{Z}_8$. The set $\mathbb{Z}_8^*$ of nonzero elements of $\mathbb{Z}_8$ is partitioned into two sets:

$$\mathbb{Z}_8^* = \{3, 4, 5\} \cup \{1, 2, 6, 7\}.$$

Call the two sets R and B respectively. Then $x \sim y$ in G if and only if $x - y \in R$. It follows that two vertices are joined in $\overline{G}$ if and only if their difference is in B. Observe that both R and B contain the additive inverses of all their elements; this is important because the differences $x - y$ and $y - x$ both correspond to the same edge xy. (This property might be relaxed for some applications to directed graphs.)

Notice that R contains no solution to the equation

$$a + b = c;$$

no element of R equals the sum of two elements of R. We say R is a *sum-free set.* By contrast, B is not sum-free; not only is $1 + 6 = 7$, but also $1 + 1 = 2$ (a, b and c need not be distinct). If xyz were a triangle in G, then $x - y$, $y - z$ and $x - z$ would all be members of R; but

$$x - y + y - z = x - z,$$

so R would not be sum-free.

In general, suppose G is any group, written additively. A nonempty subset S of G is a *sum-free set* if there never exist elements a, b of S such that $a + b \in S$. (This means that 0 cannot belong to S, since $0+0 = 0$.) S is *symmetric* if $-x \in S$ whenever x is in S. A *symmetric sum-free partition* of G is a partition of G^* into symmetric sum-free sets. As examples,

$$\mathbb{Z}_5^* = \{1, 4\} \cup \{2, 3\}$$
$$\mathbb{Z}_8^* = \{3, 4, 5\} \cup \{1, 7\} \cup \{2, 6\}$$

are symmetric sum-free partitions.

If S is a symmetric sum-free set in G, the *graph of S* is the graph with vertex-set G, where x and y are adjacent if and only if $x - y \in S$. From our earlier discussion, it follows that

Theorem 10.7. *If S is a symmetric sum-free set of order s in a group G of order g, then the graph of S is a triangle-free regular graph of degree s on g vertices.*

If there is a symmetric sum-free partition $S_1 \cup S_2 \cup \cdots \cup S_k$ of G^*, then one obtains a k-painting of $K_{|G|}$ that contains no monochromatic triangle, by applying c_i to all the edges of the graph of S_i for $i = 1, 2, \ldots, k$. So

Corollary 10.8. *If there exists a sum-free partition of a g-element group into k parts, then $R_k(3) > g$.*

For example, the partition

$$\mathbb{Z}_5^* = \{1, 4\} \cup \{2, 3\}$$

provides the well-known partition of $\mathbb{Z}_5$ into two 5-cycles that is used in proving that $R_2(3) = 6$. The partition

$$\mathbb{Z}_8^* = \{3, 4, 5\} \cup \{1, 7\} \cup \{2, 6\}$$

yields a (not very interesting) triangle-free 3-painting of K_8.

There are two abelian groups of order 16 that have symmetric sum-free partitions. The group $\mathbb{Z}_4 \times \mathbb{Z}_4$ can be written as the set of all ordered pairs xy where both x and y come from $\{0, 1, 2, 3\}$ and

$$xy + zt = (x + z)(y + t)$$

(additions modulo 4). Then

$$(\mathbb{Z}_4 \times \mathbb{Z}_4)^* = R \cup B \cup G$$

where

$$\begin{aligned} R &= \{02, 10, 30, 11, 33\}, \\ B &= \{20, 01, 03, 13, 31\}, \\ G &= \{22, 21, 23, 12, 32\}. \end{aligned} \tag{10.1}$$

$(\mathbb{Z}_2 \times \mathbb{Z}_2 \times \mathbb{Z}_2 \times \mathbb{Z}_2)^*$ has a similar partition

$$\begin{aligned} R &= \{1000, 1100, 1010, 1111, 0001\}, \\ B &= \{0010, 0011, 1011, 0111, 1101\}, \\ G &= \{0100, 0110, 0101, 1110, 1001\}. \end{aligned} \tag{10.2}$$

The existence of these partitions proves of course that $R_3(3) \geq 17$. To see that $R_3(3) = 17$, we use the following argument. Suppose K_{17} could be colored in three colors. Select any vertex x. Since there are sixteen edges incident with x, there must be at least six in one color, red say. Consider the subgraph generated by the other endpoints of those edges. If it has a red edge, then there is a red triangle; if not, the subgraph is a K_6 colored in the two remaining colors, and it must contain a monochromatic triangle.

The above argument can be used to show that $R_4(3) \leq 66$, but there is no sum-free partition of a 65-element group into four parts. In fact, we know that $51 \leq R_4(3) \leq 65$ ([27, 43, 132]). The lower bound was proven by exhibiting a triangle-free coloring of K_{50}, while the upper bound comes from a lengthy argument proving that if a triangle-free 4-painting of K_{65} existed, then the adjacency matrices of the monochromatic subgraphs would have eigenvalues of irrational multiplicities.

The method of sum-free sets can be generalized to avoid larger complete subgraphs. For example, consider the subset $B = \{1, 2, 6, 7\}$ of K_8 that arose in discussing $R(3, 4)$. This set is not sum-free, and its graph will contain triangles. However, suppose there were a K_4 in the graph, with vertices a, b, c and d. Then B would contain a solution to the following system of three simultaneous equations in six unknowns:

$$\begin{aligned} x_{ab} + x_{bc} &= x_{ac} \\ x_{ac} + x_{cd} &= x_{ad} \\ x_{bc} + x_{cd} &= x_{bd} \end{aligned}$$

(in each case, x_{ij} will be either $i - j$ or $j - i$). But a complete search shows that B contains no solution to these equations. So the graph contains no K_4. (The graph is $\overline{G}$ in Figure 10.2.)

Exercises 10.3

10.3.1 Prove that $R(3,5) > 13$ by choosing $R = \{4, 6, 7, 9\}$, $B = \{1, 2, 3, 5, 8, 10, 11, 12\}$ in Z_{13}.

10.3.2 Show that $R(4,4) > 17$, by choosing $R = \{1, 2, 4, 8, 9, 13, 15, 16\}$, $B = \{3, 5, 6, 7, 10, 11, 12, 14\}$ in Z_{17}.

10.3.3 Verify that the partitions in (10.1) and (10.2) are in fact symmetric sum-free partitions.

10.4 Bounds on Classical Ramsey Numbers

Very few Ramsey numbers are known. Consequently much effort has gone into proving upper and lower bounds.

Lemma 10.9. *If p and q are integers greater than 2, then*

$$R(p,q) \le R(p-1,q) + R(p,q-1).$$

Proof. Write $m = R(p-1,q) + R(p,q-1)$. Suppose the edges of K_m are colored in red and blue. We shall prove that K_m contains either a red K_p or a blue K_q. Two cases arise.

(i) Suppose that one of the vertices x of K_m has at least $s = R(p-1,q)$ red edges incident with it, connecting it to vertices $x_1, x_2, \ldots, x_s$. Consider the K_s on these vertices. Since its edges are colored red or blue, it contains either a blue K_q, in which case the lemma is proved, or a red K_{p-1}. Let the set of vertices of the red K_{p-1} be $\{y_1, y_2, \ldots, y_{p-1}\}$. Then the vertices $x, y_1, \ldots, y_{p-1}$ are those of a red K_p and again the lemma holds.

(ii) Suppose that no vertex of K_m has $R(p-1,q)$ red edges incident with it. Then every vertex must be incident with at least $m - 1 - [R(p-1,q) - 1] = R(p,q-1)$ blue edges. The argument is then analogous to that of part (i). □

Theorem 10.10. *For all integers, $p, q \ge 2$,*

$$R(p,q) \le \binom{p+q-2}{p-1}.$$

Proof. Write $n = p + q$. The proof proceeds by induction on n. Clearly $R(2,2) = 2 = \binom{2+2-2}{2-1}$. Since $p, q \ge 2$, we can have $n = 4$ only if $p = q = 2$. Hence the given bound is valid for $n = 4$. Also for any value of q, $R(2,q) = q = \binom{2+q-2}{2-1}$, and similarly for any value of p, $R(p,2) = p = \binom{p+2-2}{p-1}$, so the bound is valid if $p = 2$ or $q = 2$.

Without loss of generality assume that $p \ge 3$, $q \ge 3$ and that

$$R(p',q') \le \binom{p'+q'-2}{p'-1}$$

for all integers p', q' and n satisfying $p' \geq 2, q' \geq 2, p' + q' < n$ and $n > 4$. Suppose the integers p and q satisfy $p + q = n$.

We apply the induction hypothesis to the case $p' = p - 1, q' = q$, obtaining

$$R(p-1, q) \leq \binom{p+q-3}{p-2},$$

and to $p' = p, q' = q - 1$, obtaining

$$R(p, q-1) \leq \binom{p+q-3}{p-1}.$$

But by the properties of binomial coefficients,

$$\binom{p+q-3}{p-2} + \binom{p+q-3}{p-1} = \binom{p+q-2}{p-1},$$

and from Lemma 10.9

$$R(p, q) \leq R(p-1, q) + R(p, q-1),$$

so

$$R(p, q) \leq \binom{p+q-3}{p-2} + \binom{p+q-3}{p-1} = \binom{p+q-2}{p-1}. \qquad \square$$

If $p = 2$ or $q = 2$ or if $p = q = 3$, this bound is exact. But suppose $p = 3$, $q = 4$. Then $\binom{p+q-2}{p-1} = \binom{5}{2} = 10$, and the exact value of $R(3, 4)$ is 9. Again if $p = 3$, $q = 5$, then $\binom{p+q-2}{p-1} = \binom{6}{2} = 15$, whereas the exact value of $R(3, 5)$ is 14. In general, Theorem 10.10 shows that

$$R(3, q) \leq \binom{q+1}{2} = \frac{q(q+1)}{2} = \frac{q^2+q}{2}.$$

But for the case $p = 3$, this result can be improved [9]. It is shown there that for every integer $q \geq 2$,

$$R(3, q) \leq \frac{q^2+3}{2}.$$

The following lower bound for $R_n(k)$ was proved by Abbott [1].

Theorem 10.11. *For integers $s, t \geq 2$,*

$$R_n(st - s - t + 2) \geq (R_n(s) - 1)(R_n(t) - 1) + 1.$$

Proof. Write $p = R_n(s) - 1$ and $q = R_n(t) - 1$. Consider a K_p with vertices $x_1, x_2, \ldots, x_p$ and a K_q with vertices $y_1, y_2, \ldots, y_q$. Color the edges of K_p and K_q in n colors $c_1, c_2, \ldots, c_n$ in such a way that K_p contains no monochromatic K_s and K_q contains no monochromatic K_t (such colorings must be possible by the definitions of p and q).

Now let K_{pq} be the complete graph on the vertices z_{ij}, where $i \in 1, 2, \ldots, p$ and $j \in 1, 2, \ldots, q$. Color the edges of K_{pq} as follows:

(i) Edge $w_{gj}w_{gh}$ is given the color y_jy_h received in K_q.

(ii) If $i \neq g$, $w_{ij}w_{gh}$ is given the color $y_j y_h$ received in K_p.

Now write $r = st - s - t + 2$ and let G be any copy of K_r contained in K_{pq}. Suppose G is monochromatic, with all its edges colored c_1. Two cases arise:

(i) There are s distinct values of i for which the vertex w_{ij} belongs to G. Then from the coloring scheme K_p contains a monochromatic K_s, which is a contradiction;

(ii) There are at most $s - 1$ distinct values of i for which w_{ij} belongs to G. Suppose there are at most $t - 1$ distinct values of j such that w_{ij} belongs to G. Then G has at most $(s-1)(t-1) = st - s - t + 1 = r - 1$ vertices, which is a contradiction. So there is at least one value of i such that at least t of the vertices w_{ij} belong to G. Applying the argument of Case (i), K_q contains a monochromatic K_t, which is again a contradiction.

Thus K_{pq} contains no monochromatic K_r. □

In order to develop the ideas of sum-free sets and obtain some bounds for $R_n(3)$, we define the *Schur function*, $f(n)$, to be the largest integer such that the set

$$1, 2, \ldots, f(n)$$

can be partitioned into n mutually disjoint nonempty sets $S_1, S_2, \ldots, S_n$, each of which is sum-free. Obviously $f(1) = 1$, and $f(2) = 4$ where $\{1, 2, 3, 4\} = \{1, 4\} \cup \{2, 3\}$ is the appropriate partition. Computations have shown that $f(3) = 13$ with

$$\{1, 2, \ldots, 13\} = \{3, 2, 12, 11\} \cup \{6, 5, 9, 8\} \cup \{1, 4, 7, 10, 13\}$$

as one possible partition, that $f(4) = 44$ and $f(5) \geq 138$.

Lemma 10.12. *For any positive integer n*

$$f(n+1) \geq 3f(n) + 1,$$

and since $f(1) = 1$,

$$f(n) \geq \frac{3^n - 1}{2}.$$

Proof. Suppose that the set $S = \{1, 2, \ldots, f(n)\}$ can be partitioned into the n sum-free sets $S_1 = \{x_{11}, x_{12}, \ldots, x_{1\ell_1}\}, \ldots, S_n = \{x_{n1}, x_{n2}, \ldots, x_{n\ell_n}\}$. Then the sets

$$\begin{aligned}
T_1 &= \{3x_{11}, 3x_{11} - 1, 3x_{12}, 3x_{12} - 1, \ldots, 3x_{1\ell_1}, 3x_{1\ell_1} - 1\}, \\
T_2 &= \{3x_{21}, 3x_{21} - 1, 3x_{22}, 3x_{22} - 1, \ldots, 3x_{2\ell_1}, 3x_{2\ell_1} - 1\}, \\
&\vdots \\
T_n &= \{3x_{n1}, 3x_{n1} - 1, 3x_{n2}, 3x_{n2} - 1, \ldots, 3x_{n\ell_n}, 3x_{n\ell_n} - 1\}, \\
T_{n+1} &= \{1, 4, 7, \ldots, 3f(n) + 1\}
\end{aligned}$$

form a partition of $\{1, 2, \ldots, 3f(n) + 1\}$ into $n + 1$ sum-free sets. So

$$f(n+1) \geq 3f(n) + 1.$$

Now $f(1) = 1$, so equation 10.4 implies that

$$f(n) \geq 1 + 3 + 3^2 + \cdots + 3^{n-1} = \frac{3^n - 1}{2}.$$ □

Theorem 10.13. *For any positive integer*

$$\frac{3^n+3}{2} \le R_n(3) \le n(R_{n-1}(3)-1)+2.$$

Proof. (i) The proof of the upper bound is a generalization of the method used to establish $R_3(3) \le 17$.

(ii) Let $K_{f(n)+1}$ be the complete graph on the $f(n)+1$ vertices $x_0, x_1, \ldots, x_{f(n)}$. Color the edges of $K_{f(n)+1}$ in n colors by coloring $x_i x_j$ in the k-th color if and only if $|i-j| \in S_k$.

Suppose the graph contains a monochromatic triangle. This must have vertices x_a, x_b, x_c with $a.b.c$, such that $a-b, b-c, a-c \in S_k$. But now $(a-b)+(b-c)=a-c$, contradicting the fact that S_k is sum-free. Hence

$$f(n)+1 \le R_n(3)-1,$$

so that

$$\frac{3^n-1}{2}+2 = \frac{3^n-3}{2} \le R_n(3),$$

which proves the lower bound. □

Exercises 10.4

A10.4.1 Verify that, for any $p \ge 2$ and any $q \ge 2$,

$$R(2,q)=q, \qquad R(p,2)=p.$$

10.4.2 Is it possible to 2-color the edges of K_{35} so that no red K_4 or blue K_5 occurs?

A10.4.3 Is it possible to 2-color the edges of K_{25} so that no monochromatic K_5 occurs?

10.4.4 Verify that the sets $T_1, T_2, \ldots, T_{n+1}$ of Lemma 10.12 form a sum-free partition.

10.4.5 Verify the upper bound in Theorem 10.13.

10.5 The General Case of Ramsey's Theorem

In its general (finite) form, Ramsey's Theorem deals with the partition of the collection of all r-sets on a set. The graphical case is the case $r = 2$. The case $r = 1$ is the well-known pigeonhole principle: *if n objects are distributed among more than n sets, some set will contain at least two objects.* This obvious statement has some less-than-obvious applications.

We state Ramsey's Theorem in its general form as Theorem 10.14. The proof is left to the exercises.

Theorem 10.14. (Ramsey's Theorem) *Suppose S is an s-element set. Write $\Pi_r(S)$ for the collection of all r-element subsets of S, $r \ge 1$. Suppose the partition*

$$\Pi_r(S) = A_1 \cup A_2 \cup \cdots \cup A_n$$

is such that each r-subset of S belongs to exactly one of the A_i, and no A_i is empty. If the integers $p_1, p_2, \ldots, p_n$ satisfy $r \le p_i \le s$, for $i = 1, 2, \ldots, n$, then there exists an integer

$$R(p_1, p_2, \ldots, p_n; r),$$

depending only on n, $p_1, p_2, \ldots, p_n$ *and* r, *such that if* $s \geq R(p_1, p_2, \ldots, p_n; r)$, *then for at least one* i, $1 \leq i \leq n$, *there exists a* p_i*-element subset* T *of* S, *all of whose* r*-subsets belong to* A_i.

Exercises 10.5

0.5.1 Suppose p, q and r are integers satisfying $1 \leq r \leq p, q$. Prove:
(i) $R(p, q; 1) = p + q - 1$;
(ii) $R(r, q; r) = q$;
(iii) $R(p, r; r) = p$.

0.5.2 Prove Ramsey's Theorem for $n = 2$.

0.5.3 Prove Ramsey's Theorem.

11
Digraphs

11.1 Basic Ideas

Recall from Chapter 1 that a *digraph* is a finite set v of objects called *vertices* together with a finite set of directed edges, or *arcs*, which are *ordered pairs* of vertices. It is like a graph except that each edge is allocated a direction — one vertex is designated a *start* and the other is a *finish*. An arc directed from start s to finish t is denoted by (s, t), or simply st. It is important to observe that, unlike a graph, a digraph can have two arcs with the same endpoints, provided they are directed in opposite ways. But we shall not allow *multiple arcs* or *loops*.

The idea of adjacency needs further consideration in digraphs. Associated with a vertex x are the two sets

$$A(x) = \{y : (x, y) \text{ is an arc}\}$$
$$B(x) = \{y : (y, x) \text{ is an arc}\}.$$

($A(x)$ consists of the vertices *after* x, and $B(x)$ means *before* x.) A vertex x is called a start or *source* in the digraph if $B(x)$ is empty and a finish or *sink* if $A(x)$ is empty.

The *indegree* and *outdegree* of a vertex are the numbers of arcs leading into and leading away from that vertex respectively, so if multiple arcs are not allowed, then the indegree and outdegree of v equal $|B(v)|$ and $|A(v)|$ respectively.

The notation of (1.5) is extended to directed graphs in the obvious way, so that if X and Y are any sets of vertices of G, then $[X, Y]$ consists of all arcs with start in X and finish in Y. If X or Y has only one element, it is usual to omit the set brackets in this notation. Observe that, if V is the vertex set of G, then

$$[x, A(x)] = [x, V] = \text{ set of all arcs leading out of } x,$$
$$[B(x), x] = [V, x] = \text{ set of all arcs leading into } x.$$

A (directed) *walk* of length n in a directed multigraph is a sequence $(x_0, a_1, x_1, a_2, x_2, \ldots, x_{n-1}, a_n, x_n)$ of vertices and arcs such that $a_i = (x_{i-1}, x_i)$. (This is analogous to the definition of a walk in a graph, but takes into account the direction of each arc. Each arc must be traversed in its proper direction.) A *directed path* is a directed walk in which the vertices are all different. Not every path is a directed path. If a directed path is considered as a digraph, then x_0 is a source, and is unique, and a_n is the unique sink, so we call x_0 and x_n the *start* or *source* and *finish* or *sink* of the path respectively. We say x_i precedes x_j (and x_j succeeds x_i) when $i < j$.

A *directed circuit* $(x_1, x_2, \ldots, x_n)$ is a directed walk in which the first and last vertex are the same. A *directed cycle* is a directed circuit in which all of the members are distinct. A digraph is called *acyclic* if it contains no directed cycle.

Most of the above reads as though we are applying the word "directed" to many of the terms that were used in describing undirected walks, paths and cycles. However, you should notice that there can be a directed cycle of length 2, or *digon*, which does not occur in the undirected case.

Example. Consider the digraph of Figure 11.1. It has

$$\begin{aligned} A(a) &= \{b, c\}, & B(a) &= \emptyset, \\ A(b) &= \emptyset, & B(b) &= \{a, c, d\}, \\ A(c) &= \{b, d\}, & B(c) &= \{a\}, \\ A(d) &= \{b\}, & B(d) &= \{c\}; \end{aligned}$$

a is a source and b is a sink. $[\{a, c\}, \{b, d\}] = \{ab, cb, cd\}$. There are various directed paths, such as (a, c, d, b), but no directed cycle.

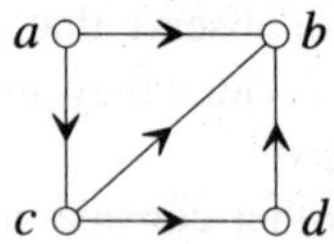

Fig. 11.1. A typical digraph

Lemma 11.1. *If a digraph contains an infinite sequence of vertices $(x_0, x_1, \ldots)$ such that $x_{i-1}x_i$ is an arc for every i, then the digraph contains a cycle.*

Proof. Any digraph has finitely many vertices, so the sequence $(x_0, x_1, \ldots)$ must contain repetitions. Suppose x_j is the first repetition — that is, $x_i = x_j$ for some $i < j$, and no vertex x_k with $k < j$ has this property. Then $(x_i, x_{i+1}, \ldots, x_j)$ is a cycle in g. □

In a similar way we can prove

Lemma 11.2. *If a digraph contains an infinite sequence of vertices $(x_0, x_1, \ldots)$ such that $x_{i+1}x_i$ is an arc for every i, then the digraph contains a cycle.*

Theorem 11.3. *Every acyclic digraph has a source and a sink.*

Proof. Suppose the digraph D contains no sink. Select any vertex x_0 of D. Since x_0 is not a sink, $A(x_0)$ is not empty; select a vertex x_1 in $A(x)$. Similarly, select x_2 in $A(x_1)$,

x_3 in $A(x_2)$, and so on. Then Lemma 11.1 implies there is a cycle in D, a contradiction. The proof in the case of a digraph with no source is similar (but uses Lemma 11.2). □

The following theorem provides an unexpected link between directed paths and chromatic numbers. Notice that the concepts relating to vertex-coloring do not depend on whether or not the edges are directed, so one can define the chromatic number of a digraph to be the chromatic number of its underlying graph.

Theorem 11.4. [104] *A digraph D contains a directed path of length $\chi(D) - 1$.*

Proof. Let A be a set of smallest possible size of arcs of the digraph D such that $D' = D - A$ is acyclic. If k is the length of the longest directed path in D', we shall color the vertices of D with colors $1, 2, \ldots, k$ as follows: if the longest directed path in D' with start x has length i, then x receives color i. We shall show that this is a proper coloring of $V(D)$.

Suppose there is a directed path P from x to y in D'. Then it is clear that x and y receive different colors: if Q is a longest directed path in D' with start y, then Q can contain no vertex of P other than y (if it contained z, then there would be a cycle in D', formed by following P from z to y and then Q from y to z), so $P \cup Q$ is a path in D' with start x, and it is longer than Q. So the endpoints of any directed path in D' receive different colors under the coloring.

We now observe that the endpoints of every arc xy of D receive different colors. If the arc is in D', then it constitutes a directed path (of length 1) in D', and the result follows from the preceding paragraph. If not, xy is in A; the minimality of A implies that $D' + xy$ must contain a cycle, and deleting xy from that cycle yields a directed path from y to x in D'. So the coloring is proper.

This means that $\chi(D) \le k+1$. So D' contains a directed path of length $\chi(D) - 1$; certainly D will contain a directed path of this length. □

The concept of a complete graph generalizes to the directed case in two ways. The *complete directed graph* on vertex set V, denoted by DK_V, has as its arcs all ordered pairs of distinct members of V, and is uniquely determined by V. On the other hand, one can consider all the different digraphs that can be formed by assigning directions to the edges of the complete graph on V; these are called *tournaments*, and will be discussed in Section 11.2.

In those cases where a directed graph is fully determined, up to isomorphism, by its number of vertices, notation is used that is analogous to the undirected case. The directed path, directed cycle and complete directed graph on v vertices are denoted by DP_v, DC_v and DK_v respectively.

We shall say vertex x is *reachable* from vertex y if there is a walk (and consequently a directed path) from y to x. (When x is reachable from y, some authors say "x is a descendant of y" and "y is an ancestor of x".) Two vertices are *strongly connected* if each is reachable from the other, and a digraph (or directed multigraph) is called strongly connected if every vertex is strongly connected to every other vertex. For convenience, every vertex is defined to be strongly connected to itself. We shall say a directed graph or multigraph is *connected* if the underlying graph is connected, and *disconnected* otherwise. However, some authors reserve the word "connected" for a digraph in which, given any pair of vertices x and y, either x is reachable from y or y is reachable from x.

It is clear that strong connectivity is an equivalence relation on the vertex set of any digraph D (see Exercise 11.1.3). The equivalence classes, and the subdigraphs induced by them, are called the *strong components* of D.

Exercises 11.1

A11.1.1 For each of the digraphs in Figure 11.2:
(i) write down the list of arcs;
(ii) write down $A(x)$ and $B(x)$ for every vertex x;
(iii) find all directed paths from s to t;
(iv) write down a directed cycle of maximum length in the digraph;
(v) write down $[X, Y]$, where $X = \{s, a, b\}$, and $Y = V \backslash X$.

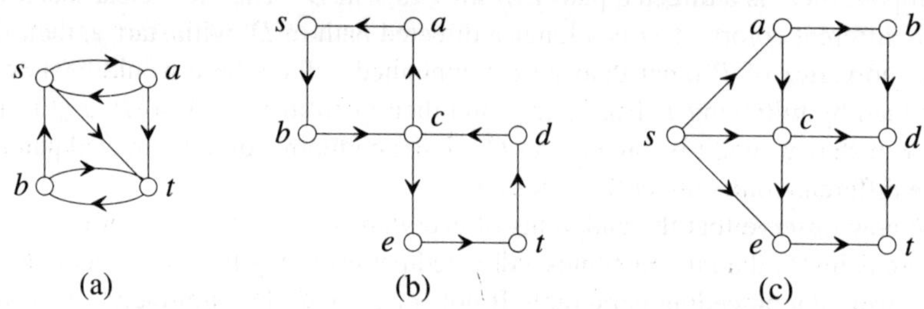

Fig. 11.2. Digraphs for analysis in Exercise 11.1.1.

11.1.2 Repeat the preceding exercise for the digraphs in Figure 11.3.

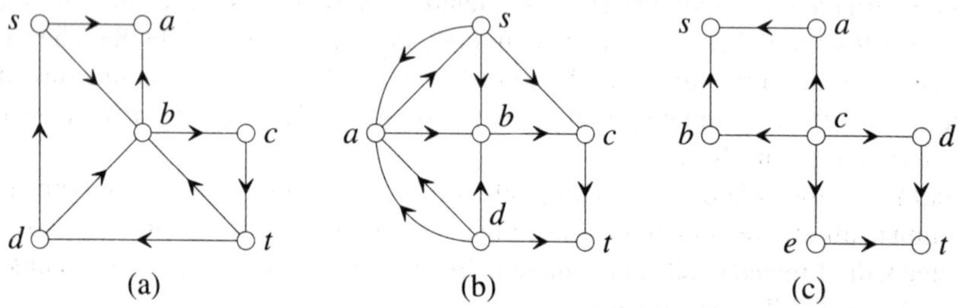

Fig. 11.3. Digraphs for analysis in Exercise 11.1.2.

H11.1.3 Prove (the directed analog of Theorem 2.1) that if there is a walk from x to y in the digraph D, then there is a directed path from x to y in D. Hence prove that strong connectivity is a transitive relation on the vertex set of a digraph.

11.1.4 The *reachability digraph* $R(D)$ of D is the digraph whose vertices are the vertices of D, in which yx is an arc if and only if x is reachable from y.

(i) Prove that the strong components of D are complete subdigraphs of $R(D)$, and are maximal (that is, they are not contained in any larger complete subdigraph).

(ii) Prove that not every digraph is the reachability digraph of some digraph.

1.1.5 For each of the digraphs in Figure 11.2:

(i) construct the reachability digraph (as defined in the preceding exercise);

(ii) list the strong components of the original digraph.

1.1.6 Do the same for the digraphs in Figure 11.3.

1.1.7 Prove that, in any digraph with vertex set V,

$$\sum_{v \in V} |A(v)| = \sum_{v \in V} |B(v)|.$$

1.1.8 Human blood comes in four main types — O, A, B and AB. A person with type O can give blood to anybody; type A or B can give to their own type or to AB; AB can give only to AB. Draw a diagram to illustrate these relationships. Is it a digraph?

11.2 Orientations and Tournaments

If G is any graph, it can be converted to a digraph by assigning a direction to each edge. This process is called *orientation*. An *oriented graph* is a digraph that can be obtained by orienting a graph; in other words, it is a digraph that never contains both arcs xy and yx for any pair of vertices x and y.

If G has e edges, it can be oriented in 2^e ways, but these orientations will not usually be nonisomorphic. For example, there are eight different orientations of K_3, but only two different ones up to isomorphism.

An important practical problem is to orient edges so that the resulting digraph is strongly connected. Such an orientation will be called *strong*. For example, consider a communications network in which messages along any given link can be sent in only one direction. If every node is to be able to send messages to every other node, the directions chosen must form a strong orientation.

Theorem 11.5. *A connected graph G has a strong orientation if and only if every edge of G belongs to at least one cycle.*

Proof. The necessity is obvious. Now suppose every edge of G belongs to at least one cycle. We give an algorithm, similar to the Euler walk algorithm, which produces a strong orientation.

Suppose S is some set of edges of G that can be oriented in such a way that the resulting digraph $D(S)$ is strongly connected. If S contains all edges of G, we are done. Otherwise, select an edge xy that has not yet been oriented, where x is a vertex of a member of S. (This must be possible, since G is connected.) Find a cycle that contains xy, and orient it consistently, except that if an edge has already been assigned a direction (that is, the edge is common to S and the new cycle), the orientation is not changed. Call the new cycle, together with its orientations, C. Then either C is a directed cycle or it consists of a number of arcs of $D(S)$ together with a number of directed paths that lead from one vertex of $D(S)$ to another.

We verify that $D(S) \cup C$ is strongly connected. Suppose z and t are two vertices of the union. If z and t are both vertices of $D(S)$, each is reachable from the other because $D(S)$ is strongly connected. If z lies in C but not in $D(S)$, say z lies on a directed path P that leads from a to b, both of which are vertices of $D(S)$; then one can get from z to any vertex t of $D(S)$ by traversing P from z to b, and then traversing $D(S)$ from b to t, and to get from t to z one travels through $D(S)$ from t to a and then along P to z. To travel between any two vertices not in $D(S)$, one can find a walk that passes through some point of $D(S)$.

If there are edges that have not yet received orientations, one replaces $D(S)$ by $D(S) \cup C$ and proceeds as above. Eventually the process halts because of the finiteness of G, and the result is a strong orientation.

It remains to show that this whole process can be started — that is, the initial set S must be found. But one can select any cycle in G, and orient its edges to form a directed cycle. □

Suppose a competition is conducted in a sport such as tennis, where every match results in a win or loss — ties and draws are impossible. Then the results can be represented graphically as follows: if x beats y, there is an arc xy. If the competition is a (single) round robin tournament, the possible result diagrams are precisely the orientations of the complete graph whose vertices are the contestants. For this reason an oriented complete graph is called a *tournament*, and the outdegree of a vertex is called its *score*. If xy is an arc in a tournament, it is usual to say x *dominates* (or *beats*) y.

Several special kinds of tournaments have been defined. A *transitive* tournament is one in which, whenever x dominates y and y dominated z, x dominates z. This implies a linear "dominance" structure from which the whole structure of a transitive tournament can be deduced, and there is exactly one transitive tournament on v vertices, up to isomorphism. A *reducible* tournament is one in which the vertex set can be partitioned into two nonempty sets S and T such that each vertex of S dominates each vertex of T; if no such partition exists, then the tournament is *irreducible*. A transitive tournament can be viewed as a very special kind of reducible tournament.

Theorem 11.6. *A tournament is irreducible if and only if it is strongly connected.*

Proof. Suppose D is a reducible tournament, and every vertex in S dominates every vertex in T. Then no vertex in S is reachable from any vertex in T, so D is not strongly connected. Now assume D is not strongly connected; say vertex y is not reachable from vertex x. Write S for the set of all vertices that are not reachable from x, and T for the set of all vertices reachable from x. Then neither set is empty ($y \in S$ and $x \in T$), and every member of S must dominate every member of T. So D is reducible. □

Corollary 11.7. *Every vertex of an irreducible tournament lies in a directed cycle of length 3.*

Proof. Say vertex x lies in no directed triangle. If $A(x)$ is empty, then every other vertex dominates x. Otherwise, $B(x) \cup \{x\}$ dominates $A(x)$. □

A tournament is called *regular* if every vertex has the same score.

Hamilton paths and cycles in digraphs are defined in the obvious way, as *directed* paths and cycles that contain all the vertices. Then we have

Theorem 11.8. [99] *Every tournament contains a Hamilton path.*

Proof. Consider a tournament D on v vertices. Since its underlying graph is K_v, $\chi(D) = v$. So by Theorem 11.4, D contains a path of length $v - 1$. □

Theorem 11.9. [88] *Every vertex of any strongly connected tournament on v vertices, $v \geq 3$, lies on a directed cycle of length k for every k, $3 \leq k \leq v$.*

Proof. We proceed by induction on k. The theorem is true for $k = 3$ by Corollary 11.7. Suppose it is true for $k \leq t$, where $t < v$. Select a vertex x_0 and a directed cycle

$$C = (x_0, x_1, \ldots, x_{t-1}, x_0)$$

containing it. Define sets of vertices R, S, T as follows: S is the set of vertices that dominate all vertices in C, T is the set of vertices that are dominated by all vertices in C, and R consists of all vertices not in C, S or T.

First suppose R is empty. If there is a vertex t in T and a vertex s in S such that t dominates s, then

$$(x_0, x_1, \ldots, x_{t-2}, t, s, x_0)$$

is a directed cycle of length $t + 1$. But such a vertex must exist: if S is empty, then T is not empty and $V(C)$ dominates T; if T is empty, or if no arc ts exists, then S dominates all other vertices. In either case, D is reducible and therefore not strongly connected.

On the other hand, suppose R contains a vertex, say r. Suppose x_i dominates r. Going around C from x_i in the direction $x_i, x_{i+1}, \ldots$, find the first vertex x_j such that r dominates x_j. (If no such vertex existed, r would not be in R.) Then

$$(x_0, x_1, \ldots, x_{j-1}, r, x_j, \ldots, x_{t-1}, x_0)$$

is a directed cycle of length $t + 1$, as required. (If $j = 0$, r is inserted between x_{t-1} and x_0.) □

Corollary 11.10. *Every strongly connected tournament contains a Hamilton cycle.*

Exercises 11.2

1.2.1 Find strong orientations for:
A(i) K_v ($v \geq 3$);
(ii) $K_{m,n}$ ($m, n \geq 2$);
(iii) the Petersen graph.

1.2.2 Prove Theorem 11.8 by induction, without using Theorem 11.4.

11.2.3 Given a reducible tournament, must the sets S and T be unique? Provide a proof or counterexample.

11.2.4 D is any tournament. Prove that one can produce an irreducible tournament by changing the direction of at most one arc of D.

A11.2.5 The *score sequence* of a tournament is formed by arranging the set of scores of its vertices in nondecreasing order. Find the score sequences of the tournaments in Figure 11.4.

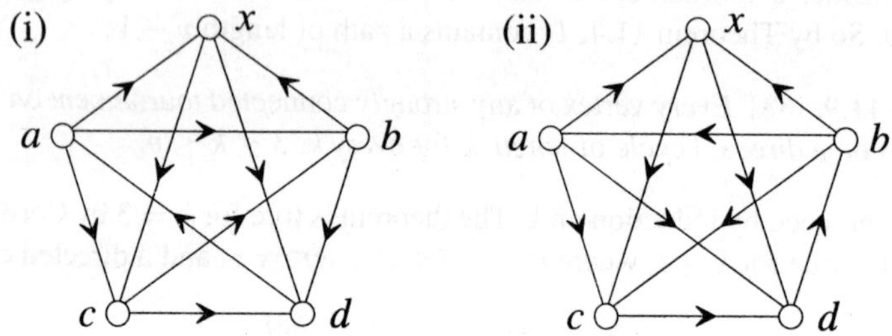

Fig. 11.4. Tournaments for Exercises 11.2.5 and 11.2.9

11.2.6 If the score sequence (as defined in Exercise 11.2.5) of the tournament D is $(a_1, a_2, \ldots, a_v)$, prove:
(i) $\sum_{i=1}^{v} a_i = \frac{1}{2}v(v-1)$;
(ii) if $k < v$, then $\sum_{i=1}^{k} a_i \geq \frac{1}{2}k(k-1)$, and the inequality is strict for every k if and only if D is strongly connected.

11.2.7 Prove by induction, without using Theorem 11.9, that every strongly connected tournament contains a Hamilton cycle.

11.2.8 C is a directed cycle in a strongly connected tournament D. Prove that D has a Hamilton cycle in which the cyclic order of the vertices of C is preserved. [37]

A11.2.9 Find cycles through x of lengths 3, 4 and 5 in the tournaments shown in Figure 11.4.

11.2.10 Repeat Exercises 11.2.5 and 11.2.9 for the following tournaments.

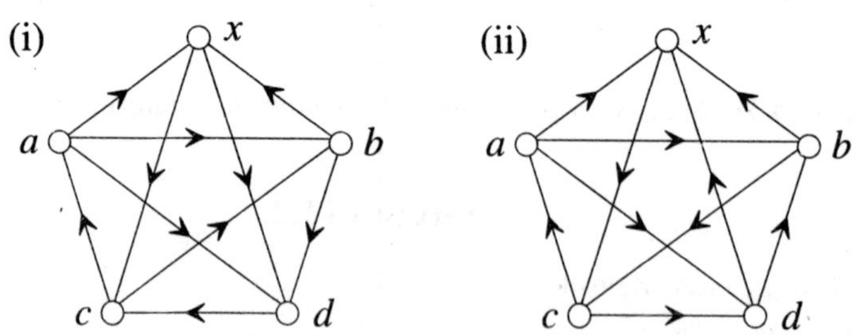

11.2.11 Prove that a tournament is transitive if and only if it contains no directed 3-cycle.

A11.2.12 A regular tournament in which the common score of all the vertices is s is called *s-regular*.

(i) Prove that an s-regular tournament must have $2s + 1$ vertices.
(ii) Prove that there exists an s-regular tournament for every positive integer s.

2.13 Prove that every regular tournament, as defined in the preceding exercise, is strongly connected.

11.3 Directed Euler Walks

We use the phrase "Euler walk" when discussing directed graphs and multigraphs in the obvious way. It means a walk (in the directed sense) that covers each arc exactly once. There is a directed analog of Theorem 2.6.

Theorem 11.11. *A directed multigraph has an Euler walk if and only if it is connected, and the indegree of every vertex equals its outdegree, with the possible exception of two vertices. For these two vertices, the indegree of one is* 1 *greater than its outdegree, and the indegree of the other is* 1 *less than its outdegree.*

Theorem 11.12. *A directed multigraph has a closed Euler walk if and only if it is connected, and the indegree of every vertex equals its outdegree.*

Good [54] gave the following application of Euler walks to the construction of an automated generalized switch. Suppose a machine carries out several different tasks in sequence: an example is a washing machine, which will first fill with water, then agitate, then spin, and so on. The switch is in the form of a rotating drum divided into a number of sectors — let us say 16. The construction is illustrated in Figure 11.5, where the shaded areas are made of conducting material and the white areas are made of nonconducting material. A small electric motor makes the drum rotate, and the position of the drum determines whether the terminals a, b, c, d are connected to the earth or insulated from it. This is the information necessary for the machine to carry out a specific type of operation. For example, when the drum position is as shown in Figure 11.5, a, c and d are connected to earth, while b is not.

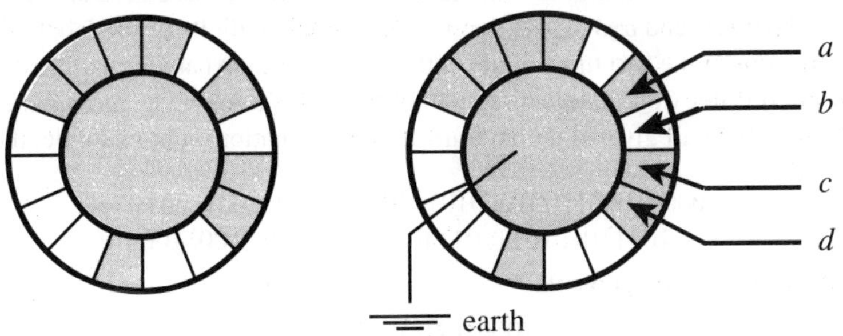

Fig. 11.5. A generalized switch

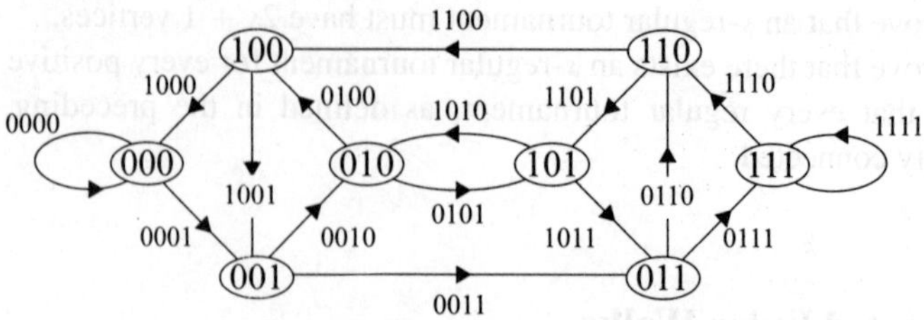

Fig. 11.6. Directed multigraph corresponding to a generalized switch

For each of the 16 positions of the drum to give a distinct signal to the machine, the sectors must be constructed so that no two conducting and nonconducting patterns of four consecutive sectors are the same. We wish to know whether such an arrangement exists, and, if so, how to construct it.

Label the earthed (conducting) sectors with 0 and the insulating (nonconducting) sectors with 1. The setting in the Figure 11.5 would read 1011. If the drum were rotated clockwise one position, it would read 1101. (Terminals a, b, c and d are fixed in position.) The problem becomes:

Arrange 16 *binary digits in a circular array such that every sequence of four consecutive digits is distinct.*

We construct a directed multigraph with eight vertices labeled with the eight 3-digit binary numbers:

$$000, \quad 001, \quad 010, \quad 011, \quad 100, \quad 101, \quad 110, \quad 111.$$

We construct the arcs of the multigraph and label each with a 4-digit binary number as follows:

We draw one arc from the vertex $\alpha_1\alpha_2\alpha_3$ to the vertex $\alpha_2\alpha_3 0$ and another to the vertex $\alpha_2\alpha_3 1$, labeling the arc from $\alpha_1\alpha_2\alpha_3$ to $\alpha_2\alpha_3 0$ by $\alpha_1\alpha_2\alpha_3 0$ and the arc from $\alpha_1\alpha_2\alpha_3$ to $\alpha_2\alpha_3 1$ by $\alpha_1\alpha_2\alpha_3 1$. Since there are eight distinct vertices labeled with eight distinct 3-digit binary numbers, this process will yield a multigraph with 16 arcs labeled with 16 distinct 4-digit binary numbers, shown in Figure 11.6.

In any walk in the multigraph, the labels for any two consecutive arcs must be of the form $\alpha_1\alpha_2\alpha_3\alpha_4$ and $\alpha_2\alpha_3\alpha_4\alpha_5$; that is, the final three digits of the label of the first arc are the same as the first three digits of the label of the next arc. Thus, corresponding to any closed Euler walk in this directed multigraph is a (circular) arrangement of 16 binary digits that will give all the 16 four-digit combinations. For example, the closed Euler walk

$$0000, 0001, 0010, 0101, 1010, 0100, 1001, 0011,$$
$$0110, 1101, 1011, 0111, 1111, 1110, 1100, 1000$$

corresponds to the arrangement

$$0000101001101111.$$

A circuit is obtained by joining the ends of this arrangement. This is the circuit that was used to construct the drum shown in Figure 11.5; the start of the sequence is shown by an arrow in that diagram.

A closed Euler walk exists because the indegree of each vertex equals its outdegree.

It is possible by a similar construction to arrange 2^n binary digits in a circular array so that 2^n sequences of n consecutive digits in the arrangement are all distinct. We construct a directed multigraph by 2^{n-1} vertices labeled with 2^{n-1} $(n-1)$-digit binary numbers and draw arcs from vertex $\alpha_1\alpha_2\alpha_3\cdots\alpha_{n-1}1$ to $\alpha_2\alpha_3\cdots\alpha_{n-1}0$ and $\alpha_2\alpha_3\cdots\alpha_{n-1}1$ respectively. This multigraph then contains a closed Euler walk.

Sequences of this type are called *de Bruijn sequences*, after the mathematician who introduced them. More recently they have been called *universal sequences*.

Exercises 11.3

.3.1 Find a digraph with four vertices that has no Euler walk, although its underlying graph has an Euler walk.

.3.2 Suppose the graph G has an Euler walk. Prove that it is possible to orient G in such a way that the resulting digraph has an Euler walk.

.3.3 Prove Theorem 11.11.

.3.4 Find a cyclic sequence of length 27 that contains the digits 0, 1, 2, nine times each, such that each of the 27 strings of length three chosen from the set $\{0, 1, 2\}$ appears exactly once.

12

Critical Paths

12.1 Activity Digraphs

One application of digraphs is in the scheduling of compound activities, ones that are made up of various tasks. One easy example is building a house. There are several different tasks — roofing, assembling walls, carpeting and so on. Sometimes there is a strict priority relationship (you cannot lay carpet until the floors are done); sometimes the tasks are independent (carpeting can be done before, after or during the exterior painting). We shall assume that these are the only possibilities: given two tasks a and b, either one *must* precede the other or they are independent.

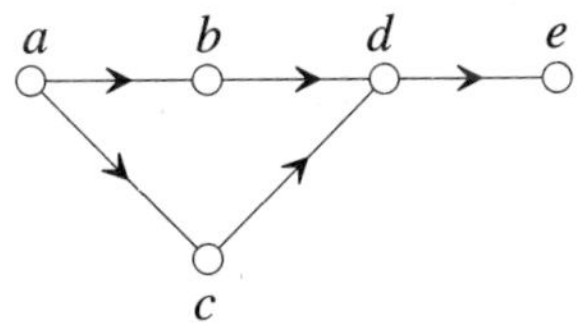

Fig. 12.1. Illustration of precedence relations

We shall set up a digraph, called an *activity* digraph, to represent the precedence requirements of a compound activity. The various tasks are represented by vertices. The digraph contains a path in which vertex a precedes vertex b if and only if task a must be completed before task b is started. For example, in Figure 12.1, tasks b and c must precede task d, but tasks b and c are independent. We assume that all the required tasks are included in the digraph — if it is possible to perform all the tasks that precede x, then task x can be performed.

In order for this representation of tasks to be a digraph, there can be no loops or multiple arcs. This is no problem. A loop will never occur, because it would represent a task that is a prerequisite for itself; such a task could never be performed. Multiple arcs have no sensible interpretation.

In constructing an activity digraph, it is not always clear that there is a "starting-point" s and an "endpoint" or "finish" t; so, for convenience, we formalize the following rules to ensure that the digraph represents a task with a well-defined start and finish:

- If the digraph contains more than one source, define a new vertex s; and add an arc sx to the digraph whenever x is a source;
- If the digraph contains more than one sink, define a new vertex t; and add an arc yt to the digraph whenever x is a sink.

We refer to this process as *initializing* the digraph. If necessary, we shall assume that any activity digraph we discuss has been initialized.

Example. In building a barn one must first build the frame. The outside walls must be erected before painting, and also before any inside fittings are installed. The roof must also be in place before the inside fittings are done. Let us denote the tasks by:

a: build frame
b: build walls
c: fit roof
d: paint
e: do inside fittings.

Then we have:

a must precede b, c, d, e;
b must precede d, e;
c must precede e.

The digraph in Figure 12.2 is easily constructed. The unique source a serves as a start; a finish t has been added.

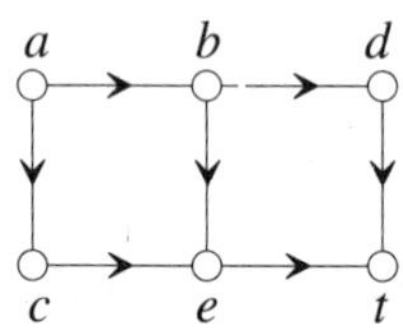

Fig. 12.2. Digraph for the barn example

Clearly some digraphs represent impossible sets of prerequisites — for example, a directed cycle ab of length 2 means that a is prerequisite for b and b is prerequisite for a. More generally, if a digraph contains a directed cycle, the corresponding project cannot be completed. We shall now prove that the converse is true.

Theorem 12.1. *A set of tasks can be completed if its activity digraph D is acyclic.*

Proof. By Theorem 11.3, D has at least one source and at least one sink. Assume we have initialized the digraph, providing a unique start s and a unique finish t for the activity. Now define sets $S_0, S_1, \ldots$ as follows:

$$S_0 = \{s\},$$

and for all positive integers i,

$$S_i = \{A(x); x \in S_{i-1}\} \cup S_{i-1}.$$

We prove by induction that it is possible to perform every task represented by a vertex in any of the sets S_i.

Trivially, s can be performed, so the statement is true for $i = 0$. Now suppose every vertex in S_{i-1} can be performed. Then the prerequisites of every task in $\{A(x); x \in S_{i-1}\}$ can be performed, so every task in S_i can be performed.

It remains to show that $t \in S_j$ for some j. Now suppose the contrary. The finiteness of digraphs tells us that there must be some integer n such that $S_n = S_{n+1} = \cdots$. By assumption $t \notin S_n$. So $t \notin Sn + 1$, whence $B(t)$ contains an element that is not in S_n. Call this element v_1; $v_1 \neq t$. Applying the same argument to v_1 instead of t, one obtains a vertex v_2; $v_2 \neq v_1$ or t. In this way an infinite sequence $(t, v_1, v_2, \ldots)$ may be constructed, and therefore D contains a cycle by Lemma 11.2 — a contradiction. □

Exercises 12.1

12.1.1 Set up a digraph that represents the priorities involved in the following set of tasks a, b, c, d, e, f and g.

a must precede b, c, e and g.

b must precede d and e.

f must precede c and c must precede d.

12.1.2 Draw a digraph to represent the complex of tasks a to k, where

a and b precede d and h.

c and f precede g.

h precedes e and f.

e and h precede i.

c, d, f and i precede j.

g and j precede k.

12.1.3 Construct a digraph for a project comprising tasks a to p that satisfy the following precedence relationships.

a, b and c, the first tasks, can start simultaneously.

d, e and f can start immediately after a is completed.

i and g can start after both b and d are completed.

h starts after both c and g are completed.

k and k both succeed i.

j succeeds both e and h.

e, f and h precede m and n.

m and i precede o.

p succeeds j, ℓ and o.

k, n and p are the final tasks of the project.

12.2 Critical Path Analysis

A major consideration in scheduling a project is the time needed to perform the various tasks. To represent this we associate with each vertex x a *weight* $w(x)$. Vertex weights are defined similarly to edge weights, so that w is a mapping from the vertices of a digraph to the nonnegative real numbers. $w(x)$ is called the *duration* of x, and represents the time needed to complete the task x. A digraph with a weight on the vertices or the arcs is often called a *network*, and the network formed by attaching a weight to an order requirement digraph is called an *activity network*.

In this section we shall define a technique, *critical path analysis*, for minimizing the total time taken to carry out a project. It will be assumed that manpower is no problem, in the following sense. Suppose a and b have the same prerequisites, and both are prerequisites for the same set of tasks. Say a takes seven days and b takes five days. It is assumed that there is enough manpower to perform both tasks simultaneously. If this is not so — if, for example, a and b must be performed by the same person — it may be necessary to regard the two as one task x, and write $w(x) = 12$.

If some parts of a project are delayed, this could cause delays in the project as a whole. For example, suppose task a must precede task b, a takes three days and b takes six days. If the rest of the activity can be carried out in nine or less days, then it is necessary to schedule a at the very beginning of the activity and b to start at the beginning of day 4, if unnecessary delays are to be avoided. On the other hand, if the other parts of the activity take more than nine days, there may be some leeway ("slack") in scheduling the start of a or of b, or in the transition from completing a to starting b.

We illustrate the technique using an example. Suppose there are five tasks, called a, b, c, d and e. Task a must precede c; b and c must precede e; b must precede d. The durations of the task are:

a takes 13 minutes.
b takes 25 minutes.
c takes 15 minutes.
d takes 22 minutes.
e takes 27 minutes.

The activity network, with start s and finish t added, is shown in Figure 12.3.

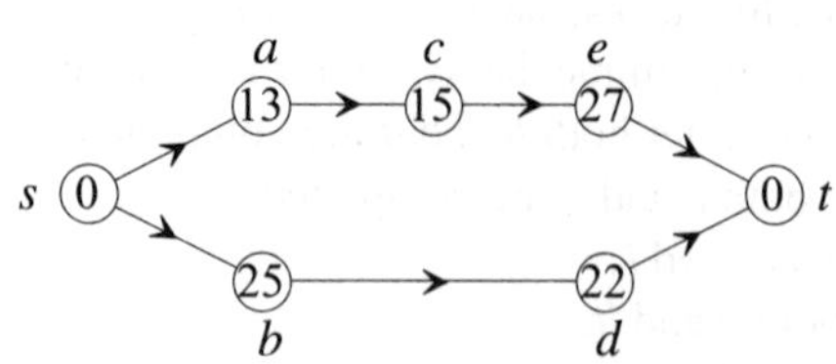

Fig. 12.3. Activity network example

We ask, "given a task, what is the earliest time that it can start and finish?" The task a, for example, could be started immediately (at time 0). It could finish 13 minutes after the start, so the earliest possible finish time would be time 13. We denote this by writing 0 before, and 13 after, the vertex a. Since c cannot start until a is completed, this "earliest finish time" for a is also an "earliest start time" for c, and c cannot be completed until a further 15 minutes have elapsed — time 28. Similarly, the earliest finish time for b is 25.

Now consider vertex e. Since b is a prerequisite, e cannot start before time 25. But c is also a prerequisite, so e cannot start before 28. Clearly the later time, 28, is the important one. So the earliest start time for e is 28, and its earliest finish time is 55.

Continuing in this way, Figure 12.4 is obtained.

The earliest start (and finish) time for t is 55, so the whole procedure must take at least 55 minutes.

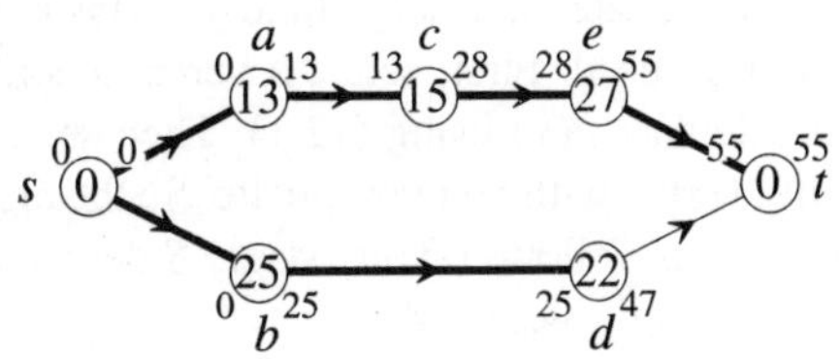

Fig. 12.4. Activity network with earliest start and finish times

A path in the order requirement digraph is a list of tasks such that each is a prerequisite of the rest. So the total time required to finish the set of tasks on a path will be at least as great as the sum of the individual durations of the tasks on the path. Let us call this the *length* of the path. The total time required to finish the project will be at least as long as the longest path.

Look at the arcs shown with heavy lines in the above diagram. In each case the arc leads from a vertex with a certain finish time to a vertex with the same start time. If there is any delay in completing the first task, or in the transition from the first to the second task, the second task will be delayed. In particular, there is a path $sacet$ from s to t, with every arc a heavy line. *If any task or transition along this path is delayed, the whole project will be held up.* Such a path is called a *critical path*. Finding critical paths is very important in the elimination of bottlenecks, the design of assembly lines, and so on.

To find all critical paths, proceed as follows. First, calculate the earliest start and finish times of each task: for S, both times equal 0; for any other task X, the earliest start time will be the maximum among all the finish times of prerequisites of X, and the finish time is the start time plus the duration. An *arc* is critical if the finish time of its first vertex equals the start time of its second vertex. A *critical path* is a path from s to t, all of whose arcs are critical.

This is easily done for small networks. In general it is necessary to express this procedure as an algorithm, which is done as follows. $e(x)$ is defined to be the earliest start time of x, the shortest time in which all tasks in the network necessary to reach x can be carried out. Then the earliest finish time for x is $f(x) = e(x)+w(x)$. It follows that:

$$\begin{aligned} e(x) &= max_{y\in B(x)} f(y) \\ &= max_{y\in B(x)}(e(y) + w(y)). \end{aligned} \tag{12.1}$$

The slack in arc xy is defined as

$$s(x, y) = e(y) - f(x).$$

In this terminology, a critical arc is one with zero slack. A *critical path* is a path in which any delay must cause a delay in the whole project, and *critical path analysis* is the process of finding all critical paths. Once every critical path has been found, one can identify the tasks most likely to cause delays and bottlenecks in the system, and devote resources to trying to speed these tasks, rather than others.

Clearly $e(x)$ cannot be calculated until $e(y)$ (and $f(y)$) is known for all y in $B(x)$. So the following algorithm is used. First, select a vertex x such that $e(x)$ is known for all $y \in B(x)$. Then calculate $e(x)$ using (12.1). Then select another vertex. The process starts at s, the only vertex with no prerequisite. So the algorithmfor calculating the values of the function e is as follows (at any stage, S denotes the set of all vertices y such that $e(y)$ has already been calculated):

1. $e(s) \leftarrow \{0\}$
2. $S \leftarrow \{s\}$
3. **select** x **such that** $B(x) \subseteq S, x \notin S$
4. $e(x) \leftarrow \max_{y\in B(x)}(e(y) + w(y))$
5. $S \leftarrow S \cup x$
6. **if** $x \neq t$ **then goto [3]**

The algorithm is illustrated in the following small example (which would, of course, be solved in practice by constructing the diagram).

Example. To paint a door. The tasks are as follows (time is shown in minutes):

j. Remove old paint (75).
k. Sand the door (30).
ℓ. Open can and stir paint (4).
m. Prepare brushes (5).
n. Clean up paint scrapings (4).
p. Paint the door (30).
q. Clean equipment and put brushes away (15).

j must precede k. j and k must precede n. ℓ, m and n must precede p. p must precede q. A suitable digraph is shown in Figure 12.5.

The algorithm proceeds:

[1] $e(s) = \{0\}$
[2] $S = \{s\}$

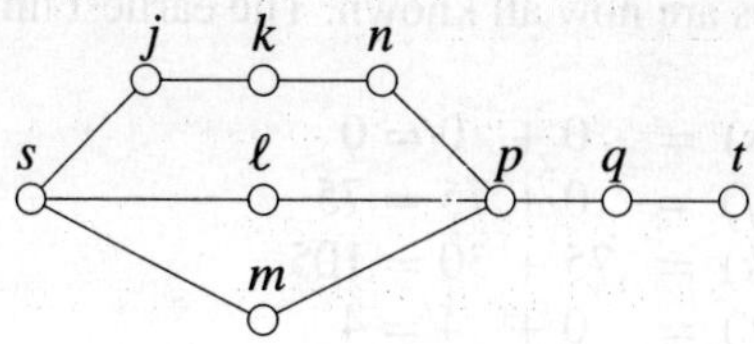

Fig. 12.5. Network for door painting.

[3] $x = j, \ell$ or m, but $x \neq n$ as $k \in B(n)$ but $k \notin S$. Say $x = j$. $B(a) = \{s\}$
[4] $e(j) = \max_{y \in B(a)}(e(y) + w(y)) = e(s) + w(s) = 0 + 0 = 0$
[5] $S = \{s, j\}$
[6] go to **[3]**
[3] $x = k, \ell$ or m. Say $x = k$. $B(k) = \{j\}$
[4] $e(k) = e(j) + w(j) = 0 + 75 = 75$
[5] $S = \{s, j, k\}$
[6] go to **[3]**
[3] Say $x = \ell$. $B(\ell) = \{s\}$
[4] $e(\ell) = e(s) + w(s) = 0 + 0 = 0$
[5] $S = \{s, j, k, \ell\}$
[6] go to **[3]**
[3] Say $x = m$. $B(m) = s$
[4] $e(m) = e(s) + w(s) = 0 + 0 = 0$
[5] $S = \{s, j, k, \ell, m\}$
[6] go to **[3]**
[3] $x = n$. $B(n) = \{k\}$
[4] $e(n) = e(k) + w(k) = 75 + 30 = 105$
[5] $S = \{s, j, k, \ell, m, n\}$
[6] go to **[3]**
[3] $x = p$. $B(p) = \{\ell, m, n\}$
[4] $e(\ell) + w(\ell) = 0 + 4 = 4$
$e(m) + w(m) = 0 + 5 = 5$
$e(n) + w(n) = 105 + 4 = 109$
$e(p) = \max\{44, 5, 109\} = 109$
[5] $S = \{s, j, k, \ell, m, n, p\}$
[6] go to **[3]**
[3] $x = q$. $B(q) = \{p\}$
[4] $e(q) = e(p) + w(p) = 109 + 30 = 139$
[5] $S = \{s, j, k, \ell, m, n, p, q\}$
[6] go to **[3]**
[3] $x = f$. $B(f) = \{q\}$
[4] $e(f) = e(q) + w(q) = 139 + 15 = 154$
[5] $S = \{s, j, k, \ell, m, n, p, q, t\}$
[6] **stop**

The earliest start times are now all known. The earliest finish times are now easily computed:

$$\begin{aligned}
f(s) &= e(s) + w(s) = 0 + 0 = 0\\
f(j) &= e(j) + w(j) = 0 + 75 = 75\\
f(k) &= e(k) + w(k) = 75 + 30 = 105\\
f(\ell) &= e(\ell) + w(\ell) = 0 + 4 = 4\\
f(m) &= e(m) + w(m) = 0 + 5 = 5\\
f(n) &= e(n) + w(n) = 105 + 4 = 109\\
f(p) &= e(p) + w(p) = 109 + 30 = 139\\
f(q) &= e(q) + w(q) = 139 + 15 = 154\\
f(t) &= e(t) + w(t) = 154 + 0 = 154
\end{aligned}$$

The slacks for the arcs are:

$$\begin{aligned}
s(s, j) &= e(j) - f(s) = 0 - 0 = 0\\
s(s, \ell) &= e(\ell) - f(s) = 0 - 0 = 0\\
s(s, m) &= e(m) - f(s) = 0 - 0 = 0\\
s(s, n) &= e(n) - f(s) = 105 - 0 = 105\\
s(j, k) &= e(k) - f(j) = 75 - 75 = 0\\
s(k, n) &= e(n) - f(k) = 105 - 105 = 0\\
s(\ell, p) &= e(p) - f(\ell) = 109 - 4 = 105\\
s(m, p) &= e(p) - f(m) = 109 - 5 = 104\\
s(n, p) &= e(p) - f(n) = 109 - 109 = 0\\
s(p, q) &= e(q) - f(p) = 139 - 139 = 0\\
s(q, t) &= e(f) - f(q) = 154 - 154 = 0
\end{aligned}$$

The critical arcs are sj, $s\ell$, sm, jk, kn, np, pq, qf, and the (unique) critical path is (s, j, k, n, p, q, t).

Exercises 12.2

2.2.1 Draw a network to represent a compound activity that is made up of tasks a, b, c, d, e with the requirements

a must precede c and d,
b must precede c and d,
c must precede e,
d must precede e.

Find a critical path in the network on the assumption that a, b, c, d, e take 1, 3, 2, 1, 1 hours respectively.

2.2.2 For each of the following problems, construct an activity network. Find the shortest time for completion, and identify all critical paths and tasks.

(i) To shine a pair of boots. (Time in minutes.)
Clean mud from boot A (2); clean mud from boot B (2); apply polish to boot A (1); apply polish to B (1); wait for polish on A to dry (1); wait for polish on B to dry (2); rub A (1); rub B (1).

A(ii) To send a letter. (Time in minutes.)
Letter dictated to secretary (10); letter typed by secretary (10); envelope ad-

dressed by clerk (2); envelope stamped by clerk (1); clerk places letter in envelope and seals (1); letter taken to mailbox (8).

(iii) To make a wooden box. (Time in minutes.)
Mark out pieces on sheet of wood (15); cut out pieces (12); glue body together (10); screw hinges to lid (5); wait for glue to dry (15); screw hinges (with lid) to box (6).

(iv) To replace a flat tire. (Time in seconds.)
Get jack and tools (60); remove hubcap (30); loosen wheelnuts (120); place jack under car (25); lift car (80); get spare (25); remove wheelnuts and wheel (40); place spare on studs (30); put wheelnuts (half-tight) on studs (60); lower car and remove jack (45); put jack away (20); tighten wheelnuts on studs (30); put away tools and punctured tire (50); replace hubcap (20).

2.3 Construct a network to represent the following data, find out how early you can finish the project, and find all critical paths.

Task	Time	Prerequisites
a	4	nil
b	3	a
c	3	a
d	3	b
e	3	a
f	3	b, c
g	3	e, f
h	3	d, g

2.4 The following networks represent projects (times for the tasks are shown on the arcs). Find all critical paths.

(i)

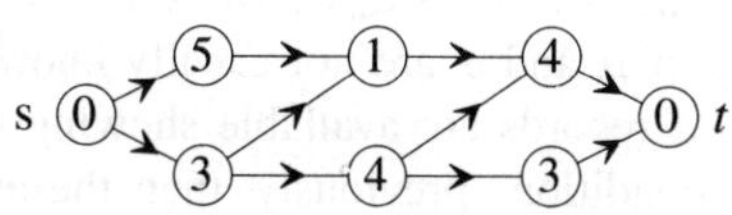

(ii)

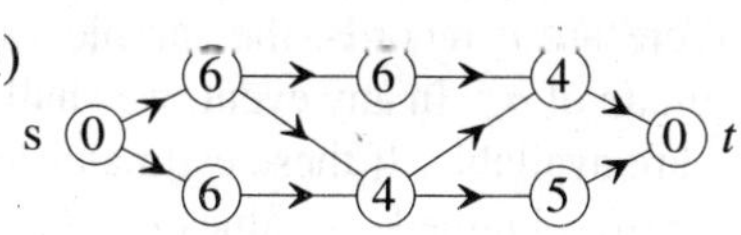

(iii)

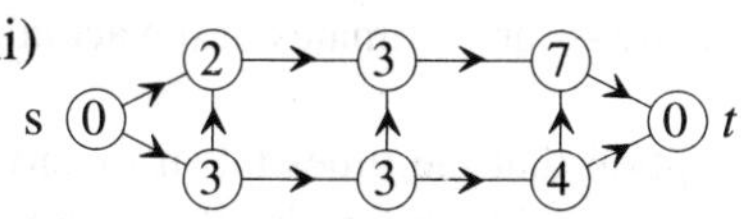

2.5 Activities r, u, v, x, y, z are necessary to produce a certain product. r must precede u and y, v must precede x, x and y must precede z. The numbers of days necessary to carry out the activities are:

$$w(r) = 4, w(u) = 4, w(v) = 4, w(x) = 4, w(y) = 4, w(z) = 4.$$

The cost of carrying out all the activities in this way is \$3,500. It is possible to reduce the time for x by one, two or three days, at a cost of \$150 per day's reduction. You are offered a contract to produce the product in 17 days, for a price of \$4,000. There is a bonus of \$200 for each day by which you finish before the deadline, but your product will not be accepted if you take more than 17 days. Should you accept the contract? If so, how long should you take over activity x?

12.2.6 The following tasks are to be subcontracted in building a house; the times quoted by the subcontractors are shown in brackets. Construct the activity network and find all critical paths.

a. Site preparation; foundations (5 days)
b. Underground plumbing; sewer and water connections (7 days)
c. Internal plumbing; hot and cold water pipes and faucets (4 days)
d. Brickwork and piers to floor level (5 days)
e. Timber frame, flooring, walls, roof (10 days)
f. Roof tiling and gutters (5 days)
g. Electrical: all exterior connections and internal wiring (8 days)
h. Exterior construction: siding, glazing, doors (10 days)
i. Exterior painting (10 days)
j. Internal drywall and plastering (3 days)
k. Internal finishing, painting etc. (8 days)

12.3 Critical Paths Under Uncertainty

Often the tasks represented by the arcs of a digraph do not have a fixed duration; there are variables x_a, the "time it takes to carry out task a," and one cannot predict with certainty the value of x_a in the particular case.

The x_a are random variables in the usual statistical sense. Each x_a will have a mean, μ_a, the (theoretical) expected value of x_a, and a variance, $var(x_a)$ or σ_a^2, the (theoretical) expected value of $(x_a - \mu_a)^2$. We write σ_a to denote the positive square root of σ_a^2. The values of μ and σ are not usually known, but often they can be estimated. For example, if records are available showing the time it has taken to perform a under similar conditions previously, then the average $\overline{x}_a$ of those times is an estimate of μ_a. If there are n records, the sample variance $s_a^2 = \Sigma(x_a - \overline{x}_a)^2/(n-1)$ provides an estimate of σ_a^2. In any event, we shall assume that some estimates $\overline{x}_a$ and s_a^2 of μ_a and σ_a^2 are available. If these estimates are used in critical path computations, the values that are obtained — values of $e(x)$ and $f(x)$ for each vertex x and of the slack for each arc — are estimates of the actual values of these quantities in the project.

We make the assumption that the times taken to carry out the different tasks are independent. If $a, b, \ldots, f$ are different tasks, this assumption implies that the time to carry out the sequence $a, b, \ldots, f$ will have mean $\mu_a + \mu_b + \cdots + \mu_f$ and variance $\sigma_a^2 + \sigma_b^2 + \cdots + \sigma_f^2$. Estimates can be found by replacing each μ_a by $\overline{x}_a$ and each σ_a^2 by s_a^2.

The particular model to be discussed is called Program Evaluation and Review Technique, or **PERT**. This makes allowance for the fact that variances are nearly always hard to estimate and that means are often difficult also. One first estimates the shortest and longest durations that are likely for the task a. These are called "optimistic" and "pessimistic" times for a, and are denoted by α_a and β_a respectively. In general, a fair guide is that the probabilities of x_a being less than α_a or greater than β_a should be about .01. Then $s_a^2 = ((\beta_a - \alpha_a)/6)^2$ is a reasonable approximation to σ_a^2. If no estimate of x_a is available, we use an estimate m_a of the *mode* of x_a, the value that x_a takes most frequently. Although modes and means can differ considerably (see Exercise 12.3.1), we can compensate by putting

$$x_a = (4\mu_a + \alpha_a + \beta_a)/6.$$

Given estimates x_a of each μ_a, we proceed to write down an activity network with x_a shown as the duration of a and find the (estimated) earliest start and finish times and the slacks. The estimated variance of the earliest start and finish times can then be calculated.

Once an earliest time $e(a)$, and an optimal path to a (the path to a that actually takes the earliest time), have been discovered, then the estimated variance of the earliest time is $\sum \sigma_b^2$, where the sum is taken over the vertices b in the maximal path. (If there are two such paths, use the larger variance.)

In the case where task durations were fixed, one could calculate whether or not a scheduled deadline could be met. In the case of uncertainty one can ask what is the *probability* of meeting a schedule. To calculate this, we must make certain assumptions — essentially, we assume that the earliest times have a normal distribution. In most cases this is reasonably close to the truth, particularly for the later tasks, and typically the most important case is the finish state t.

Suppose there is a deadline d. We wish to calculate the probability that $e(t) < d$. We assume that $e(t)$ is normally distributed with true mean $e(t)$ and variance var[e(t)]. Then $\delta = (d - e(t))/\sqrt{(\text{var}[e(t)])}$ will be a normal variable with mean 0 and standard deviation 1, which is called a unit (or standardized) normal variable, and denoted by $N(0, 1)$. So the probability that $e(t) < d$ equals the probability that a unit normal variable should be less than δ. This probability function is tabulated, and is readily available in any volume of statistical tables.

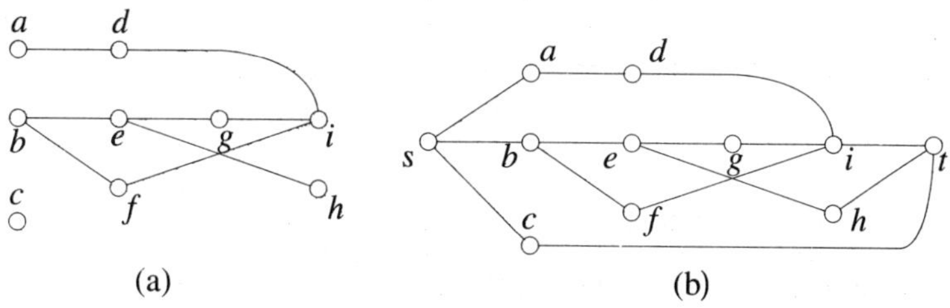

Fig. 12.6. Digraph for the example

Example. Consider the digraph of Figure 12.6(a). (Arcs are directed from left to right.) After a source and sink are added, the resulting digraph is shown in Figure 12.6(b). Suppose estimates have been made as shown in the first three columns of Table 12.1.

task	a	b	c	d	e	f	g	h	i
α	10	15	20	10	8	5	1	10	8
m	15	18	26	14	13	8	2	15	10
β	26	33	56	24	30	23	9	26	24
$\bar{x}$	16	20	30	15	15	10	3	16	12

Table 12.1. Estimated times for the example

Then the mean and variance estimates can be calculated: for example, in the case of a, the mean estimate is

$$\frac{10 + (4 \times 15) + 26}{6} = \frac{96}{6} = 16.$$

Similar calculations yield the other means shown in the last column of the table. Inserting these into the digraph and solving as an ordinary critical path problem we obtain the earliest and latest times as shown in Figure 12.7. The unique critical path is s, b, e, h, t, so the critical tasks are b, e and h.

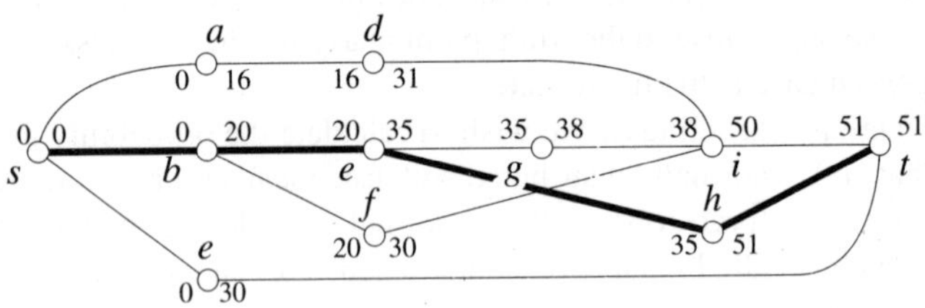

Fig. 12.7. Digraph for the example with earliest and latest times

We need the variance estimates for b, e and h. They are

$$\begin{aligned}\sigma_b^2 &= \frac{(33-15)}{6}^2 &= \frac{324}{36},\\ \sigma_e^2 &= \frac{(30-8)}{6}^2 &= \frac{484}{36},\\ \sigma_h^2 &= \frac{(26-10)}{6}^2 &= \frac{256}{36},\end{aligned}$$

and the estimated variance of the completion time is

$$\frac{324 + 484 + 256}{36} = \frac{1064}{36} = 29.55... = (5.437)^2.$$

So the estimated completion time for the project is distributed approximately as a normal variable with mean 51 days and variance 5.437^2 days. Suppose there is a penalty if the project is not completed within 55 days. To calculate the probability of completion within that time, we use the conversion

$$\delta = (d - 51)/5.437;$$

$d = 55$ yields $\delta = 0.7357$ approximately, so the probability of completion before 55 days equals the probability that $N(0,1) \leq 0.736$, which is found from tables to be 0.769. There is a 77% chance of completion without penalty.

Exercises 12.3

2.3.1 A random variable can attain the values 1, 2 and 3; it equals 1 in 60% of trials, 2 in 30% and 3 in 10%.

(i) Find the mean m and the mode m of this variable.

(ii) Using $\alpha = 1$ and $\beta = 3$, calculate $(\alpha + 4m + \beta)/6$. How does this compare to m as an estimate of μ?

2.3.2 Consider the project described in Exercise 12.2.1. Suppose the durations given in that exercise are actually means, and the variance estimates are 1.0 for task b and 0.5 for each other task. What is the probability of completing the task within 5 hours?

2.3.3 Suppose the times given in Exercise 12.2.6 are optimistic estimates; the most likely times are 25% longer than those given, and pessimistic estimates are twice the suggested times. What is the probability that the building will be completed within 65 days?

2.3.4 A project consists of tasks a, b, c, d, e, f, g, h and i. a must precede c, b must precede d and e, c and d must each precede f and g, e and f must precede h, and g must precede i. The optimistic, most likely and pessimistic times (in days) of the tasks are

task	a	b	c	d	e	f	g	h	i
α	2	1	3	2	2	6	2	5	1
m	5	4	8	7	6	9	5	6	2
β	6	7	13	12	10	12	11	7	3

(i) Construct a network and find all critical paths.

(ii) What is the expected length of the project?

(iii) What is the probability of completion within 30 days? Within 24 days?

2.3.5 A project has tasks a, b, c, d, e, f, g, h, i and j with estimated durations as follows:

task	a	b	c	d	e	f	g	h	i	j
α	14	10	13	7	16	20	7	8	13	6
m	16	14	18	8	16	26	8	10	17	9
β	18	15	23	9	16	38	12	12	21	15

The precedence relations are that a must precede c, b must precede d, e and f, c, f and h must precede j, d must precede g, and e and g must precede h and i.

(i) Construct a network and find all critical paths. What is the expected length of the project?

(ii) The project is allocated 52 days. What is the probability of completion on or before the deadline?

13

Flows in Networks

13.1 Transportation Networks and Flows

Graphs are used to model situations in which a commodity is transported from one location to another. A common example is the water supply, where the pipelines are edges, vertices represent water users, pipe joins, and so on. In the example of an airline, given in Section 2.3, we can interpret freight or passengers as commodities to be transported. Highway systems can be thought of as transporting cars. In many examples it is natural to interpret some or all edges as directed (some roads are one-way, water can flow only in one direction at a time in a given pipe, and so on). A common feature of transportation systems is the existence of a *capacity* associated with each edge — the maximum number of cars that can use a road in an hour, the maximum amount of water that can pass through a pipe, and so on.

Example. A communications network connects the n centers $x_1, x_2, \ldots, x_n$. The maximum number of messages, c_{ij}, that can be sent from x_i to x_j per minute depends on the number of lines between x_i and x_j. Given the $n \times n$ matrix $C = (c_{ij})$, it is important to to know the maximum amount of information that can be transmitted between a given pair of centers per minute. The appropriate model is a graph with vertices representing centers, edges representing direct communication lines and capacities representing the maximum rates of information transfer. This is an example where directions are not attached to the edges.

Normally there will exist several places where new material can enter the system. These will be modeled by vertices called *sources*. Material leaves at vertices called *sinks*. The usual model has one source and one sink; it will be shown that this involves no loss of generality. We are primarily concerned with the amount of material that flows through the system; the classical problem is to maximize this quantity.

We define a *transportation network* to be a digraph with two distinguished vertices called the *source* and the *sink*, and with a weight c defined on its arcs. $c(x, y)$ is called the *capacity* of the arc xy.

Example. An oil company pumps crude oil from three wells w_1, w_2 and w_3 to a central terminal t. The oil passes through a network of pipelines that has pumping stations at all three wells and also four intermediate stations p_1, p_2, p_3 and p_4. The digraph of this network, along with the capacities of the pipelines (in units of ten thousand barrels) is shown in Figure 13.1. There are three sources, w_1, w_2, w_3, and one sink, t.

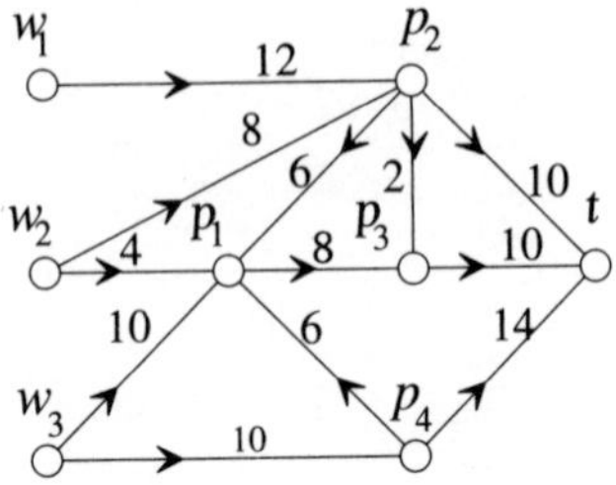

Fig. 13.1. Network for the oil pipeline problem

This example is not a transportation network, because of the multiple sources. The usual way to deal with this case is to add a new (dummy) vertex, say s, together with arcs of infinite capacity from s to every source. The new network has one source, s. Multiple sinks are handled similarly. The resulting single-source, single-sink network is often called the *augmented* or *completed* network.

Sometimes there is a capacity constraint on the vertices of a system. For example, say vertex x can process at most c units of a commodity per day. To embody this constraint, replace x by two vertices x_1 and x_2. All arcs that previously led into x now lead into x_1, all those that previously were directed out of x now leave x_2, and there is an arc from x_1 to x_2 with capacity $c(x_1, x_2) = c$. The most common type of vertex capacity is when the vertex x is a source, and in that case one can simply set $c(s, x) = c$ when defining the new source s.

Example (continued). Suppose the maximum possible amounts that oil wells w_1, w_2 and w_3 can produce are 120,000, 100,000, and 105,000 barrels per day respectively, and p_2 can process at most 150,000 barrels each day. Then an appropriate network is shown in Figure 13.2.

A *flow* f of *value* $v = v(f)$ on a transportation network with source s and sink t is a weight f satisfying

$$0 \leq f(x, y) \leq c(x, y) \text{ for every arc } xy, \tag{13.1}$$

and

$$\sum_{y \in A(s)} f(s, y) - \sum_{z \in B(s)} f(z, s) = v, \tag{13.2}$$

$$\sum_{y \in A(t)} f(t, y) - \sum_{z \in B(t)} f(z, t) = -v, \tag{13.3}$$

$$\sum_{y \in A(x)} f(x, y) - \sum_{z \in B(x)} f(z, x) = 0 \text{ for other vertices } x. \tag{13.4}$$

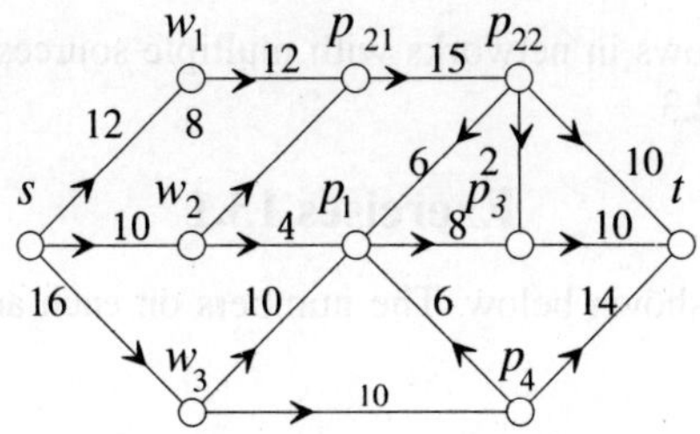

Fig. 13.2. Augmented network for the oil pipeline problem

The flow f is called a *flow from s to t*.

The quantity $\sum_{y\in A(x)} f(x, y) - \sum_{z\in B(x)} f(z, x)$ is called the *net flow at x*. The source and sink of a transportation network are often called *terminal* vertices, and the other vertices are *interior*, so (13.4) says that the net flow at any interior vertex is zero.

Recall that $[A, B]$ denotes the set of all edges of a graph with one endpoint in A and the other in B. The same notation is used in digraphs. If f is a function defined on edges, it is common to write $f[A, B]$ for $\sum_{x\in A}\sum_{y\in B} f(x, y)$. In that notation, the net flow at x is

$$F(x) = f[x, V] - f[V, x],$$

where V is the vertex-set of the network. (One could replace V by $V - x$, but as there is no flow from x to itself, this makes no difference.)

Example. The following diagram shows a real-valued function f on a network with source s and sink t. For convenience we assume that the capacity of each arc is sufficiently large that (13.1) is satisfied.

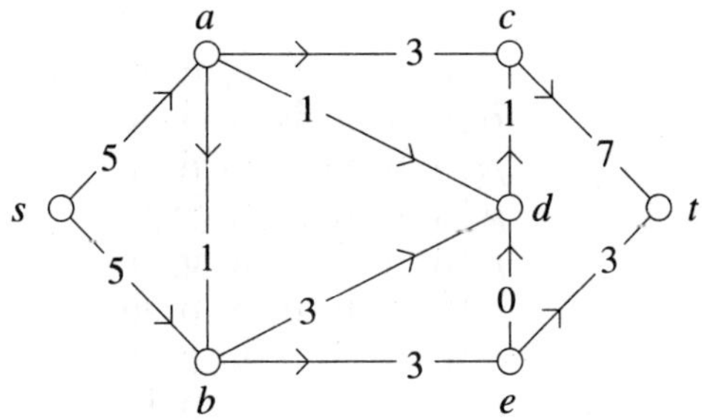

To verify that f is a flow, it suffices to check the net flow at each vertex:

$$\begin{array}{lclclcr}
F(s) & = & f(s,a) + f(s,b) & = & 5+5 & = & 10 \\
F(a) & = & f(a,b) + f(a,c) + f(a,d) - f(s,a) & = & 1+3+1-5 & = & 0 \\
F(b) & = & f(b,d) + f(b,e) - f(s,b) - f(a,b) & = & 3 = 3-5-1 & = & 0 \\
F(c) & = & f(c,t) - f(a,c) - f(d,c) & = & 7-3-4 & = & 0 \\
F(d) & = & f(d,c) - f(a,d) - f(b,d) - f(e,d) & = & 4-1-3-0 & = & 0 \\
F(e) & = & f(e,d) - f(e,t) - f(b,e) & = & 0+3-3 & = & 0 \\
F(t) & = & -f(c,t) - f(e,t) & = & -7-3 & = & -10
\end{array}$$

so f is a flow from s to t of value $v(f) = 10$.

Generalization to flows in networks with multiple sources and sinks will be considered in Exercise 13.2.3.

Exercises 13.1

A13.1.1 The network N is shown below. The numbers on each arc give the value of the function g.

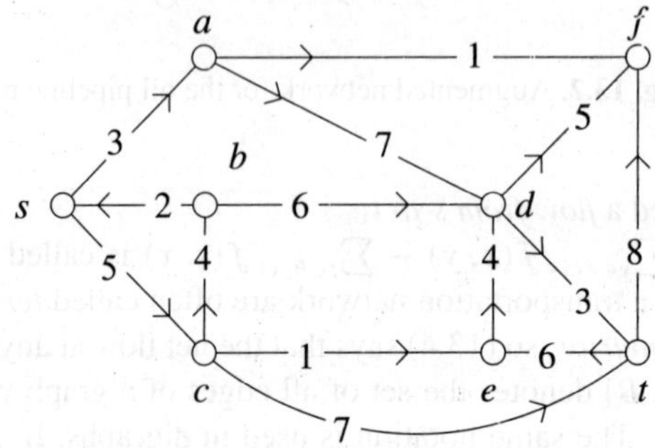

(i) Give an example of a directed (s, t)-path.
(ii) Give an example of an (s, t)-path that is not a directed (s, t)-path.
(iii) Consider the subsets $X = \{s, a, d\}$, $Y = \{c, e, f\}$ and $Z = \{c, t\}$. Find:

(a) $[X, f]$; (b) $[X, e]$; (c) $g[X, Y]$;
(d) $g[X, Z]$; (e) $g[X, Y \cap Z]$; (f) $g[X, Y \cup Z]$.

13.1.2 Water is sent from the main dam D to a metropolitan reservoir R through a pipeline network containing five pumping stations P_1, P_2, P_3, P_4 and P_5. The maximum number of gallons that can flow between the various stations per day is given by the table

	D	P_1	P_2	P_3	P_4	P_5	R
D	0	50	40	35	0	0	0
P_1	0	0	20	0	40	0	0
P_2	0	0	0	0	0	33	39
P_3	0	0	0	0	0	30	0
P_4	0	0	0	0	0	0	40
P_5	0	0	0	0	0	0	50
R	0	0	0	0	0	0	0.

The problem is to maximize the amount of water flowing from the dam to the reservoir per day.
(i) Construct the network.
(ii) Suppose a new reservoir Q is added to the network, and pipelines are constructed joining Q to P_3, P_4 and P_5 with maximum daily flow capacities of 10, 15 and 20 million gallons respectively. How does the network in Part (i) change? What optimization problems arise?
(iii) Suppose P_4 has a maximum capacity of 55 million gallons per day. How can the network in (i) be changed to reflect this?

.1.3 In the following networks two numbers are assigned to each arc: the capacity is followed by the value, in brackets, of a function f. In each case, is f a flow from s to t? If f is a flow, what is its value?

(i)

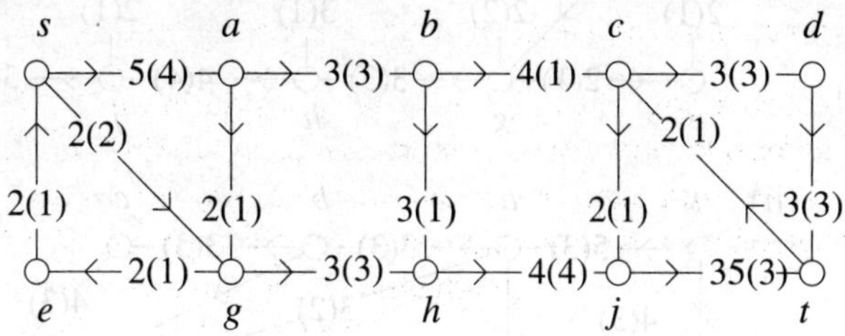

(ii)

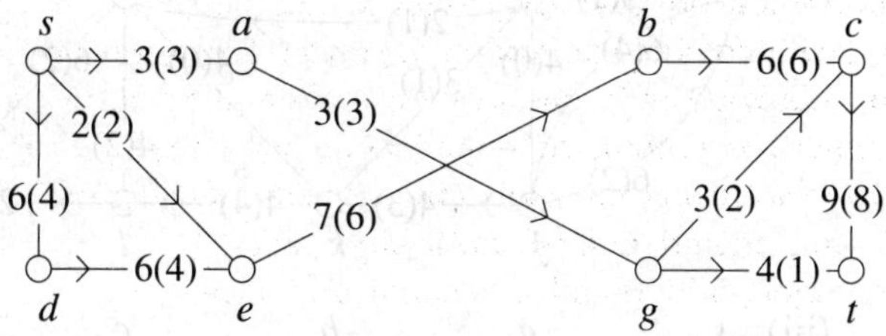

.1.4 Suppose g is a real-valued function defined on the arcs of a network N, and X, Y and Z are subsets of the vertex-set of N. Show that

$$g[X, Y \cup Z] = g[X, Y] + g[X, Z] - g[X, Y \cap Z].$$

.1.5 In the following network, the two numbers on an arc again represent the capacity and the value, in brackets, of a nonnegative function f.

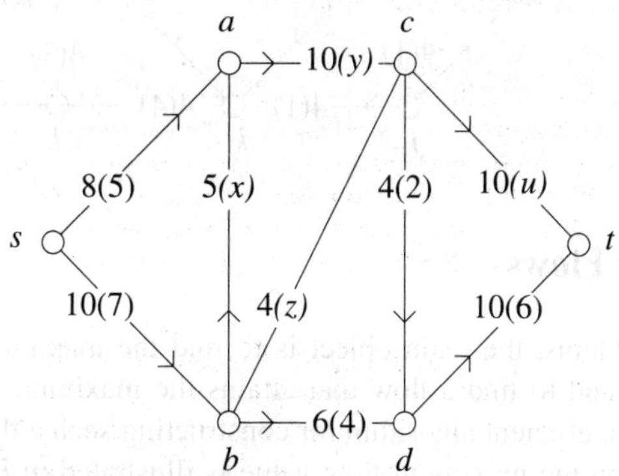

(i) List all directed paths from s to t.

(ii) Suppose $u = 5$. Can f define a flow from s to t?

(iii) Describe the possible values of x, y, z and u such that f is a flow from s to t.

.1.6 Repeat Exercise 13.1.3 for these networks:

(i)

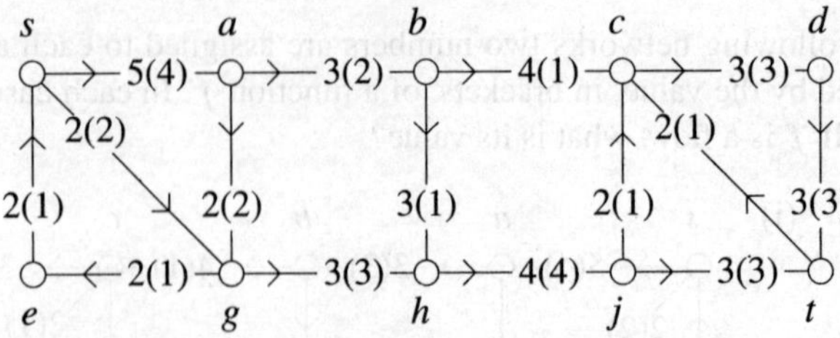

(ii)

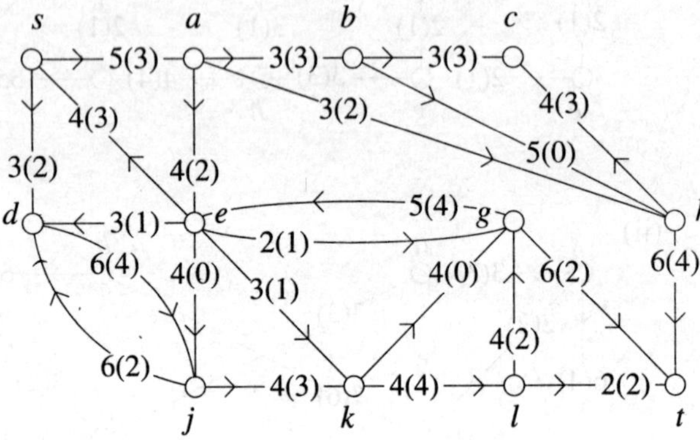

(iii)

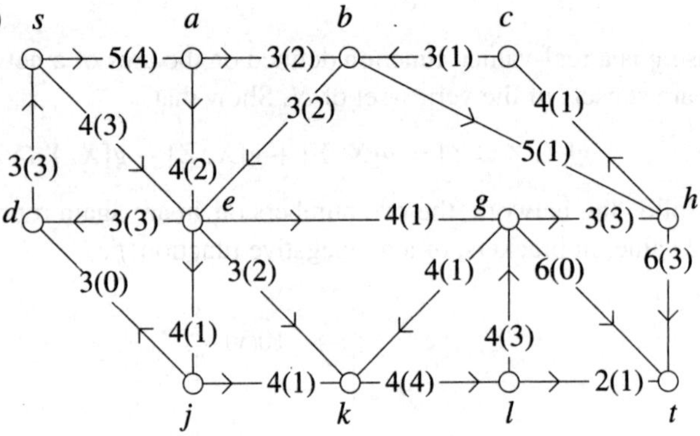

13.2 Maximal Flows

In most flow problems, the main object is to find the maximum value of a flow in a given network, and to find a flow that attains the maximum value. It is moreover desirable to find an efficient algorithm for constructing such a flow.

A restriction on the maximum flow value is illustrated in Figure 13.3. In Figure 13.3(a), if an amount v is input at s and output at t, then the flow in the arc ab must be v also. A similar situation arises in Figure 13.3(b); the total flowing through the network must pass through ab or cd, and the flow through those arcs must also counterbalance any flow in da. In other words,

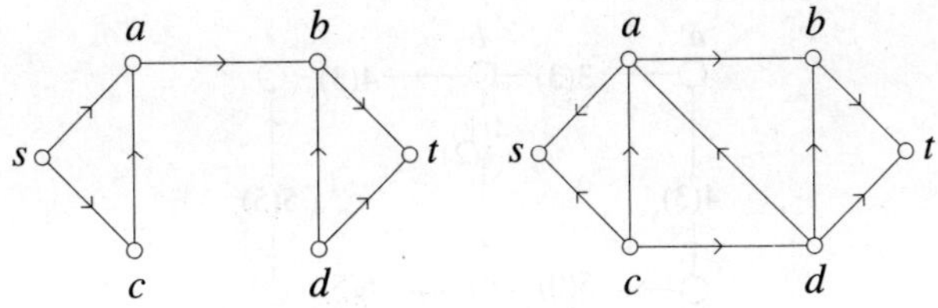

(a) (b)

Fig. 13.3. Restrictions on the maximum flow.

$$v = f(a, b) + f(c, d) - f(d, a).$$

To generalize these examples, observe that in each case the arcs discussed were those in the set $[S, T]$, where $S = \{s, a, c\}$ and $T = \{b, d, t\}$. The observation was that the value of the flow equaled the net flow from S to T. This applies in general.

If A and B are disjoint sets containing all vertices of a graph or digraph G, the set $[A, B]$ is often called a cut. We define a *separating cut* in a transportation network to be one in which the source and sink are in different parts; conventionally, if $[S, T]$ is a separating cut, then $s \in S$ and $t \in T$ (that is, the two parts of the cut are written in order, with the part containing the source being written first).

Lemma 13.1. *If $[S, T]$ is a separating cut in a transportation network, and f is a flow of value v from s to f, then*

$$v = f[S, T] - f[T, S], \tag{13.5}$$

and

$$v \le c[S, T]. \tag{13.6}$$

Proof. Denote the set of vertices of the network by V. Since t is not in S, then $f[x, V] - f[V, x] = 0$ for every $x \in S$, except for $f[s, V] - f[V, s] = v$. So

$$\sum_{x \in S} f[x, V] - f[V, x] = v,$$

i.e., $f[S, V] - f[V, S] = v$, and since $V = S \cup T$,

$$f[S, S \cup T] - f[S \cup T, S] = v. \tag{13.7}$$

Since S and T are disjoint, $f[S, S \cup T] = f[S, S] + f[S, T]$, and similarly $f[S \cup T, S] = f[S, S] + f[T, S]$. So (13.7) becomes

$$f[S, T] - f[T, S] = v. \tag{13.8}$$

Now f is nonnegative, so $f[T, S] >\geq 0$. Also $f[S, T] \ge c[S, T]$. So

$$f[S, T] - f[T, S] \ge c[S, T],$$

establishing the lemma. □

It is clear that a finite network contains only a finite number of separating cuts, so there will be a well-defined minimum among the capacities of separating cuts. Any

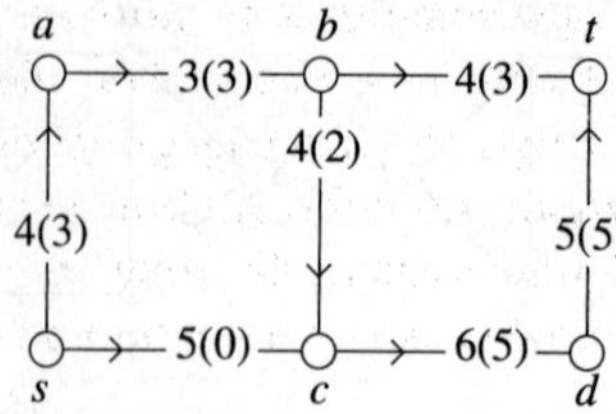

Fig. 13.4. A network with a flow

separating cut realizing this capacity will be called *minimal*. Similarly, if there is a maximum flow value, any flow attaining that value will be called *maximal*.

Example. Figure 13.4 shows a transportation network with a flow f on it. On each arc is shown the capacity, followed by the flow in parentheses. The flow f has value 6. This is not the maximum possible, because the flow g with

$$g(s,c) = 4,\ g(b,c) = 1,\ g(b,t) = 2,$$

and $g = f$ elsewhere, has value 7.

The network contains a cut of value 7, namely $[\{s, a\}, \{b, c, d, t\}]$. By Lemma 13.1, no flow can have value greater than 7. So the maximum has been attained, and g is a maximal flow.

It should be observed that g is not unique. Figure 13.5 shows a set of values on the arcs of the network that form a flow of value 7 for any real x, $0 \le x \le 1$. So the network has infinitely many maximal flows.

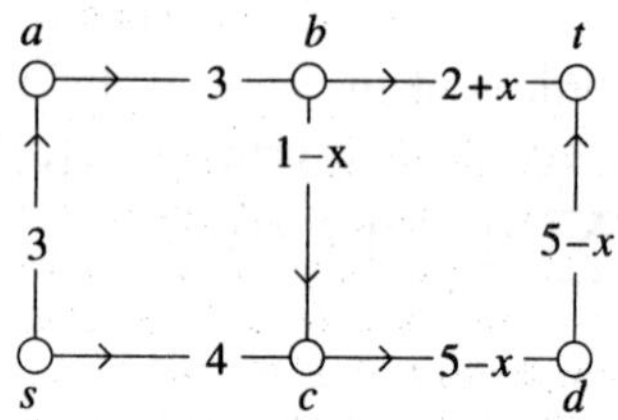

Fig. 13.5. Maximal flows in the network of Figure 13.4

In the preceding example, the flow g was obtained from the flow f by changing the flow in the three arcs sc, cb and bt, and those arcs form a path from s to t. We shall now generalize this example.

Suppose f is a flow from s to t in a transportation network. Consider a path P,

$$P = (x_0, x_1, \ldots, x_n),$$

where $s = x_0$ and $t = x_n$. This path is *not* directed, so the edge joining x_i to x_{i+1} might be the arc $x_i x_{i+1}$ or the arc $x_{i+1} x_i$; say the arc is *forward* if its direction is from

x_i to x_{i+1}, and *backward* otherwise. Loosely speaking, forward arcs are those in the direction from s to t along the path.

The path P is called an *augmenting path* for P if it has the following properties:

1. If $x_i x_{i+1}$ is a forward arc, then $f(x_i, x_{i+1}) < c(x_i, x_{i+1})$;
2. If $x_i x_{i+1}$ is a backward arc, then $f(x_i, x_{i+1}) > 0$.

One commonly calls an arc "full" if the flow in it equals its capacity, and "empty" if its flow is zero. So an augmenting path is one that contains no full arcs in the forward direction (the direction from s to t), and no empty backward arcs.

Example (continued). The flow shown in Figure 13.4 has value 6. Since arc ab is operating at capacity, it cannot be in any augmenting path, and neither can dt, but there is one augmenting path, namely $scbt$.

Lemma 13.2. *If a transportation network with a flow f of value v has an augmenting path, then it has a flow whose value is greater than v.*

Proof. Suppose $(x_0, x_2, \ldots, x_n)$ is an augmenting path, with $x_0 = s$ and $x_n = t$. If the arc joining x_i to x_{i+1} is $x_i x_{i+1}$, then define

$$\delta_i = c(x_i, x_{i+1}) - f(x_i, x_{i+1});$$

if it is $x_{i+1}x_i$, then

$$\delta_i = c(x_{i+1}, x_i).$$

Finally define $\delta = \min_{1 \le i < n} \delta_i$.

A new flow g is now constructed: if $x_i x_{i+1}$ is a forward arc of the augmenting path, then

$$g(x_i, x_{i+1}) = f(x_, x_{i+1}) + \delta;$$

if $x_i x_{i+1}$ is a backward arc of the augmenting path, then

$$g(x_{i+1}, x_i) = f(x_{i+1}, x_i) - \delta;$$

and in all other cases, when xy does not lie on the augmenting path,

$$g(x, y) = f(x, y).$$

It is clear that

$$0 \le g(x, y) \le c(x, y)$$

for every arc xy. If x is not a vertex on the augmenting path, then the net flow at x is unchanged, so it is still zero. Now consider the effect of the change in flow in the arc $x_i x_{i+1}$. If the arc is directed *out of* x_i, then the flow in the arc is *increased* by δ, and if the direction is *into* x_i, the flow is *decreased* by δ. In either case, the net flow out of x_i is increased by δ, and the net flow out of x_{i+1} is decreased by the same amount. If $i \neq 0$ or n, the net effect on x_i of the flow changes in $x_{i-1}x_i$ and $x_i x_{i+1}$ is zero. The net flow out of the source $x_0(= s)$ is increased by δ, and the net flow into the sink $x_n(= t)$ is decreased by δ. So g is a flow, of value $v + \delta$. □

Exercises 13.2

13.2.1 Suppose a network has v vertices, a single source and a single sink. How many separating cuts does it have?

A13.2.2 On the following network, arc capacities shown and values of an arc weight f in parentheses.

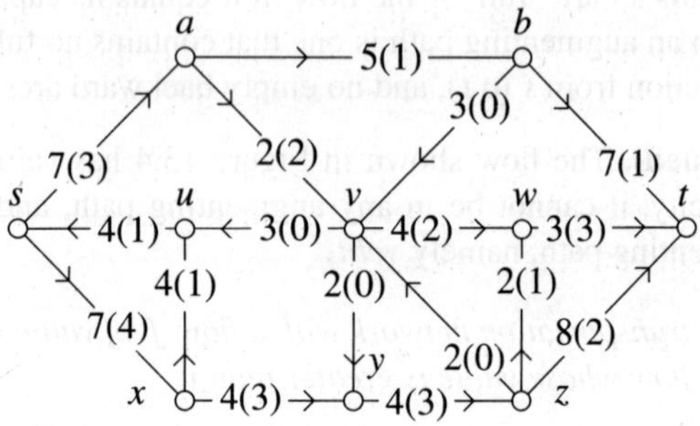

(i) Verify that f is a flow.
(ii) Find the value of the flow.
(iii) Find an augmenting path in the network.
(iv) Find a new flow of greater value than the original.
(v) If the flow is not maximal, find a better flow.
(vi) Find a maximal flow in the network.

13.2.3 Repeat the preceding exercise for the networks shown below.

(i)

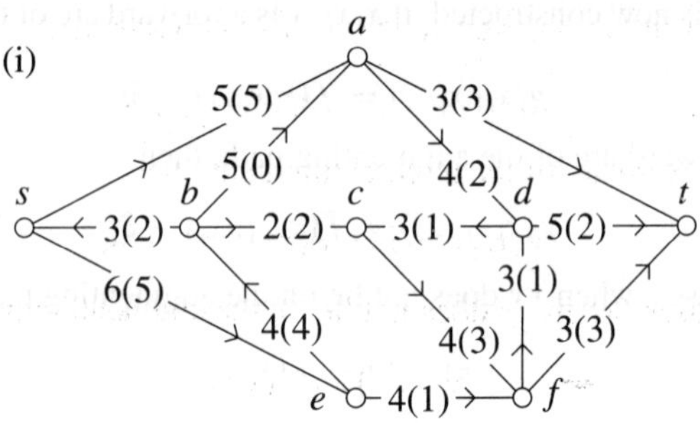

(ii)

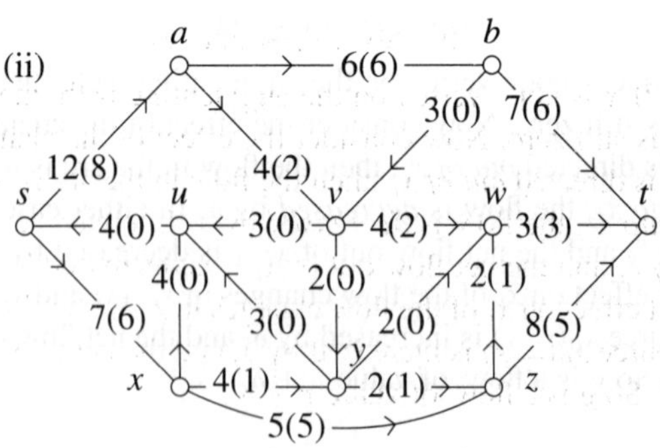

13.2.4 Consider a network N with a capacity function c whose vertex-set V contains a set S of p sources and a set T of q sinks, where S and T are disjoint; say

$$S = \{s_1, s-2, \ldots, s_p\},$$
$$T = \{t_1, t_{,2}, \ldots, t_q\}.$$

Define a *flow* g on N to be a function that satisfies (13.1), satisfies (13.4) for every vertex x not in $S \cup T$, and for which

$$g[S, V] - g[V, S] = v,$$
$$g[T, V] - g[V, T] = -v$$

for some nonnegative real number v. In the completed network (with source s and sink t), a flow f is defined from g as follows. If x and y are vertices of the original network, then $f(x, y) = g(x, y)$.

$$g(s, s_i) = f[s_i, V] - f[V, s_i] \text{ for every } i,$$
$$g(t_j, t) = f[V, t_j] - f[t_j, V] \text{ for every } j.$$

Assuming that $c(s, s_i) \geq c[s_i, V]$ and $c(t_j, t) \geq c[V, t_j]$ for every i and j, prove that f is a flow in the completed network, and that f is maximal if and only if g is maximal in N.

13.2.5 In the following networks, the number on each arc represents the capacity of that arc. In each case find all cuts separating s and t and their capacities. What is the minimum cut capacity of the network? Determine the maximum flow in the network.

(i)
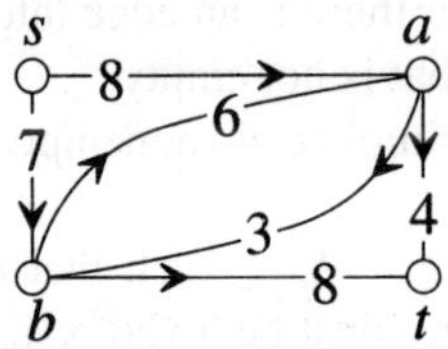

(ii)
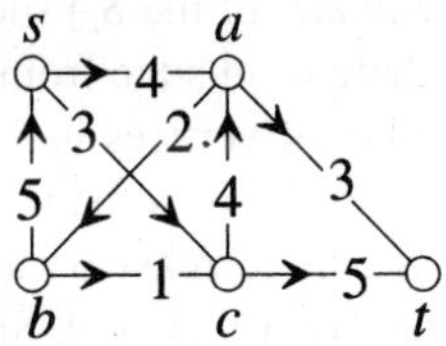

13.2.6 Repeat the preceding exercise for these networks:

(i)
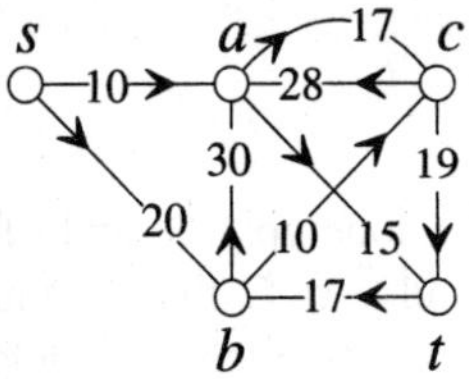

(ii)
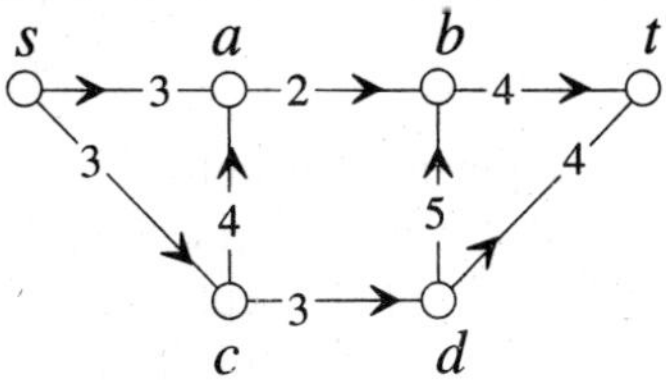

(iii)
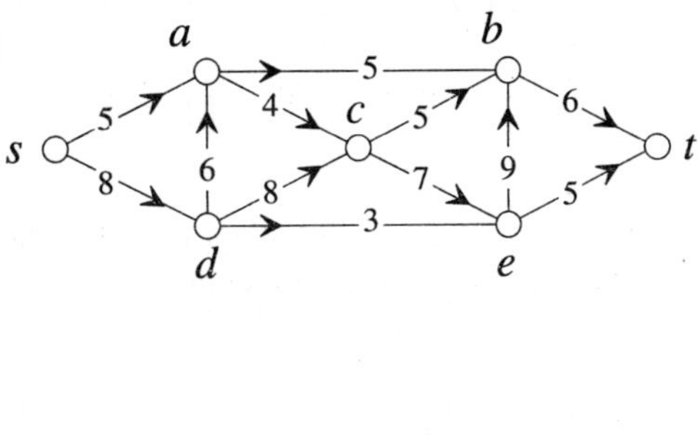

A13.2.7 Suppose vertex x in a transportation network N has the property that no more than d units of material can flow through x per unit time. (The vertex might represent a pump in a sewage system, for example.) How could you model this feature in the network?

13.3 The Max Flow Min Cut Theorem

Lemma 13.1 tells us that the value of any flow in a network must be equal to or less than the capacity of any separating cut. So the maximum flow value is no greater than the minimum capacity of a separating cut. Ford and Fulkerson [45] showed that equality can be attained. We prove this result, using Lemma 13.2 and the following, which is essentially its converse.

Lemma 13.3. *If a transportation network has a flow f of value v from source s to sink t, then either the network contains an augmenting path for f or else it contains a separating cut whose capacity is v.*

Proof. We construct a series $S_0, S_1 \dots$ of sets of vertices of the network

$$S_0 = \{s\}.$$

When $k > 0$, S_k consists of all vertices y that do not belong to $S_0, S_1, \dots$ or S_{k-1} such that either there is an arc xy such that $f(x, y) < c(x, y)$, or else there is an arc yx such that $f(y, x) > 0$, for some vertex $x \in S_{k-1}$. In words, y must be a new vertex (not already in any of the S_i) such that either there is an edge into y from S_{k-1} that is not full, or there is an edge from y to S_{k-1} that is not empty.

The number of vertices is finite, so the number of (nonempty) sets constructed is finite. Either $t \in S_n$ for some n or not.

If $t \in S_n$, there must be a vertex $x_{n-1} \in S_{n-1}$ such that either $f(x_{n-1}, t) < c(x_{n-1}, t)$, or $f(t, x_{n-1}) > 0$. Similarly there must be a vertex $x_{n-2} \in S_{n-2}$ such that either $f(x_{n-2}, x_{n-1}) < c(x_{n-2}, x_{n-1})$, or $f(x_{n-1}, x_{n-2}) > 0$. Continuing in this way, s is eventually reached, and the sequence

$$s, x_1, x_2, \dots, x_{n-1}, t$$

forms an augmenting path for f.

If $t \notin S_n$ for any n, write S for the union of the nonempty S_i, and t for its complement. Then $[S, T]$ is a separating cut. If $x \in S$ and $y \in T$, then any arc xy satisfies $f(x, y) < c(x, y)$, and any arc yx satisfies $f(y, x) > 0$, or else y would be in S. So

$$f[S, T] - f[T, S] = c[S, T] - 0 = c[S, T],$$

and thus

$$c[S, T] = v$$

from (13.5). □

Theorem 13.4. [45] *In any transportation network, the maximum flow value equals the minimum capacity of any separating cut.*

Proof. Suppose v is the maximum flow value, and suppose f is a flow of value v. By Lemma 13.2, f has no augmenting path. So Lemma 13.3 shows that the network contains a separating cut of capacity v. Lemma 13.1 says that no separating cut can have a capacity less than v. So v equals the minimum capacity. □

Since a flow that attains the theoretical maximum value is called a *maximal flow*, and a separating cut that has the minimum capacity is called a *minimal cut*, Theorem 13.4 is usually called the *Max Flow Min Cut Theorem*.

In proving this result, we have essentially used the following characterization of maximal flows:

Theorem 13.5. *A flow is maximal if and only if it has no augmenting path.*

The following characterization of minimal cuts is left as an exercise:

Theorem 13.6. *A separating cut $[S, T]$ is minimal if and only if every maximal flow makes every edge of $[S, T]$ full and every edge of $[T, S]$ empty.*

Exercises 13.3

3.3.1 Prove Theorem 13.6.

3.3.2 Suppose $[S, T]$ and $[X, Y]$ are two minimal cuts in a network N. Prove that both $[S \cup X, T \cap Y]$ and $[S \cap X, T \cup Y]$ are also minimal cuts of N.

3.3.3 Refer to Exercise 13.2.4. A *separating cut* in a network with a set S of sources and a set T of sinks is defined by a partition $\{X, Y\}$ where $S \subseteq X$ and $T \subseteq Y$. State and prove the appropriate version of the Max Flow Min Cut Theorem for such networks.

3.3.4 Consider a transportation network N with vertex-set and arc-set $V(N)$ and $A(N)$. A set D of arcs is called a *blocking set* if every directed path from s to t must contain an arc of D.

(i) Prove that every separating cut in N is a blocking set.

(ii) Given a blocking set D in N, a set S is constructed as follows:

1. $s \in S$;
2. if $x \in S$ and $xy \in A$ but $xy \notin D$, then $y \in S$;
3. every member of S can be found using rules 1 and 2.

Prove that $[S, \overline{S}]$ is a subset of D, and is a cut in N.

(iii) A blocking set D is called *minimal* if no proper subset D' of D is a blocking set. Prove that every minimal blocking set is a separating cut.

13.4 The Max Flow Min Cut Algorithm

Suppose one wishes to find a maximal flow in a network. One technique is to start with any flow (if necessary, use the trivial case of zero flow in each arc). If the given flow admits of an augmenting path, then by Lemma 13.2 it can be improved. Find such a path and find the improved flow. Then repeat the procedure for the new flow. Continue until no augmenting path exists. Theorem 13.5 says that the resulting flow is maximal.

Example. Consider the network shown in Figure 13.6(a). As usual, the capacity is shown on each arc, followed by a flow f, of value 7, in parentheses. It will be observed that this flow has an augmenting path, s, a, b, c, t, with $\delta = 1$. If the flow is augmented accordingly, adding 1 to $f(s, a)$, $f(a, b)$ and $f(c, t)$ and subtracting 1 from $f(c, b)$, the new flow (shown in Figure 13.6(b)) has value 8. It can be shown that this flow is maximal.

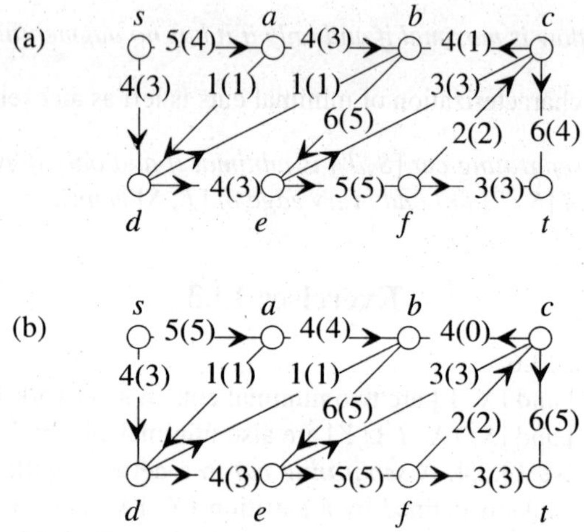

Fig. 13.6. Augmenting a flow

Clearly it is desirable to have an algorithm that will either find an augmenting path or prove that no such path exists. Such an algorithm was constructed by Ford and Fulkerson [46]. Given a transportation network N and a flow f on N, it either produces a flow g on N with value greater than that of f, or proves that f is maximal.

To each vertex u, the algorithm assigns a label of the form (z^+, δ) or (z^-, δ), where δ is a positive real number or ∞. If one of these labels is assigned to a vertex u, this means that we can construct an (undirected) (s, u)-path P in which:

(i) $c(x, y) - f(x, y) \geq \delta$ for every forward arc of P;
(ii) $f(y, x) \geq \delta$ for every backward arc of P.

The z in the label is a vertex adjacent to u; $+$ means that the forward arc zu is the last edge in the path; $-$ means it is the backward arc uz.

The labeling process ends when either the sink t is labeled or no further labels can be assigned. (Labeling of the sink is called *breakthrough*.) If termination occurs because t has been labeled, then f has an augmenting path, as described in (i) and

(ii) above. If t does not receive a label, there is no possible augmenting path and f is maximal.

To simplify the description of labeling, we use the following definition: to *scan* a labeled vertex z means to label every unlabeled vertex y that is adjacent to x and satisfies either $f(x, y) < c(x, y)$ or $f(y, x) > 0$.

Labeling Algorithm:

1. Label the source vertex s with $(-, \infty)$.
2. Select any labeled, unscanned vertex x. Suppose it is labeled $(z^+, \varepsilon(x))$ or $(z^-, \varepsilon(s))$. (In this notation, we could say $\varepsilon(s) = \infty$.) Scan x and assign labels according to the rules:
   - if xy is an arc in which $f(x, y) < c(x, y)$ and y is unlabeled, assign y the label $(x^+, \varepsilon(y))$, where $\varepsilon(y) = \min\{\varepsilon(x), c(x, y) - f(x, y)\}$;
   - if yx is an arc with $f(y, x) > 0$ and y is unlabeled, assign y the label $(x^-, \varepsilon(y))$, where $\varepsilon(y) = \min\{\varepsilon(x), f(y, x)\}$.
3. Repeat Step 2 until either t is labeled (breakthrough), or until no more labels can be assigned and t is unlabeled. In the latter case there is no augmenting path. If breakthrough occurs, then f admits a flow augmenting path, which can be constructed by backtracking from t.

Example. The first diagram in Figure 13.4 shows a network with a flow f of value 8 from s to t. As usual, on each edge the number in parentheses indicates the edge flow and the other number indicates the edge capacity. The labeling algorithm will be used to find an augmenting path.

The construction of an augmenting path is illustrated in the remainder of Figure 13.4. The labeling algorithm terminates in breakthrough. Backtracking from t obtains the augmenting path; the arcs of this path are indicated in the final diagram by heavy lines.

When the labeling routine ends in breakthrough, an improved flow is constructed. This is done using another algorithm, the Flow Augmentation Algorithm. This algorithm takes a flow augmenting path P and processes the vertices along the path sequentially. It increases the flow along each forward edge of P by $\varepsilon(t)$ and decreases the flow along each reverse edge of P by $\varepsilon(t)$.

Flow Augmentation Algorithm:

1. First process t:
   - if t is labeled $(y^+, \varepsilon(t))$, define $g(y, t) = f(y, t) + \varepsilon(t)$;
   - if t is labeled $(y^-, \varepsilon(t))$, define $g(y, t) = f(t, y) - \varepsilon(t)$;
   - next, process vertex y.

2. To process vertex u, where $u \neq t$:
   - if u is labeled $(x^+, \varepsilon(u))$, define $g(x, u) = f(x, u) + \varepsilon(t)$;
   - if u is labeled $(x^-, \varepsilon(u))$, define $g(u, x) = f(u, x) - \varepsilon(t)$;
   - next, process vertex x.

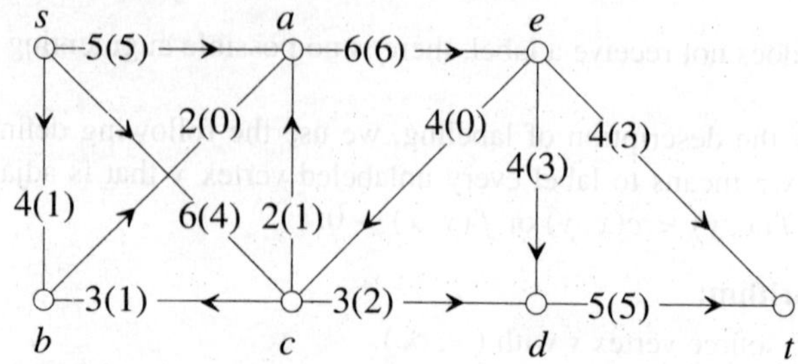

Network for which the maximal flow is to be found

Initial configuration:

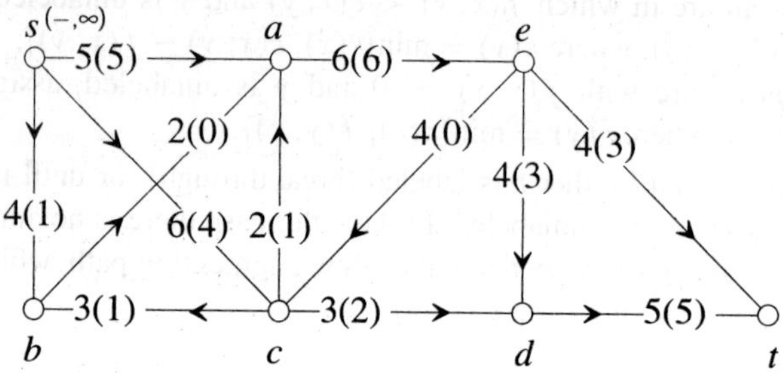

After scanning s:

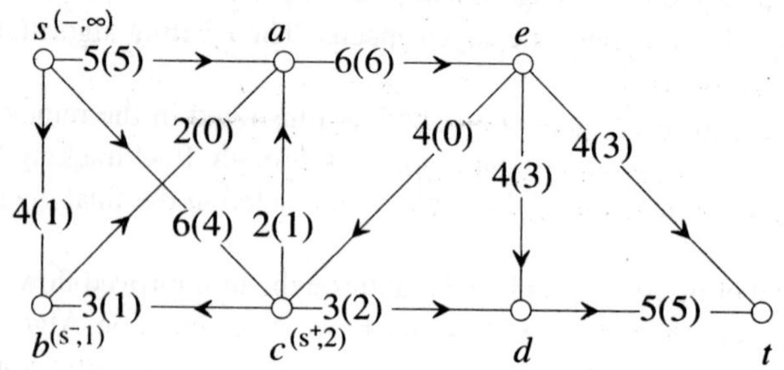

After scanning b:

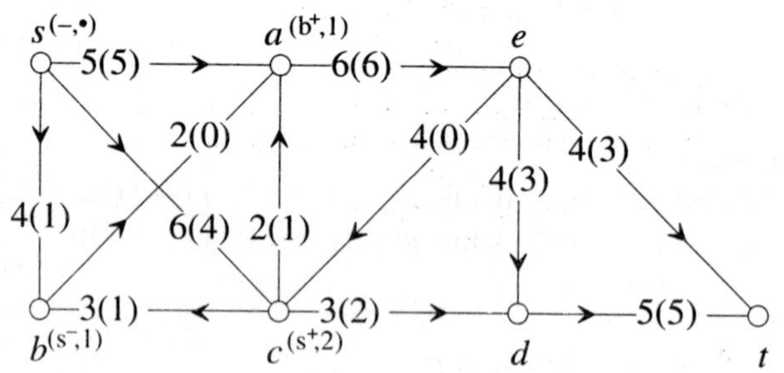

Fig. 13.7. The Labeling Algorithm

After scanning a (no result) and c:

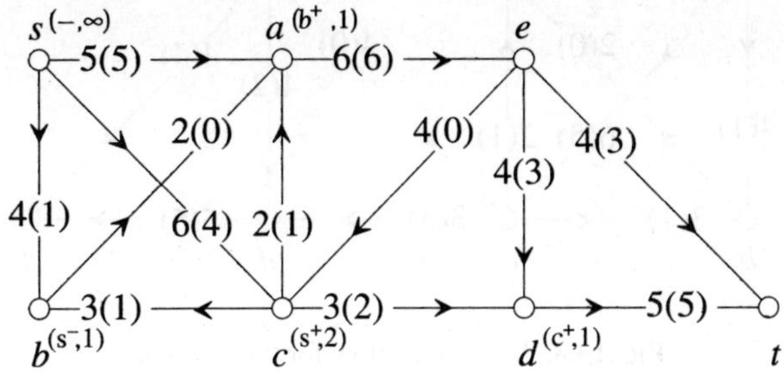

After scanning d:

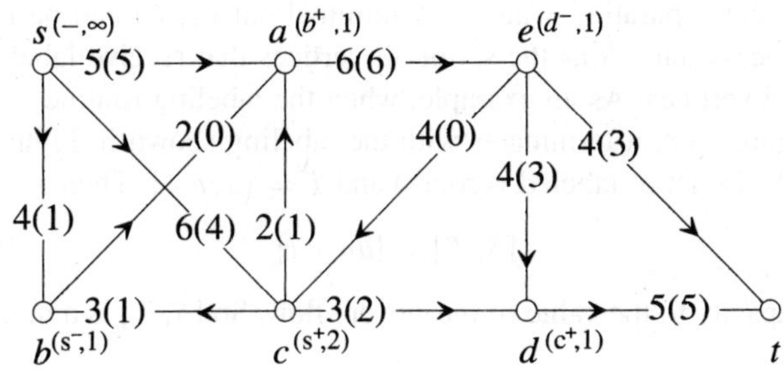

After scanning e; breakthrough occurs:

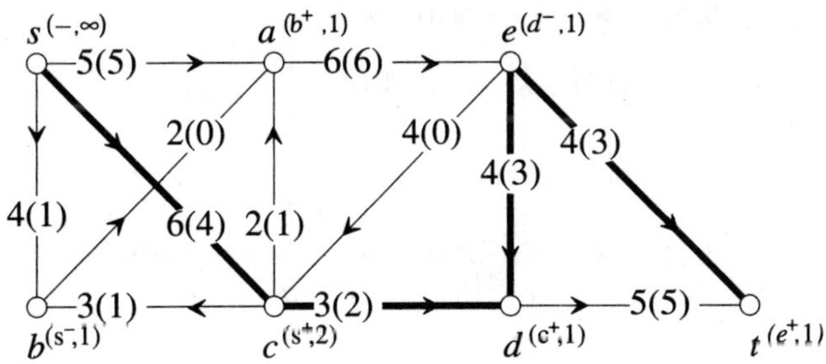

Figure 13.7, continued

3. Repeat Step 2 until the source vertex s is reached. If we take $g(x, y)$ to be equal to $f(x, y)$ for all edges not on the augmenting path, then g defines a flow from s to t of value $v(f) + \varepsilon(t)$.

Example (continued). Applying the flow augmentation algorithm to the flow augmenting path found in our earlier example, we get the revised flow indicated in Figure 13.8.

If the labeling algorithm does not reach breakthrough, then the current flow f is maximal. According to Theorem 13.4 the value v of f is equal to the capacity

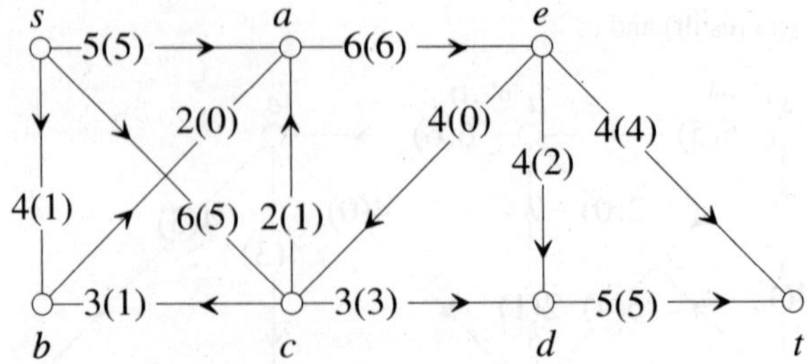

Fig. 13.8. Augmented flow for the example

of the minimal cut separating s and t. A minimal cut (S, T) can be identified from the labeling process: take S as the set of all vertices that receive labels and T as the set of unlabeled vertices. As an example, when the labeling routine is applied to the network of Figure 13.8, it terminates with the labeling shown in Figure 13.9. Define $S = \{s, a, b, c\}$ (the set of labeled vertices) and $T = \{d, e, t\}$. Then

$$[S, T] = \{ae, cd\}$$

is a cut with capacity 9, the value of the current flow. So $[S, T]$ is a minimal cut.

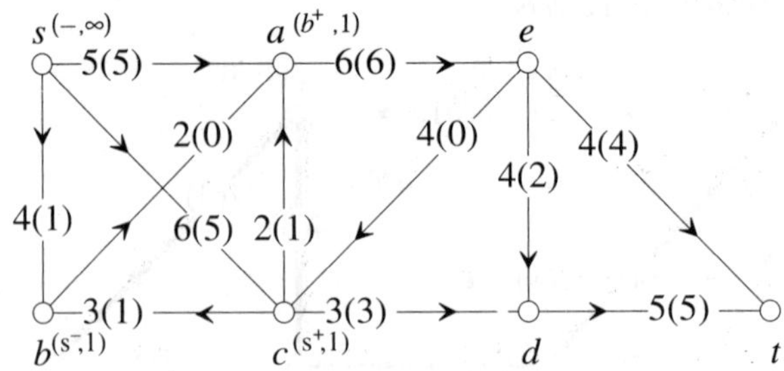

Fig. 13.9. Finding the minimal cut

Exercises 13.4

A13.4.1 Prove that the flow in Figure 13.6(b) is maximal.

A13.4.2 In Exercise 13.1.1, if g is interpreted as a capacity function, find a maximal flow in the network.

13.4.3 Apply the labeling algorithm to find maximal flows in the networks in Exercises 13.2.2, 13.2.3, 13.2.5 and 13.2.6. (Compare your results with the results for those exercises.)

H13.4.4 In the following street network, the numbers are the traffic flow capacities. The problem is to place one-way signs on the streets not already oriented (the streets marked ?) so as to maximize the traffic flow from s to t. Solve this problem using the labeling algorithm.

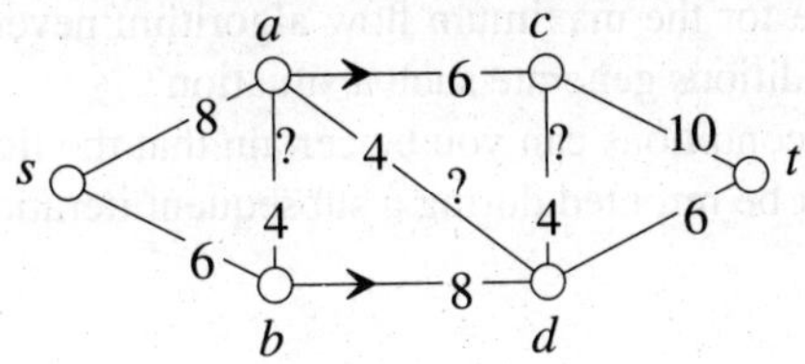

3.4.5 For each network, find a maximal flow from s to t, starting from a zero flow.

(i)

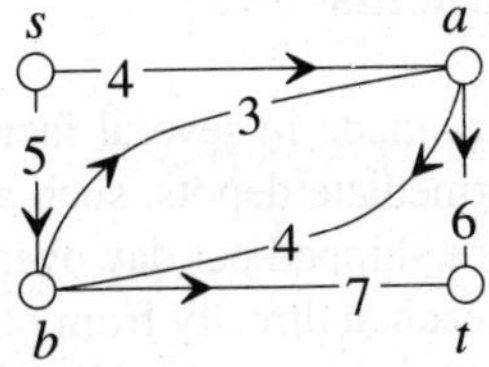

(ii)

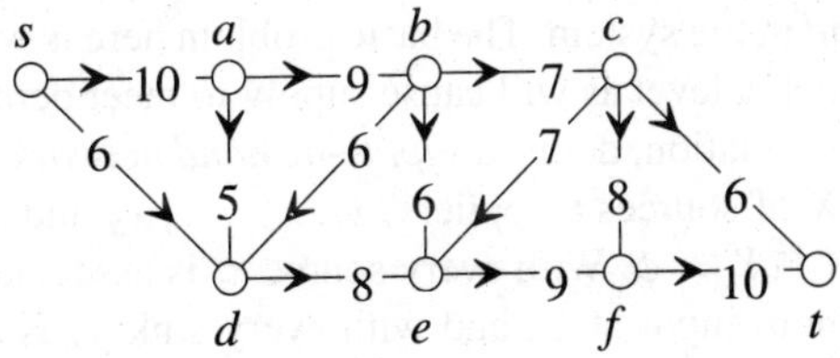

3.4.6 Repeat the preceding exercise for these networks.

(i)

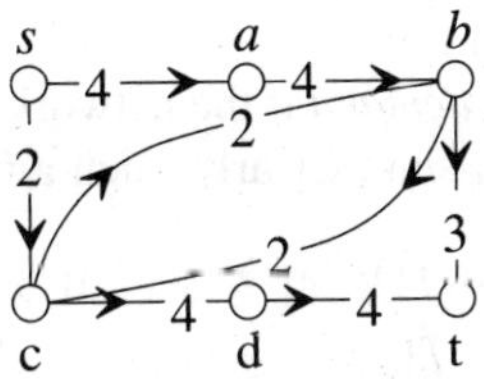

(ii)

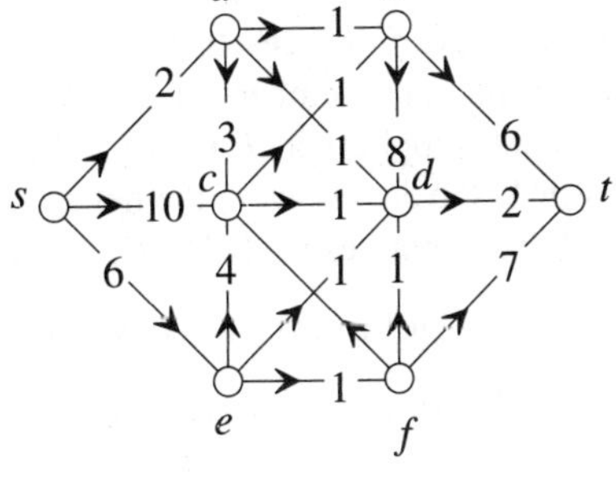

13.4.7 If the maximum flow algorithm discovers a flow augmenting path that contains an edge directed back along the path, then some flow will be removed from this backward edge and rerouted.

(i) Is is possible for the maximum flow algorithm never to reroute any flow? If so, what conditions generate such a situation?

(ii) Under what conditions can you be certain that the flow assigned to a specific edge will not be rerouted during a subsequent iteration of the maximum flow algorithm?

13.5 Supply and Demand Problems

Consider the situation where goods are made in several factories and shipped to retailers. Often they will be sent to intermediate depots, such as distributors and warehouses. The number of items that can be shipped per day over part of the route will be restricted and some places cannot be reached directly from others.

Every factory will have a certain maximum amount that it can produce, and every retailer will have a certain minimum amount that it needs. The total of these needs is called the *demand* in the system. The basic problem here is whether or not the factories can produce at such a level as will cause supply to meet demand.

To model this situation, define a *supply-demand network* to be a transportation network with a set X of sources (suppliers) $x_1, x_2, \ldots, x_\alpha$ and a set Y of sinks (retailers) $y_1, y_2, \ldots, y_\beta$; $X \cap Y = \phi$. With every source x_i is associated a positive real number $a(x_i)$, the maximum input at x_i, and with every sink y_j is associated a positive real number $b(y_j)$, the demand at y_j. If $x \notin X$, then $a(x) = 0$ and if $y \notin Y$, then $b(y) = 0$. (For convenience it has been assumed that no factory also acts as a retailer. In practice this is not an important restriction.)

It is said that *supply can meet demand* if the network has a flow whose output is at least equal to the demand at each sink. Clearly, such a flow must meet the following constraints:

$$f[(x, V] - f[V, x] \leq a(x) \text{ if } x \in X; \tag{13.9}$$

$$f[V, x] - f[x, V] \geq b(x) \text{ if } x \in Y; \tag{13.10}$$

$$f[x, V] - f[V, x] = 0 \text{ for other } x; \tag{13.11}$$

$$0 \leq f(x, y) \leq c(x, y). \tag{13.12}$$

Suppose S and T are sets of vertices that partition the vertex-set V. The total demand from the retailers that are members of T is $b(T)$, and the maximum amount that can be produced by the suppliers in T is $a(T)$. The rest of the demand must be supplied from vertices in S. Given a flow f, the net amount that flows from S to T is $f[S, T] - f[T, S]$, and this cannot exceed $c[S, T]$. So, if supply is to meet demand,

$$b(T) - a(T) \leq c[S, T]$$

for every such partition S, T. We now prove that this necessary condition is also sufficient.

Theorem 13.7. *If N is a supply-demand network with vertex-set V, then supply can meet demand if and only if*

$$b(T) - a(T) \leq c[S, T] \tag{13.13}$$

for every partition $\{S, T\}$ of V.

Proof. We know the condition is necessary. To prove sufficiency, suppose N satisfies (13.13) for every subset S. We construct a new network N', whose vertices and arcs are the vertices and arcs of N together with two new vertices s and t and arcs sx for all x in S and yt for all y in T. The capacity function c' is defined by

$$\begin{array}{lcll} c'(s, x) & = & a(x) & \text{if } x \in S, \\ c'(y, t) & = & b(y) & \text{if } x \in T, \\ c'(x, y) & = & c(x, y) & \text{if } x, y \in V. \end{array}$$

N' is treated as a network with one source s and one sink t.

Define $Q = \{t\}$, and $P = \{s\} \cup V$. Then $[P, Q]$ is a separating cut in N'. Select any separating cut $[G, H]$ in N' and write S for $G \backslash \{s\}$ and T for $H \backslash \{t\}$ (so that T is the complement of S *as a set of vertices of the original network* N). We evaluate $c'[G, H] - c'[P, Q]$. We use equations (13.1) to (13.4) and various other facts: $c'[s, Q]$ is zero because there is no edge st; $c'[S, Q] = b(S)$, and $c'[P, Q] = b(V)$; $c'[s, T] = a(T)$.

$$\begin{aligned} c'[G, H] - c'[P, Q] &= c'[S \cup \{s\}, T \cup Q] - b(V) \\ &= c'[S, T] + c'[s, T] + c'[S, Q] - b(V) \\ &= c[S, T] + a(T) + b(S) - b(V). \end{aligned} \tag{13.14}$$

Now S and T are disjoint, so $b(S) + b(T) = b(S \cup T) = b(V)$, since all members of T are in $S \cup T$. Therefore $b(S) - b(V)$ equals $-b(T)$, and (13.14) is

$$c[S, T] + a(T) - b(T).$$

This expression is nonnegative by (13.13), so

$$c'[G, H] \leq c'[P, Q].$$

Thus the separating cut $[P, Q]$ is minimal.

Suppose f' is a maximal flow in N'. Then f' must make every edge in $[P, Q]$ full and $f'(x, t) = c'(x, t) = b(x)$. The function f, defined on N by putting $f(x, y) = f'(x, y)$, is a flow that satisfies (13.11) and (13.12). If x is in T then (13.13) yields

$$f'[x, V] = f'(x, t) - f'[V, x] = 0,$$

since there is no edge xs, sx or tx; so

$$f[V, x] - f(x, V)] = f'(x, t) = b(x).$$

Similarly, one can prove that

$$f[x, V] - f[V, x] = f'(s, x) \leq a(x)$$

when x lies in S. So supply meets demand under the flow f. □

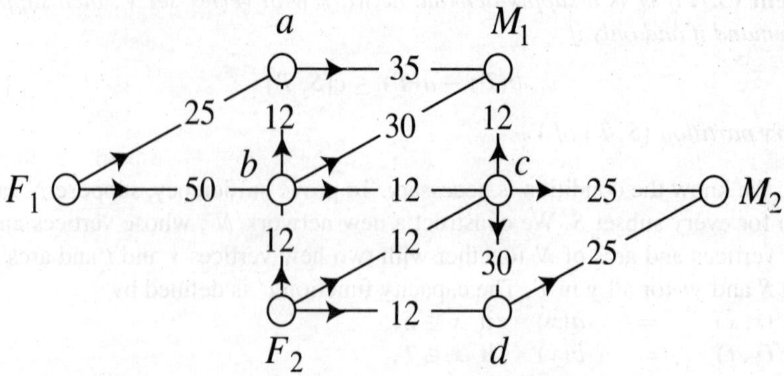

(a) The distribution network

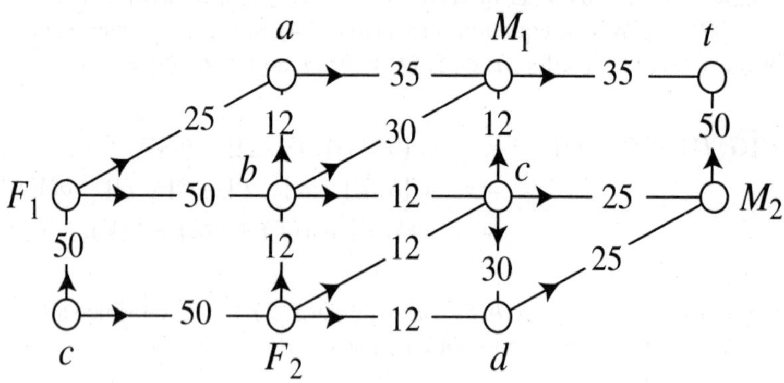

(b) The transportation network

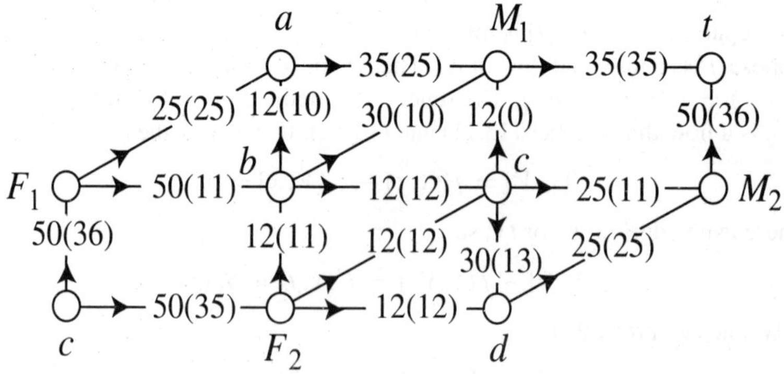

(c) A maximal flow

Fig. 13.10. Figures for the example

This theorem provides a method to find out whether supply can meet demand in a network. First set up the network N', and then find a maximal flow in it. If the flow fills all the edges into t, then it induces a solution to the supply-demand problem; if not, then the problem is insoluble.

Example. A company has two factories F_1 and F_2 producing a commodity sold at two retail outlets M_1 and M_2. The product is marketed by four distributors a, b, c and d. Each factory can produce 50 items per week. The weekly demand at M_1 and M_2 are 35 and 50 units respectively. The distribution network is given in Figure 13.10(a); the number on an arc indicates its weekly capacity. Can the weekly demands at the retail outlets be met?

The problem amounts to maximizing the flow from s to t in the network of Figure 13.10(b). Applying the algorithm yields the maximum flow f given in Figure 13.10(c). (Verifying this is left as an exercise.)

Since $v(f) = 71$ is less than the requirement, the demand cannot be met. The labeled vertices at the conclusion of the algorithm are s, F_1, F_2, a, b and M_1 If $S = \{F_1, F_2, a, b, M_1\}$ and $T = \{c, c, M_2\}$, then $a(T) = 0$, $b(T) = 85$ and $c(S, T) = 36$. Thus (13.13) is violated.

Note that when the problem is not feasible a violation of (13.13) can always be found, as occurred in this example.

Exercises 13.5

3.5.1 Verify that the flow f given in Figure 13.10(c) is a maximal flow for the network.

3.5.2 Two factories F_1 and F_2 produce a commodity that is required at three markets M_1, M_2 and M_3. The commodity is transported from the factories to the markets through the network shown below. (Capacities are indicated.) Use the maximal flow algorithm to determine the maximum amount of the commodity that can be supplied to the markets from the factories.

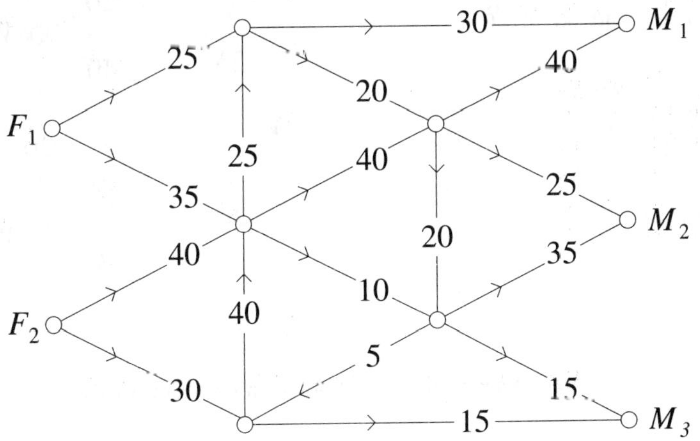

3.5.3 A supply-demand network is shown below. It has two factories x_1 and x_2 and two retailers y_1 and y_2. Capacities are shown in thousands of units per week. x_1 can output 6000 units per week, x_2 can output 8000 units per week, y_1 needs 7000 units per week and y_2 needs 4000 units per week. Can supply meet demand?

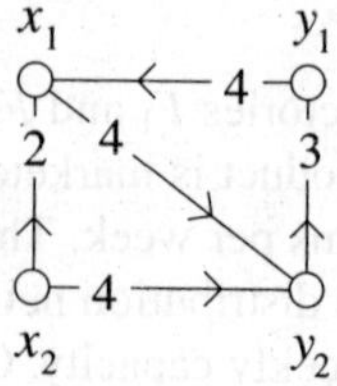

13.5.4 A manufacturing company has two factories F_1 and F_2 producing a certain commodity that is required at three retail outlets or markets M_1, M_2 and M_3. Once produced, the commodity is stored at one of the five company warehouses W_1, W_2, ..., W_5 from where it is distributed to the various retail outlets. Because of location, it is not feasible to move the commodity from any factory to any warehouse, and from any warehouse to any outlet. Information is given in Figure 13.11; the maximum weekly amount of the commodity that can be moved from F_i to W_j, and from W_j to M_k, are given by the appropriate entries in the matrices, and movement of the commodity is possible through the network shown.
Factory 1 has a weekly production capacity of 60 units and Factory 2 has a weekly production rate of 40 units. Using an appropriate flow algorithm, determine the maximum amount of the commodity that can be supplied to the markets. (Assume that the demand is unlimited.)

	W_1	W_2	W_3	W_4	W_5
F_1	25	35	0	0	0
F_2	0	40	30	0	0

	M_1	M_2	M_3
W_1	30	0	0
W_2	0	0	0
W_3	0	0	15
W_4	40	25	0
W_5	0	35	15

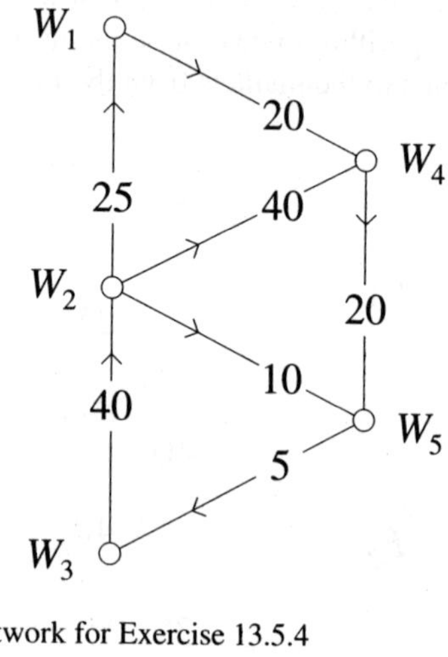

Fig. 13.11. Matrices and network for Exercise 13.5.4

14

Computational Considerations

14.1 Computation Time

In this chapter we very briefly discuss the theory of computability and the relationship between graphs and algorithms. In this book we can only introduce the ideas very sketchily. For a more thorough introduction to complexity and NP-completeness, see [52]; an excellent treatment of algorithmic graph theory is found in [53].

We first consider the question: How long does it take to carry out a computation? An answer in terms of seconds or minutes is of no use: not only does this depend on the computer used, but it does not take into account the difference between apparently similar instances of a problem.

Suppose one needs to input n pieces of data to define a particular case (an *instance*) of a problem. The number of steps required to produce a solution will be a function of n. We shall call this function the *complexity* of the solution process. In many graph algorithms, a graph G can be described by listing its vertices and edges, so a common measure of the size of the problem is $n(G) = |V(G)+|E(G)|$. The number of vertices, v, is also a useful measure of input size for some problems.

For example, suppose one wants to find the length of the shortest path between two specified vertices in a graph G. G can be described by listing its vertices and edges, so the size of the input is $n(G)$. To find the complexity, one could count the number of arithmetical or logical steps required for the computation.

Complexity is ordinarily used to discuss algorithms, not specific instances. Rather than calculating the complexity of one computation, it is usual to ask what is the largest number of steps required to apply a certain algorithm. For example, if Dijkstra's algorithm is applied to the above problem, we ask what is the largest number of steps required to solve the problem for any instance with a given value of $n = |V(G) + |E(G)|$. This is called the *worst-case complexity* of the algorithm, and is expressed as a function of n. Different algorithms for the same problem may have different complexities.

We are not usually interested in the precise value of the complexity. Suppose, for example, that one algorithm takes n^2 steps in the worst case, another takes $2n^2$ steps and a third takes $n^2 + n$. All of these algorithms are comparable, in the sense that if a problem can be solved using any one, it can be solved using the others — maybe it will take twice as long to run, but this is usually feasible. On the other hand, suppose two algorithms take n^2 and $n^3 - 99n^2$ steps respectively. A problem of size 100 can be solved in the same time using either algorithm, but a problem of size 10^6 will take a thousand times as long using the second algorithm — three years, perhaps, compared to one day.

Suppose there exist a value n_0 and a positive constant K such that functions $f(n)$ and $g(n)$ satisfy $f(n) \le Kg(n)$ whenever $n \ge n_0$. (It does not matter whether n is restricted to the integers, the rational numbers or the reals; but we shall usually be interested in integer n.) Then we say f *is of order less than or equal to that of* g, and write

$$f = \mathrm{O}(g).$$

When both $f = \mathrm{O}(g)$ and $g = \mathrm{O}(f)$, we say f and g are of the same order, and write $f = \Theta(g)$. (More precisely, one should define two sets of functions, $\mathrm{O}(g)$ being the set of all functions of order less than or equal to that of g, and $\Theta(g)$ the set of all functions of the same order as g, and write things like $f \in \mathrm{O}(g)$ and $f \in \Theta(g)$, but the "=" notation is customarily used.) For example, $2n^2 = \Theta(n^2)$ and $n^2+n = \Theta(n^2)$, and $n^3 - 99n^2 \neq \Theta(n^2)$; $n^2 = \mathrm{O}(n^3 - 99n^2)$ but not conversely. We say n^2 is of *lower* order than $n^3 - 99n^2$, and write $n^2 = \mathrm{o}(n^3 - 99n^2)$.

Alternatively, if there exists a value n_0 and a positive constant K such that functions $f(n)$ and $g(n)$ satisfy $f(n) \ge Kg(n)$ whenever $n \ge n_0$, then we say f *is of order greater than or equal to that of* g, and write

$$f = \Omega(g).$$

It is easy to see that $f = \Omega(g)$ if and only if $g = \mathrm{O}(f)$ (the proof is left as an exercise), and we could define $f = \Theta(g)$ to mean that both $f = \Omega(g)$ and $g = \Omega(f)$.

To test the order of functions f and g, it is often most convenient to examine $\lim_{n\to\infty} f(n)/g(n)$. If the limit is a nonzero constant, then $f = \Theta(g)$; if it is zero, then $f = \mathrm{o}(g)$, and if it is ∞, then $g = \mathrm{o}(f)$.

The following theorem provides a hierarchy of orders of functions.

Theorem 14.1. *The relations $f = \mathrm{O}(g)$ and $f = \mathrm{o}(g)$ are transitive. The relation $f = \Theta(g)$ is an equivalence relation. Moreover:*

(i) $\log n = \mathrm{o}(n)$;

(ii) $n^x = \mathrm{o}(n^y)$ *whenever* $x < y$;

(iii) *If f and g are polynomials in n of the same degree, then* $f = \Theta(g)$;

(iv) $n^k = \mathrm{o}(x^n)$ *for any constants k and x both greater than 1;*

(v) $x^n = \mathrm{o}(n!)$ *if x is a positive constant.*

Proof. The first part is left as an exercise.

(i) One can use L'Hôpital's rule to evaluate $\frac{\log n}{n}$, since both terms tend to ∞. So

$$\lim_{n\to\infty} \frac{\log n}{n} = \lim_{n\to\infty} \frac{\frac{1}{n}}{1} = \lim_{n\to\infty} \frac{1}{n} = 0.$$

(ii) Since $y - x > 0$,

$$\lim_{n\to\infty}\frac{n^x}{n^y}=\lim_{n\to\infty}\frac{1}{n^{y-x}}=0.$$

(iii) Say $f = ax^k +$ (*terms of degree lower than k*) and $g = bx^k +$ (*terms of degree lower than k*), where a and b are nonzero. Then

$$\lim_{n\to\infty}\frac{f}{g}=\lim_{n\to\infty}\frac{a + \textit{terms that} \to 0}{b + \textit{terms that} \to 0}=\frac{a}{b},$$

which is a nonzero constant.

(iv) Write $h(n) = \frac{n^k}{x^n}$. Then

$$\frac{h(n+1)}{h(n)}=\frac{(n+1)^k x^n}{n^k x^{n+1}}=\frac{(1+\frac{1}{n})^k}{x}.$$

Now k is constant, so

$$\lim_{n\to\infty}\frac{h(n+1)}{h(n)}=\frac{1}{x}<1$$

so $h(n) \to 0$.

(v) Write $h(n) = \frac{x^n}{n!}$. Then

$$\frac{h(n+1)}{h(n)}=\frac{x^{n+1}n!}{x^n(n+1)!}=\frac{x}{n+1}<1 \text{ for sufficiently large } n.$$

Again, $h(n) \to 0$. □

To make the orders of complexity more concrete, suppose a computer can carry out 2^6 steps per second. Table 14.1 shows how long it would take to carry out n^2, 2^n and $n!$ steps, for $n = 2$, 10, 20 and 50. To put these numbers into perspective, remember that the universe is about 1.5×10^{10} years old.

n	Time for n^2	Time for 2^n	Time for $n!$
2	.000004 seconds	.000004 seconds	.000002 seconds
10	.0001 seconds	.001 seconds	3.6 seconds
20	.0004 seconds	1 second	almost 1000 centuries
50	.0025 seconds	35 years	about 10^{52} years

Table 14.1. Running times at one million steps per second

For obvious reasons, an algorithm of complexity $f(n)$ is called *linear* if $f(n) = \Theta(n)$, *polynomial* if $f(n) = \Theta(p(n))$ for some polynomial function $p(n)$, and so on. If $f(n) = o(n)$, f is *sublinear*. Because of the explosion in running-times illustrated in Table 14.1, an algorithm whose complexity f satisfies $p(n) = o(f(n))$ for every polynomial function p is called *hard*; hard algorithms include exponential and factorial algorithms.

Exercises 14.1

4.1.1 Prove that if $f = O(g)$ and $g = O(h)$, then $f = O(h)$.

4.1.2 Prove that if $f = o(g)$ and $g = o(h)$, then $f = o(h)$.
4.1.3 Prove that $f = \Theta(g)$ is an equivalence relation.
4.1.4 Prove that $f = \Omega(g)$ if and only if $g = O(f)$.
4.1.5 Suppose your present computer can solve a problem of size $n = 10$ in one hour. You are given a computer that is 100 times as fast as your present machine. Calculate the maximum (integer) value of n for which your new computer can solve the problem in one hour, if the size of the computation is a constant multiple of:
(i) n^2;
(ii) 2^n;
(iii) $n!$.
4.1.6 Two algorithms for the same problem have running times 2^n and n^2. For which values of n is the second algorithm faster?
4.1.7 Give an example of functions f and g with the property that $f = \Theta(g)$ but $\lim_{n\to\infty} f(n)/g(n)$ does not exist.
4.1.8 "Testing whether n is prime is supposed to be hard, but one needs only to test potential divisors up to $\sqrt{n}$, so the complexity of the problem is $\sqrt{n}$." What is wrong with this argument?

14.2 Data Structures

Common data structures used to specify a graph in a computer program include the adjacency and incidence matrices, which were defined in Section 1.2. If the vertex-set and edge-set of G are ordered in some way, say

$$V(G) = \{x_1, x_2, \ldots, x_v\}$$

and

$$E(G) = \{y_1, y_2, \ldots, y_e\},$$

then the *adjacency matrix* M_G with respect to this ordering is the $v \times v$ matrix with entries m_{ij}, where

$$m_{ij} = \begin{cases} 1 & \text{if } x_i \sim x_j, \\ 0 & \text{otherwise,} \end{cases}$$

and the *incidence matrix* N_G is the $v \times e$ matrix with entries n_{ij},

$$n_{ij} = \begin{cases} 1 & \text{if vertex } x_i \text{ is incident with edge } a_j, \\ 0 & \text{otherwise.} \end{cases}$$

In many cases, one of these matrices will be the most convenient way to describe a graph. However, observe that checking all elements of the adjacency matrix of a graph G with v vertices and e edges requires v^2 steps, and checking the incidence matrix requires ve steps. If nearly all pairs of vertices of G are adjacent, then e will be close to $\frac{1}{2}v^2$, so that $n = \Theta(v^2)$ and processing the adjacency and incidence matrices will take $\Theta(n)$ and $\Theta(n^{\frac{3}{2}})$ steps respectively. But if e is $\Theta(v)$ (such graphs are called *sparse*), then processing either matrix takes $\Theta(n^2)$ steps. This seems to suggest that no graph algorithm can be carried out in $o(n^2)$ steps.

Less complex algorithms are in fact possible, but a different data structure must be used. With each vertex x of G is associated a list $AL(x)$ of vertices adjacent to x. $AL(x)$ is called the *adjacency list of* x. The *adjacency list of* G is a list $AL(G)$ made up of v sublists, each consisting of a vertex x followed by $AL(x)$. $AL(x)$ has $d(x)$ elements, so $AL(G)$ has $\sum_x d(x) + 1 = v + \sum_d (x) = v + 2e$ elements (using Theorem 1.1). So the complexity of specifying G is of order $v + 2e$. Since $v + e \leq v + 2e \leq 2(v + e)$, this is $\Theta(n)$.

Many algorithms involve graphs with weighted edges. Weights are easily represented in matrices. If $x_i x_j$ has weight w_{ij}, then the *weighted adjacency matrix* has (i, j) entry w_{ij}. The *weighted incidence matrix* has an additional row, whose entry in a given column equals the weight of the edge represented by that column.

In the place where the adjacency list of vertex x_i has entry x_j (to indicate edge $x_i x_j$), the weighted adjacency list of x_i has the ordered pair (x_j, w_{ij}). The weighted adjacency list of G is formed from these lists in the same way as in the unweighted case.

It is easy to see that the complexity of describing a graph is not significantly increased by the addition of weights (see Exercise 14.2.1).

Exercises 14.2

14.2.1 For each of the three data structures discussed in this section (adjacency matrix, incidence matrix, adjacency list), prove that the complexity of describing a weighted graph is of the same order as describing an unweighted graph.

14.2.2 Suggest a definition of an adjacency list for a digraph. What is the complexity of describing a digraph using your definition?

14.3 Some Graph Algorithms

In this section we shall study a few graph algorithms, which we find it convenient to write in a pseudocode. We shall present implementations by writing "programs" in a simplified language rather like *Pascal* or C, but without any details. An experienced programmer will find it easy to construct programs from these outlines, and they will be an aid in working out the complexity of the algorithms.

Example. Dijkstra's algorithm

Dijkstra's algorithm for finding shortest paths in a weighted graph was discussed in Section 2.3. Edge weights are interpreted as lengths, and the length of a path is the sum of its edge lengths. For a given source vertex s, the algorithm produces the length of the shortest weighted path from s to each other vertex. In order to implement it we introduce a function $P(x)$. At any time during the running of the algorithm, $P(x)$ is the length of the shortest known path from s to x. We define a function $W(x, y)$ on pairs of vertices as follows: $W(x, x) = 0$ for every vertex; if x_i and x_j are adjacent, then $W(x_i, x_j) = w_{ij}$, the weight of edge $x_i x_j$; and if x_i and x_j are not adjacent, then $W(x_i, x_j) = \infty$ (on the computer, ∞ is replaced by some large number). At any stage,

F will be the set of vertices whose processing is finished. A pseudocode program for implementing the algorithm is given in Figure 14.1.

```
algorithm
   1.  for all vertices x, P(x) ← W(s, x)
   2.  F ← {u}
   3.  while T ≠ V(G) do
   4.        begin
   5.              find y in V(G)\F such that P(y) is minimal
                   (i. e. for all x ∉ F, P(y) ≤ P(x))
   6.              F ← F ∪ {y}
   7.              for all x ∉ F
   8.                    if P(x) > P(y) + W(y, x)
                         then P(x) ← P(y) + W(y, x)
   9.        end
```

Fig. 14.1. Pseudocode for Dijkstra's algorithm

The complexity of Dijkstra's algorithm is easily calculated. Step 1 is of order v. Both Step 5 and Step 7 are also of order v, and because Step 7 is nested within Step 5, the complexity is v^2. So the complexity is $O(v^2 + v)$, which is $O(v^2)$.

Example. All shortest paths

Suppose one needs the lengths of the shortest paths between every pair of edges in a graph. One possible method would be to run Dijkstra's algorithm v times, once for each vertex. The following algorithm [53, p.19] appears more efficient.

Assume the vertices have been labeled $x_1, x_2, \ldots, x_v$ in some order. Write W for the matrix of all weights of edges: w_{ij} is the weight of $x_i x_j$, and if there is no edge joining x_i and x_j, then $w_{ij} = \infty$ (in practice, it is given a very large value). $w_{ii} = 0$ for every i. Then a sequence of matrices $W_0, W_1, \ldots, W_v$ is defined recursively. $W_0 = W$, and if $w_{k;ij}$ is the (i, j) entry in W_k, then

$$w_{k;ij} = \min\{w_{k-1;ij}, w_{k-1;ik} + w_{k-1;kj}\}. \tag{14.1}$$

It is easy to prove that $w_{v;ij}$ is the length of the shortest path from x_i to x_j (see Exercise 14.3.1). An implementation is given is Figure 14.2. Its complexity is easily seen to be $O(v^3)$, because each of the nested Steps 4, 5 and 6 have complexity v. (Steps 1 and 2 , between them, are quadratic, and can be ignored.)

Example. Depth-first search

In many graph problems, it is necessary to go through all the vertices of the graph and to process them. An algorithm to visit all the vertices of a graph in a systematic manner is called a *search*. The output from a search is a list of all the vertices of the graph, in some order. Two important kinds of search are the *depth-first search* or *DFS* and the *breadth-first search* or *BFS*. We examine *DFS*s; *BFS*s are left to the exercises.

```
algorithm
for i = 1 to v do
    for j = i + 1 to v do
        w_{0;ij} ← w_{ij}
for k = 1 to v do
    for i = 1 to v do
        for j = 1 to v do
            w_{k;ij} ← min{w_{k−1;ij}, w_{k−1;ik} + w_{k−1;kj}}
```

Fig. 14.2. Pseudocode for the all paths algorithm.

To define depth-first searches, consider a search algorithm being applied to a connected graph G, and suppose it is currently visiting vertex x. In a *DFS*, the next step if possible is to visit a vertex that is adjacent to x and has not yet been visited. If no such vertex exists, the algorithm returns to the vertex visited just before x and looks for a new vertex adjacent to it; and so on. This returning process is called *backtracking*. Whenever the algorithm goes from vertex x to a new vertex y, we say it *traverses* the edge xy; similarly, if it backtracks from x to z, we say it traverses xz. If a *DFS* visits vertex x twice, then the edge traversed in going to x the second time must be one that was already traversed earlier, in moving from x. For this reason the edges traversed by a *DFS* form a spanning tree in G, called a *depth-first spanning tree*. More generally, if a *DFS* is applied to a disconnected graph, it produces a *depth-first spanning forest*. Edges not in this forest (or tree) are called *back-edges*.

Suppose a *DFS* is applied to G. The algorithm will assign to each vertex x a label $L(x)$; if $L(x) = i$, this means that x will be the i-th element in the output list. $L(x)$ is the *depth-first index* of x. The edges of the spanning forest will be denoted by F, and the set of all back-edges will be B. Initially vertex x receives label $\ell(x) = 0$; by the end of the run, $\ell(x) = L(x)$ for each x.

We describe the algorithm using a procedure that we call *process*. Suppose $i - 1$ vertices have so far been processed, and the algorithm is examining vertex x for the first time. There will be a counter that holds the value i. To process x, assign to $\ell(x)$ the value i, and increment the counter by 1. Then, for each vertex y adjacent to x: if $\ell(y) \neq 0$, do nothing, but if $\ell(y) = 0$, then add xy to F and process y. It will be seen that this gives label $i + 1$ to y, label $i + 2$ to the first new vertex encountered that is adjacent to y (if there are any), and so on. When y and the new vertices that came from it are exhausted, the next new neighbor of x will be processed.

The algorithm starts by setting $i \leftarrow 1$, $F \leftarrow \emptyset$ and $\ell(x) \leftarrow 0$ for each x. Then, while there is any vertex y with $\ell(x) = 0$, *process*(y) is carried out. Observe that this ensures that all components are searched, in the case of a disconnected graph.

A pseudocode program is given in Figure 14.3. The input is the set of all adjacency lists $A(x)$ of vertices x of G.

To calculate the complexity of this algorithm, observe that execution of line 3 requires v steps, and so does execution of line 4. The output is also of this order.

The procedure *process* is called v times, once for each vertex (after the first call for a given x, $\ell(x)$ is nonzero). Running $process(x)$ requires $O(d(x))$ steps, so the total complexity of calling it once for each vertex is $O(\sum d(x))$, or $O(2e)$. So the whole algorithm is $O(n)$. (In fact, if v and e are of different orders, the complexity equals the greater of $\Theta(v)$ and $\Theta(e)$.)

This is one case where it was very important to have a linear way of representing the graph. For sparse graphs, if the adjacency matrix were used, the resulting algorithm would be quadratic.

All of the above algorithms are polynomial. There are, however, many important problems where no polynomial algorithm is known. Two of the most important examples are the Traveling Salesman Problem and the problem of deciding whether two given graphs are isomorphic. It is not known for certain whether these problems are hard — that is, no polynomial algorithm exists — but in both cases the majority of researchers believe a polynomial algorithm is impossible.

Exercises 14.3

14.3.1 Consider the algorithm to find all shortest paths. Prove, by induction, that $w_{k;ij}$ is the weight of the shortest path from x_i to x_j among all paths that contain no intermediate vertices other than members of $\{x_1, x_2, \ldots, x_k\}$. Hence show that the algorithm does in fact provide the shortest paths.

A14.3.2 As an alternative to the "all shortest paths" algorithm, one could construct a sequence $W_1, W_2, \ldots, W_v$ where $w_{k;ij}$ is the length of the shortest path from x_i to

```
procedure process(x)
    begin
        ℓ(x) ← i
        i ← i + 1
        for all y ∈ A(x) do
            if ℓ(y) = 0 then
            begin
                F ← F ∪ {xy}
                process(y)
            end
    end

algorithm
    i ← 1
    F ← ∅
    for all x ∈ V(G) do ℓ(x) ← 0
    while there exists y ∈ V(G), ℓ(y) ← 0 do
        process(y)
    for all x ∈ V(G) output {ℓ(x)}
```

Fig. 14.3. Pseudocode for depth-first search

x_j among all paths that contain at most k edges: in this case $W_1 = W$ (again, if there is no edge joining x_i and x_j, then $w_{ij} = \infty$), and (14.1) is replaced by

$$w_{k;ij} = \min\{w_{k-1;ij}, \min_h\{w_{k-1;ih} + w_{hj}\}\}.$$

(i) Prove that W_v is the matrix of shortest paths;
(ii) Give a pseudocode implementation of this algorithm.
(iii) What is the complexity of this algorithm?

14.3.3 Use the result of Exercise 2.1.10 to construct an algorithm to find the number of walks of length at most k between each pair of vertices of a graph. What is the complexity of:
(i) finding all paths of length equal to or less than v;
(ii) finding all paths of length less than 5.

14.3.4 A *breadth-first search* of the vertices of a tree is carried out as follows. First, a counter i is set to 1 and some vertex x is chosen. An ordered set Q (called a *queue*) is then constructed. The members are the vertices that are adjacent to x. When this is done, x is assigned label 1, i is incremented to 2 and x is considered to have been processed. Then the first member of the queue is chosen to be processed (and all other members move one place nearer to the beginning of the queue). To process a vertex y, one selects neighbors of y that are not already in the queue or already processed, and puts them at the end of the queue. Then y receives label i, i is implemented and the first member of the queue is chosen for processing.
Produce the pseudocode for this algorithm. Prove that the *BFS* and *DFS* algorithms are of the same complexity.

14.3.5 Write a pseudocode algorithm to implement Prim's algorithm for finding a minimal spanning tree (see Section 4.3). What is the complexity of your implementation? (A quadratic algorithm is easily found.)

14.3.6 Write pseudocode for the following algorithms, and calculate their complexity:
(i) the nearest neighbor algorithm for the Traveling Salesman Problem in Section 2.6;
(ii) the sorted edges algorithm for the Traveling Salesman Problem in Section 2.6;
(iii) the algorithm used to prove that $\chi(G) \le \Delta G + 1$ in Section 7.1;
(iv) the max flow min cut algorithm in Section 13.4.

14.4 Intractability

Our description of intractability and of the classes ***P*** and ***NP*** will be somewhat informal. A more rigorous development can be found in [52] or [53, Chapter 8].

The study of intractability arises from the following consideration. Suppose an algorithm takes $f(n) = n^3$ steps. Then a small increase in the problem size produces a relatively small increase in the computation time. For example, increasing from $n = 1000$ to $n = 1005$ increases f from 10^6 to 1.015×10^6, an increase of 1.5%. On the other hand, if $f(n)$ were 2^n, the increase would be 3100%. In general, if your present computer can handle a certain instance of a polynomial problem, then a slightly larger instance of the problem can also be solved with at worst a slightly faster machine. On the other hand, even a small increase in the size of a problem can render its solution by a hard algorithm impractical. A problem with no polynomial algorithm is therefore called *intractable*.

For convenience, we initially restrict our attention to decision problems, problems with a yes-or-no answer. This is not an important restriction, as most problems can be recast as decision problems. For example, the problem of finding the shortest path from s to t can be recast in terms of the decision problem: is there an st-path of length less than k?

The set ***P*** of decision problems consists of those decision problems for which there is a polynomial time algorithm. For example, we saw in the preceding section that the shortest path algorithm is in ***P***. However, the Traveling Salesman Problem and the Graph Isomorphism Problem are not known to be in ***P***.

Even if a problem cannot be solved in polynomial time, it may be possible to *check* a solution very quickly. For example, no polynomial algorithm is known that will answer the question, "Is G isomorphic to H?", where G and H are arbitrary graphs. On the other hand, if you are given a mapping $\phi: V(G) \mapsto V(H)$ that purports to be a graph isomorphism, to test whether it is really an isomorphism can be done in linear time. One might say that the Graph Isomorphism Problem is *polynomial time verifiable*. This process is *nondeterministic*; the solution is given, not determined. So the Traveling Salesman Problem is *nondeterministically polynomially verifiable*. We write ***NP*** for the set of all *nondeterministically polynomially verifiable* problems.

Obviously, every member of ***P*** is in ***NP***. The fundamental question in the study of the complexity of algorithms is: Does ***P*** = ***NP***? Many theorists believe that ***NP*** is a much larger set than ***P***, but this has not been proved.

Suppose ***P*** ≠ ***NP***. Then there may be problems that can be verified in polynomial time — problems that *look* easy — that cannot be solved in polynomial time. Identifying such problems is important.

To discuss this, we introduce the idea of a *polynomial transformation*. If P_1 and P_2 are two decision problems, we say that there is a polynomial transformation from P_1 to P_2, and write $P_1 \prec P_2$, if there exists a function F, which can be computed in polynomial time, with the property that, if I is any instance of P_1, then $F(I)$ is an instance of P_2, and the answer to I is "yes" if and only if the answer to $F(I)$ is "yes." If $P_1 \prec P_2$, then there is a polynomial algorithm for P_1 if there is a polynomial algorithm for P_2.

The set ***NPC*** consists of all members Q of ***NP*** such that, if $P \in$ ***NP***, then $P \prec Q$. Members of ***NPC*** are called ***NP****-complete*. The importance of this concept follows from the following observations:

(i) if Q is ***NP***-complete and $Q \in$ ***P***, then ***NP*** = ***P***;
(ii) if P is ***NP***-complete, $Q \in$ ***NP*** and $P \prec Q$, then Q is ***NP***-complete.

The fundamental paper on intractability was [30].

Graph theory has provided many examples of ***NPC*** problems. We list some examples, without proof. (For proofs, see [52] and [53, Chapter 8].)

Example. A dominating set of vertices of a graph G is a subset S of $V(G)$ such that every vertex of G is adjacent to at most one member of S. The domination problem is: does G have a dominating set with k or fewer vertices? This problem is ***NPC***.

Example. The clique problem is: does G contain a clique of k or more vertices? This problem is ***NPC***.

Example. The problem: does G have a proper (vertex)-coloring in k or fewer colors? is ***NPC***.

Example. The Traveling Salesman Problem is ***NPC***.

Exercises 14.4

14.4.1 Prove that the relation $\prec$ is transitive.

15

Communications Networks and Small-Worlds

15.1 Preliminaries

A communications network consists of a number of users, or *nodes*, a number of links through which they can communicate, and some special links that facilitate communication. There are usually various *weights* associated with the nodes or links.

As an example, you and all your friends are the nodes of a network, and acquaintanceship forms the links. This sort of network is an amalgam of local clusters, such as cliques of friends or coworkers, and random long range links that might occur through a vacation or when close friends work for different companies.

Another example is the internet. One special class of user is the *internet service provider*, or ISP; an oversimplified model is that all ISPs are always connected to each other, while other users must connect to an ISP in order to participate in the network. Another class of user is the *server*; often many users are connected by one server, which in turn connects to the ISP. This is particularly common in office settings; nodes on a server form a local network, and messages between them do not necessarily go through the ISP. Similarly the World Wide Web is a communications network, with pages as nodes and links existing when one page references another. (In fact, such references are commonly called "links.")

If nodes are interpreted as vertices and links as edges, a communications network defines a graph, the *underlying graph* of the network. (Sometimes a digraph is more appropriate, as for example in the World Wide Web.) However, there may be more than one underlying graph. For example, in the case of the internet, one could take servers, or ISPs, as vertices. For many purposes one can assume that the underlying graph of a communications network is connected (at least when the network is functioning properly; breakdowns in the network may cause it to be disconnected).

In 1929, the Hungarian author Frigyes Karithny [74] published a short story, "Láncszemek" ("Chains") based on the idea that at most six intermediaries are needed to link any two people in the world by a chain of acquaintances. This idea was called "six degrees of separation," which has been used as the title of a play and the subsequent movie.

Milgram [87] describes an experiment in which people in Omaha, Nebraska, were given a letter intended for delivery to a stranger in Massachusetts, and asked to pass it along to someone they thought might know the intended recipient, with instructions for that person to do the same. Only about 20% of the letters ever reached the "target." For those that did, though, it rarely took more than six intermediate steps, just as Karithny's story predicts. Milgram's phenomenon — that the number of edges in the shortest path between two nodes is on average quite short — is called the *small-world effect*, and networks that exhibit it are called *small-world networks;* the underlying graph of a small-world network is a *small-world graph.*

Subsequent studies have revealed that other networks based on acquaintanceship or collegiality also exhibit the small-world effect. For example, there is the "Kevin Bacon" game, in which actors are the nodes and two are linked if they have appeared together in a movie [134]; the actor Bacon was the first one chosen because "six degrees of separation" rhymes with "six degrees of Kevin Bacon" (almost). Similarly, mathematicians have invented the "Erdős number" [57]: mathematicians are the nodes, links mean that the two have coauthored a paper, and your Erdős number is the length of your shortest path to the late mathematician Paul Erdős, whose 1500 publications is the largest of any mathematician. (Coincidentally, Erdős was Hungarian.)

The theory of small-world networks was developed in order to track how information and ideas spread through a community, but it serves to model many other situations, such as the spread of infectious diseases. The mathematical discussion of the small-world phenomenon has focussed on the underlying graph of the network. This study received impetus in 1998 in a paper by Watts and Strogatz [129].

15.2 Functions on Graphs

In order to study the small-world phenomenon we shall introduce two functions on graphs: the *characteristic path length* and the *clustering coefficient*.

We wish to know the typical distance between two vertices in a graph. One obvious parameter is the *mean path length*, the average value of $D(x, y)$ where x and y range through the vertices. While this is a useful parameter when known, its calculation is impractical for large graphs. It is necessary to estimate the average. Such estimation must be carried out by sampling. As calculating a median is much faster than calculating a mean, there are better sampling techniques available for accurately estimating medians than for means, so Watts and Strogatz [129] considered the median path length as a measure.

One might think that the median would be a significantly less reliable measure of average distance than the mean. However, Huber [72] studied the sampling technique for estimating medians, and estimated its accuracy. The results are summarized in [127, pp. 29–30] and show that the median is quite a good estimator.

Another theoretical advantage is that medians can be calculated even if the graph is disconnected, provided the number of disconnected pairs of vertices is not too large. (A disconnected pair corresponds to an "infinite" path, which is longer than any finite path. Provided fewer than half the paths are "infinite," the median can be calculated.)

Newman [89] suggests ignoring disconnected pairs, but we believe the "infinite" distance technique is more appropriate.

To approximate a median of a set is straightforward, because the sample median is an unbiased estimator of the median (and of the mean, in normal or near-normal data). One simply takes a number of readings, and finds their median.

If S is a set with n elements, its median is defined as that value $M(S)$ such that $n/2$ members of S have value $\leq M(S)$ and $n/2$ members have value $\geq M(S)$. Let us say that an estimate M of $M(S)$ is "of accuracy " δ if at least $\delta n/2$ members have value $\leq M$ and $\delta n/2$ members have value $\geq M$. Huber proved:

Theorem 15.1. [72] *Suppose δ (usually near 1) and ϵ (usually small) are positive constants. Approximation of the median $M(S)$ of a set S by sampling s readings yields a value of accuracy at least δ with probability $1 - \epsilon$, where*

$$s = 8 \ln \frac{2}{\epsilon} \left(\frac{\delta}{1 - \delta} \right)^2.$$

The median of a finite set of readings is a discrete-valued function: it either takes one of the reading values or the average of two of them. This leads to some inaccuracies. Consider for example two sets of readings: A contains 49 readings 0 and 50 readings 1, while B contains 50 readings 0 and 49 readings 1. These sets are very similar, but the medians are 0 and 1 respectively. To avoid such chaotic leaps in value, the following definition is used.

The *characteristic path length* $L(G)$ of a graph G is calculated as follows. First, for each vertex x, the median $\overline{D}_x$ of all the values $D(x, y)$, where y is a vertex, is calculated. Then $L(G)$ is the mean of the values $\overline{D}_x$, for varying x. In the sampling procedure suggested by Huber, one calculates a number of values of $\overline{D}_x$ and then takes their median.

People do not make friends completely at random; the people you know tend to know each other. This illustrates a phenomenon called *clustering*. One distinctive property of small-world graphs is the clustering behavior.

The *neighborhood* (or *open neighborhood*) $N(x)$ of a vertex x was defined in Chapter 1 to consist of all the vertices adjacent to x. The graph generated by $N(x)$, denoted by $\langle N(x) \rangle$, has vertex-set $N(x)$, and its edges are all edges of the graph with both endpoints in $N(x)$. We write $d(x)$ and $e(x)$ for the numbers of vertices and edges respectively, in $\langle N(x) \rangle$. ($d(x)$ is of course the degree of x.) Then the *clustering coefficient* γ_x of x is

$$\gamma_x = \frac{e(x)}{\binom{d(x)}{2}} = \frac{2e(x)}{d(x)(d(x) - 1)}.$$

In other words, it equals the number of connections between the neighbors of x, divided by the maximum possible number of connections.

The *clustering coefficient* of a graph G equals the mean of the clustering coefficients of all vertices of G, and is denoted by $\gamma(G)$ or simply γ.

The value $\gamma(G) = 1$ occurs if and only if G consists of a number of disjoint complete graphs of the same order (each has $k + 1$ vertices, where every vertex has the same degree k); otherwise $\gamma(G) < 1$. The value $\gamma(G) = 0$ is attained if and only if the graph G contains no triangles; otherwise $\gamma(G) > 0$.

The theory of both these measures, the characteristic path length and the clustering coefficient, is much the same if directed graphs are considered. However, when calculating the clustering coefficient in a digraph, the formula

$$\gamma_x = \frac{e(x)}{2\binom{d(x)}{2}} = \frac{e(x)}{d(x)(d(x)-1)}$$

should be used. In other words the result should be halved, corresponding to the fact that there are $k(k-1)$ possible directed edges between k vertices.

Exercises 15.2

15.2.1 Find the characteristic path length and clustering coefficient of the cycle C_v.

15.2.2 Find the characteristic path length and clustering coefficient of the wheel W_n.

15.2.3 Find the characteristic path length and clustering coefficient of $K_{m,n}$.

15.2.4 Find the characteristic path length and clustering coefficient of the Petersen graph.

A15.2.5 Consider the cycle C_8. Write $C_8(n)$ for the graph derived from C_8 by adding one edge, joining two vertices whose distance in C_8 is n.

(i) Find the mean path length of $C_8(n)$ for $n = 2, 3, 4$.

(ii) Find the characteristic path length $L(C_8(n))$ for $n = 2, 3, 4$.

(iii) Find $\gamma(C_8(n))$ for $n = 2, 3, 4$.

15.2.6 Find the characteristic path lengths and clustering coefficients of the graphs in Figure 15.1.

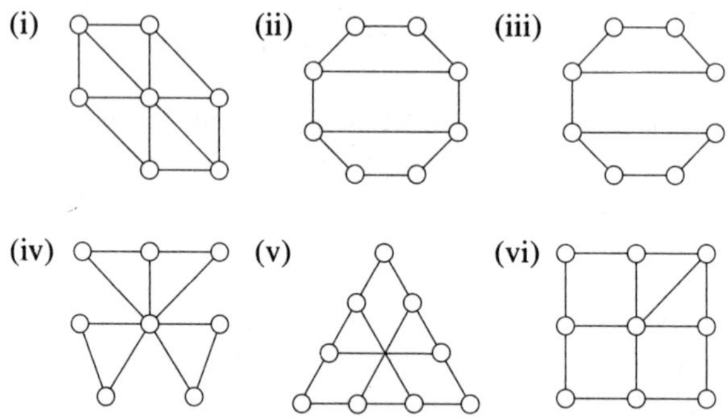

Fig. 15.1. Find the characteristic path lengths

15.3 Classes of Graphs

Two classes of "standard" graphs have been used in small-world studies. They are lattice graphs and random graphs.

A *d-lattice* is analogous to a Euclidean lattice of dimension d. The v vertices are labeled with the integers modulo v, and where x is joined to all the vertices $(x - i^{d'})$ and $(x + i^{d'})$ for $1 \le i \le \frac{1}{2}k$, $1 \le d' \le d$, where k is some integer ($k \ge 2d$). In particular, many properties of 1-lattices have been studied in connection with the small-world phenomenon. An example of a 1-lattice with $k = 4$ is shown in Figure 15.2.

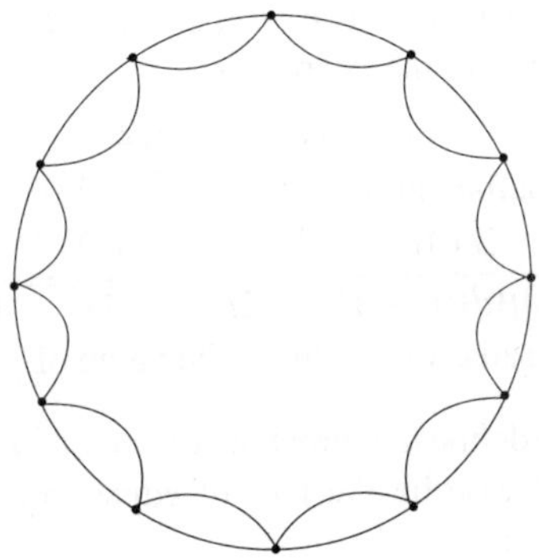

Fig. 15.2. A 1-lattice with $k = 4$

Theorem 15.2. *Suppose G is a 1-lattice with v vertices, for which k is even and at least 2. Then*

$$\frac{v + 2k - 1}{2k} \le L(G) < \frac{v + 4k - 1}{2k}.$$

Proof. We write $k = 2t$, and identify the vertices of G with the labels $0, 1, \ldots, v - 1$ modulo v. From the definition, $\overline{D}_x$ is the same for every vertex, so $L(G) = \overline{D}_0$.

Suppose $v = ak+1+d$, where $0 \le d < k$. Then $D(0, ai+j) = D(0, -ai-j) = i + 1$ for the t values $j = 1, 2, \ldots, t$, while $D(0, ak + j) = D(0, -ak - j) = i + 1$ for the d values $j = 1, 2, \ldots, d$. When $d = 1$, such that k divides $v - 1$, the average of these $v-1$ values equals the average of $1, 2, \ldots, a$, or $\frac{1}{2}(a+1) = (v+2k-1)/2k$. In any case, it is smaller than the average of $1, 2, \ldots, a + 1$, or $(v + 4k - 1)/(2k)$. □

Theorem 15.3. *Suppose G is a 1-lattice with v vertices, for which k is even and at least 2. Then*

$$\gamma(G) = \frac{3(k-2)}{4(k-1)},$$

independently of the value of v.

Proof. We use the same notation as in the preceding theorem. Each vertex of G has the same clustering coefficient, so we consider vertex 0.

The neighborhood of 0 consists of all the vertices $\{t, t-1, \ldots, 1, v-1, v-2, \ldots, v-t.\ d(0) = k = 2t\}$. If $0 < i \leq t$, vertex i is adjacent in $N(0)$ to $t, t-1, \ldots, 1, v-1, \ldots, v-t+i$, with i itself omitted, except that t is not adjacent to 0 ($0 \notin N(0)$). So i has degree $2t - i - 1$. Thus the sum of degrees of vertices $1, 2, \ldots, t$ in $N(0)$ is

$$\begin{aligned}\sum_{i=1}^{t} 2t - 1 - i &= t(2t-1) - \sum_{i=1}^{t} i \\ &= t(2t-1) - \tfrac{1}{2}t(t+1) = 3t(t-1)/2.\end{aligned}$$

The sum of degrees of vertices $v-1, v-2, \ldots, v-t$ is the same. The number of edges in $N(0)$ is half the sum of the degrees, so

$$e(0) = 3t(t-1)/2.$$

Therefore the clustering coefficient is

$$\gamma(0) = \frac{2e(0)}{d(0)(d(0)-1)} = \frac{3t(t-1)/2}{2t(2t-1)} = \frac{3t(t-1)}{4t(2t-1)}.$$

Since all vertices have the same value, this is the value of $\gamma(G)$. □

In the directed case we define the directed version of a d-lattice by replacing every edge xy in the undirected case by the pair of edges $\{xy, yx\}$. The given formulae remain true in the directed case.

The second class of graph is the so-called "random graph." There are two standard models of random graphs in the mathematical literature (see, for example, [15]), and others are possible.

A graph of type $G(v, e)$ has v vertices and e edges. The e edges are chosen from the $\binom{v}{2}$ possibilities in such a way that any of the possible e-sets is equally likely to be the one chosen. In statistical terminology, the edges are a simple random sample of size e from the set of possible edges.

A graph of type $g(v, p)$ has v vertices. The probability that the vertex-pair xy is an edge is p, and the $\binom{v}{2}$ events "xy is an edge" are independent. This model was proposed by Erdős and Rényi [39], so they are called *Erdős–Rényi graphs*.

In studying most properties — in particular, the properties of interest for the small-world phenomenon — these two models of random graphs are interchangeable, provided e is approximately $p\binom{v}{2}$. When this is true, it is more usual to study $g(v, p)$, as the variability of the number of edges makes mathematical discussions simpler and more realistic.

The expected average degree of vertices in a $g(v, p)$ is clearly $p(v-1)$. The expected value of the clustering coefficient is p; so, for a reasonably large random graph of average degree d, the clustering coefficient will be approximately d/v.

15.4 Small-World Graphs

The clustering coefficient in completely random graphs with average degree d is approximately d/v. In a lattice graph of degree d, on the other hand, it is $\frac{3(d-2)}{4(d-1)}$, independent of v. So — if one looks only at the extremes — it would seem that tight local clusters correspond to long average distances, and that reducing average distance requires abandoning local clustering.

However, Watts and Strogatz [129] discovered a class of graphs that combine strong local clustering with relatively short average distance. They started from a highly regular graph (a 1-lattice), and randomly changed a small fraction of the connections: certain edges were deleted and replaced by edges with one endpoint the same as in the original and the other chosen randomly. In these graphs the clustering coefficient stays near $3/4$, while the average distance between vertices is proportional to $\log(v)$ rather than to v.

Small world networks evolve when links are made randomly in a sparse structured network. The underlying graph of such a network is a blend of local clusters and random long range links. For example, in a network representing acquaintanceship, the clusters might be cliques of friends or coworkers, and long range links might occur through a vacation or when close friends work for different companies.

Watts and Strogatz conjectured that the underlying graphs of small-world networks would behave like their artificially constructed graphs, which they accordingly called *small-world graphs*.

Three major examples of small-world graphs were studied in [129].

One example was the Internet Movie Database (IMDb), in which the vertices are movie actors, with a connection between any two who have appeared in the same movie. Analyzing the IMDb network, which then had an average of $d = 61$ connections per actor for $v = 225,226$ actors (the IMDb has increased considerably in size since their study), Watts and Strogatz found a clustering coefficient $C = 0.79$ — surprisingly close to the theoretical value for a highly regular network — and an average distance between actors of 3.65, which is much closer to $\ln(v)/\ln(d) = 2.998$ than it is to $v/2d = 1846$.

The other two networks analyzed were electrical in nature: the power grid for the western United States and the neural system of the nematode *C. elegans*. The power grid has $v = 4941$ vertices, consisting of generators, transformers, and substations, with an average of $k = 2.67$ high-voltage transmission lines per vertex. *C. elegans* has only $v = 282$ neurons (not counting 20 neurons in the worm's throat, which biologists have not yet completely mapped), with an average of $d = 14$ connections — synapses and gap junctions — for each neuron.

The power grid has clustering coefficient $\gamma = 0.08$, 160 times what would be expected for a random network of the same size. As the power grid is inherently 2-dimensional, it is reasonable to compare its distances to those of a hexagonal honeycomb, for which most vertices have three links. The average distance between two vertices in an $m \times m$ hexagonal honeycomb, which has $v = 2m(m+2)$ vertices altogether, is approximately $2m$. When $v = 4941$, that is approximately 98, but the average distance between two vertices in the power grid is 18.7.

For *C. elegans*, the clustering coefficient is 0.28, against 0.05 for the random model, and the average distance is 2.65, while 2.25 is the value calculated for the random model and 10 for the corresponding lattice graph with $v = 14$.

All of these examples exhibited another interesting property. One can consider the set of all degrees of vertices in a graph as if it were a statistical distribution. In the case of the Erdős–Renyi random graphs, this approximates a Poisson distribution: for a large v, the fraction of vertices with degree d is proportional to e^{-d}. However, in large small-world graphs, the degrees follow a *power law*: the number of vertices of degree d is proportional to d^{-k} for some constant k.

Barabási *et al* [5] investigated the World Wide Web and found that the probability that a given page has links to d other pages is approximately $d^{-2.45}$. So the graph of the World Wide Web follows a power law.

Exercises 15.4

15.4.1 A company has a CEO and three Departments. Each Department has a Head and three junior employees. An e-mail system is set up in which the CEO communicates with each Department Head, the Department Head communicates with the three juniors in that Department, and the three juniors in a Department communicate with each other. Find the characteristic path length and clustering coefficient of the graph representing this email system. Then repeat this exercise if:
(i) the three Department Heads also communicate with each other;
(ii) three of the juniors — one from each Department — also communicate with each other.

15.4.2 Repeat the preceding exercise if there are n junior employees in each Department. If n is large, does this graph exhibit any small-world properties?

References

1. H. L. Abbott, Lower bounds for some Ramsey numbers. *Discrete Math.* **2** (1972), 289–293.
2. W. S. Andrews, *Magic Squares and Cubes.* Dover (1960).
3. K. Appel and W. Haken, Every planar graph is four colorable. *Bull. Amer. Math. Soc.* **82** (1976), 711–712.
4. K. Appel and W. Haken, *Every planar graph is four colorable.* Contemporary Math. **98**, Amer. Math. Soc., 1989.
5. R. Albert, H. Jeong and A. Barabási, Diameter of the World Wide Web. *Science* **401** (1999), 130–131.
6. A.-L. Barabási, *Linked.* Perseus, Cambridge MA, 2002.
7. E. T. Baskoro, M. Miller, Slamin and W. D. Wallis, Edge-magic total labelings. *Austral. J. Combin.* **22** (2000), 177–190.
8. L. W. Beineke and R. J. Wilson, On the edge-chromatic number of a graph. *Discrete Math.* **5** (1973), 15–20.
9. M. Behzad and G. Chartrand, *Introduction to the Theory of Graphs.* Allyn and Bacon, 1971.
10. L. W. Beineke and R. J. Wilson, *Selected Topics in Graph Theory*. Academic Press, London, 1978.
11. L. W. Beineke and R. J. Wilson, *Selected Topics in Graph Theory 2*. Academic Press, London, 1983.
12. L. W. Beineke and R. J. Wilson, *Selected Topics in Graph Theory 3*. Academic Press, London, 1988.
13. C. Berge, *Graphs and Hypergraphs*. North-Holland, 1973.
14. N. L. Biggs, E. K. Lloyd and R.J. Wilson, *Graph Theory, 1736–1936*. Clarendon Press, Oxford, 1976.
15. B. Bollobás, *Random Graphs*. Academic Press, London, 1985.
16. J. A. Bondy and V. Chvátal, A method in graph theory. *Discrete Math.* **15** (1976), 111–136.
17. J. A. Bondy and U. S. R. Murty, *Graph Theory with Applications*. Macmillan, London, 1976.
18. R. L. Brooks, On colouring the nodes of a network. *Proc. Camb. Phil. Soc.* **37** (1941), 194–197.
19. R. A. Brualdi, *Introductory Combinatorics.* (3rd ed.) Prentice-Hall, 1999.

20. S. A. Burr, Generalized Ramsey theory for graphs — a survey. In *Graphs and Combinatorics*, Lecture Notes in Mathematics **405**, Springer-Verlag, 1974, 52–75.
21. S. A. Burr, Diagonal Ramsey numbers for small graphs. *J. Graph Theory* **7** (1983), 57–69.
22. L. Caccetta and S. Mardiyono, On maximal sets of one-factors. *Austral. J. Combin.* **1** (1990), 5–14.
23. L. Caccetta and S. Mardiyono, On the existence of almost-regular graphs without one-factors. *Austral. J. Combin.* 9 (1994), 243–260.
24. G. Chartrand and F. Harary, Graphs with prescribed connectivities. In *Theory of Graphs*, Akadémiai Kiadó, Budapest, 1968, 61–63.
25. A. G. Chetwynd and A. J. W. Hilton, Regular graphs of high degree are 1-factorizable. *Proc. London Math. Soc.* (3) **50** (1985), 193–206.
26. A. G. Chetwynd and A. J. W. Hilton, 1-factorizing regular graphs of high degree – an improved bound. *Discrete Math.* **75** (1989), 103–112.
27. F. R. K. Chung, On triangular and cyclic Ramsey numbers with k colors. In *Graphs and Combinatorics*, Lecture Notes in Mathematics **405**, Springer-Verlag, 1974, 236–242.
28. V. Chvátal, Tree-complete graph Ramsey numbers. *J. Graph Theory* **1** (1977), 93.
29. V. Chvátal and F. Harary, Generalized Ramsey theory for graphs. *Bull. Amer. Math. Soc.* **78** (1972), 423–426.
30. S. A. Cook, The complexity of theorem-proving procedures. *Proceedings of the Third Annual ACM Symposium on the Theory of Computing*, ACM, 1971, 151–158.
31. E. Cousins and W. D. Wallis, Maximal sets of one-factors. In *Combinatorial Mathematics III*, Lecture Notes in Mathematics **452**, Springer-Verlag, 1975, 90–94.
32. D. Craft and E. H. Tesar, On a question by Erdös about edge-magic graphs. *Discrete Math.* **207** (1999), 271–276.
33. A. B. Cruse, A note on one-factors in certain regular multigraphs. *Discrete Math.* **18** (1977), 213–216.
34. E. W. Dijkstra, A note on two problems in connexion with graphs. *Numerische Math.* **1** (1959), 269–271.
35. G. A. Dirac, Some theorems on abstract graphs. *Proc. London Math. Soc.* (3) **2** (1952), 69–81.
36. G. A. Dirac, In abstracten Graphen verhandene vollständige 4-Graphen und ihre Unterteilungen. *Math. Nachrichten* **22** (1960), 61–85.
37. R. J. Douglas, Tournaments that admit exactly one Hamiltonian circuit. *Proc. London Math. Soc.* (3) **3** (1970), 716–730.
38. H. Enomoto, A. S. Llado, T. Nakamigawa and G. Ringel, Super edge-magic graphs. *SUT J. Math.* **2** (1998), 105–109.
39. P. Erdös and A. Rényi, On random graphs. *Publ. Math. Debrecen* **6** (1959), 290–297.
40. P. Erdös and P. Turán, On a problem of Sidon in additive number theory, and on some related problems. *J. London Math. Soc.* **16** (1941), 212–215; Addendum, **19** (1944), 242.
41. L. Euler, Solutio Problematis ad geometriam situs pertinentis. *Comm. Acad. Sci. Imp. Petropolitanae* **8** (1736), 128–140.
42. S. Fiorini and R. J. Wilson, *Edge-Colourings of Graphs*. Pitman, London, 1977.
43. J. Folkman, *Notes on the Ramsey Number* $N(3, 3, 3, 3)$, Manuscript, Rand Corporation, 1967.
44. J. Folkman and J. R. Fulkerson, Edge colorings in bipartite graphs. In *Combinatorial Mathematics and its Applications*, University of North Carolina Press, 1969, 561–577.
45. L. R. Ford and D. R. Fulkerson, Maximal flow through a network. *Canad. J. Math.* **8** (1956), 399–404.
46. L. R. Ford and D. R. Fulkerson, A simple algorithm for finding maximal network flows and an application to the Hitchcock problem. *Canad. J. Math.* **9** (1957), 210–218.

47. J.-C. Fournier, Colorations des arêtes d'un graphe. *Cahiers du CERO* **15** (1973), 311–314.
48. J. E. Freund, Round robin mathematics. *Amer. Math. Monthly* **63** (1956), 112–114.
49. D. R. Fulkerson, *Studies in Graph Theory Part I*, M.A.A., 1975.
50. D. R. Fulkerson, *Studies in Graph Theory Part 2*, M.A.A., 1975.
51. T. Gallai, Elementare relationen bezüglich der glieder und trennendenpunkte von graphen. *Mag. Tud. Akad. Mat. Kutató Int. Közlmenyei* **9** (1964), 235–236.
52. M. R. Garey and D. S. Johnson, *Computers and Intractability*, Freeman, New York, 1979.
53. A Gibbons, *Algorithmic Graph Theory*, Cambridge University Press, 1985.
54. I. J. Good, Normal recurring decimals. *J. London Math. Soc.* **21** (1946), 167–172.
55. R. D. Godbold and P. J. Slater, All cycles are edge-magic. *Bull. Inst. Combin. Appl.* **22** (1998), 93–97.
56. A. W. Goodman, On sets of acquaintances and strangers at any party. *Amer. Math. Monthly* **66** (1959), 778–783.
57. J. Grossman, `http://www.oakland.edu/enp`.
58. R. P. Gupta, The chromatic index and the degree of a graph. *Notices Amer. Math. Soc.* **13** (1966), 719 (Abstract 66T-429).
59. F. Guthrie, Note on the colouring of maps. *Proc. Roy. Soc. Edinburgh* **10** (1880), 727–728.
60. R. K. Guy, The decline and fall of Zarankiewicz's Theorem. In *Proof Techniques in Graph Theory*, Academic Press, 1969, 63–69.
61. S. L. Hakimi, On the realizability of a set of integers as degrees of the vertices of a graph. *SIAM J. Appl. Math.* **10** (1962), 496–506.
62. W.R. Hamilton, *The Icosian Game* (leaflet, Jaques and Son, 1859). Reprinted in [14], 33–35.
63. F. Harary, An elementary theorem on graphs. *Amer. Math. Monthly* **66** (1959), 405–407.
64. F. Harary, A characterization of block-graphs. *Canad. Math. Bull.* **6** (1963), 1–6.
65. F. Harary, Recent results on generalized Ramsey theory for graphs. In *Graph Theory and Applications*, Lecture Notes in Mathematics **303**, Springer-Verlag, 1972, 125–138.
66. F. Harary, *Graph Theory*, Addison-Wesley, 1972.
67. G. H. Hardy and E. M. Wright, *The Theory of Numbers*, Oxford University Press, New York, 1938.
68. V. Havel, A remark on the existence of finite graphs (in Czech). *Česk. Akad. Věd. Časop. Pěst. Mat.* **80** (1955), 477–480.
69. T. W. Haynes, S. T. Hedetniemi and P. J. Slater, *Fundamentals of Domination in Graphs*, Marcel Dekker, New York, 1998.
70. P. J. Heawood, Map-colour theorem. *Quart. J. Pure Appl. Math.* **24** (1890), 332–338.
71. F. K. Hwang, How to design round robin schedules. In *Combinatorics, Computing and Complexity (Tianjing and Beijing, 1988)*, Kluwer, Dordrecht, 1989, 142–160.
72. M. Huber, Estimating the average shortest path length in a graph. *Technical Report, Cornell University* (1996).
73. C. Jordan, Sur les assemblages de lignes. *J. reine ang. Math.* **70** (1869), 185–190.
74. F. Karithny, Láncszemek. In *Minden másképpen van*, Atheneum Press, Budapest, 1929.
75. A. B. Kempe, On the geographical problem of the four colours. *Amer. J. Math.* **2** (1879), 193–200.
76. T. P. Kirkman, On the representation of polyedra. *Phil. Trans. Roy. Soc. London* **146** (1856), 413–418.
77. D. J. Kleitman, The crossing number of $K_{5,n}$. *J. Combinatorial Theory* **9** (1970), 315–323.
78. A. Kotzig and A. Rosa, Magic valuations of finite graphs. *Canad. Math. Bull.* **13** (1970), 451–461.
79. A. Kotzig and A. Rosa, Magic valuations of complete graphs. *Publ. CRM*–**175** (1972).

80. A. Kotzíg and A. Rosa, Nearly Kirkman systems. *Congressus Num.* **10** (1974), 371–393.
81. J. B. Kruskal Jnr., On the shortest spanning subtree and the traveling salesman problem. *Proc. Amer. Math. Soc.* **7** (1956), 48–50.
82. K. Kuratowski, Sur le problème des courbes gauches en topologie. *Fund. Math.* **15** (1930), 271–283.
83. L. Lovász, Three short proofs in graph theory. *J. Combinatorial Theory* **19B** (1975), 269–271.
84. L. Lovász and M. D. Plummer, *Matching Theory*. North-Holland, Amsterdam, 1986.
85. J. MacDougall, M. Miller, Slamin and W. D. Wallis, Vertex-magic total labelings. *Utilitas Math.* **61** (2002), 3–21.
86. S. Mardiyono, *Factors in Regular and Almost-Regular Graphs*. Ph.D. Thesis, Curtin University of Technology, Australia, 1995.
87. S. Milgram, The small world problem. *Psychology Today* **2** (1967), 60–67.
88. J. W. Moon, *Topics on Tournaments*, Holt, Rinehart and Winston, New York, 1968.
89. M. E. J. Newman, The structure and function of complex networks. *SIAM Review* **45**(2003), 167–256.
90. O. Ore, Note on Hamilton circuits. *Amer. Math. Monthly* **67** (1960), 55.
91. E. T. Parker, Edge-coloring numbers of some regular graphs. *Proc. Amer. Math. Soc.* **37** (1973), 423–424.
92. T. D. Parsons, Ramsey graph theory. In *Selected Topics in Graph Theory*, Academic Press, 1977, 361–384.
93. M. D. Plummer, On minimal blocks. *Trans. Amer. Math. Soc.* 134 (1968), 85–94.
94. J. Petersen, Die Theorie der regulären Graphs. *Acta Math.* **15** (1891), 193–220.
95. J. Pila, Connected regular graphs without one-factors. *Ars Combinatoria* **18** (1983), 161–172.
96. R. C. Prim, Shortest connection networks and some generalizations. *Bell Syst. Tech. J.* **36** (1957), 1389–1401.
97. L. Pósa, A theorem concerning Hamilton lines. *Mag. Tud. Akad. Mat. Kutató Int. Közleményei* **7** (1962), 225–226.
98. R. Read, An introduction to chromatic polynomials. *J. Combinatorial Theory* **4** (1968), 52–71.
99. L. Rédei, Ein Kombinatorischer Satz. *Acta Litt. Sci. Univ. Hung. Francisco-Josephinae, Sect. Sci. Math.* **7** (1934), 39–43.
100. G. Ringel, Problem 25. *Theory of graphs and its applications* (Smolenice, 1963), 162, Publ. House Czechoslovak Acad. Sci., Prague, 1964.
101. G. Ringel and A. S. Llado, Another tree conjecture. *Bull. Inst. Combin. Appl.* **18** (1996), 83–85.
102. A. Rosa, On certain valuations of the vertices of a graph. In *Theory of Graphs, International Symposium, Rome, July 1966).* Gordon & Breach, 1967, 349–355.
103. A. Rosa and W. D. Wallis, Premature sets of 1-factors, or, how not to schedule round-robin tournaments. *Discrete Appl. Math.* **4** (1982), 291–297.
104. B. Roy, Nombre chromatique et plus longs chemins d'une graphe. *Revue Franç. d'Inf. Rech. Op.* **1** (1967), 127–132.
105. K. G. Russell, Balancing carry-over effects in round robin tournaments. *Biometrika* **67** (198), 127–131.
106. T. L. Saaty and P. L. Kainen, *The Four Colour Problem* (2nd Ed). Dover, 1986.
107. S. S. Sane and W. D. Wallis, Monochromatic triangles in three colours. *J. Austral. Math. Soc.* **37B** (1988), 197–212.
108. T. Schönberger, Ein Beweis des Petersenschen Graphensatzes. *Acta Univ. Szeged. Acta Sci. Math.* **7** (1934), 51–57.

109. A. J. Schwenk, Acquaintance graph party problem. *Amer. Math. Monthly* **79** (1972), 1113–1117.
110. J. Sedláček, Problem 27. *Theory of graphs and its applications* (Smolenice, 1963), 163–164, Publ. House Czechoslovak Acad. Sci., Prague (1964).
111. B. M. Stewart, Magic Graphs, *Canad. J. Math.* **18** (1966), 1031–1059.
112. D. P. Sumner, Graphs with 1-factors. *Proc. Amer. Math. Soc.* **42** (1974), 8–12.
113. D. P. Sumner, On Tutte's factorization theorem. In *Graphs and Combinatorics*, Lecture Notes in Mathematics **405**, Springer-Verlag, 1974, 350–355.
114. G. Szekeres and H. S. Wilf, An inequality for the chromatic number of a graph. *J. Combinatorial Theory* **4** (1968), 1–3.
115. W. T. Tutte, The factorizations of linear graphs. *J. London Math. Soc.* **22** (1947), 459–474.
116. W. T. Tutte, A theory of 3-connected graphs. *Indag. Math. 23* (1961), 441–455.
117. W. T. Tutte, *Connectivity in graphs*, University of Toronto Mathematical Expositions **15**, Oxford University Press, London, 1966.
118. L. Valdés, Edge-magic K_p. *Congressus Num.* **153** (2001), 107–111.
119. V. G. Vizing, On an estimate of the chromatic class of a p-graph [Russian]. *Discret. Anal.* **3** (1964), 25–30.
120. W. D. Wallis, The number of monochromatic triangles in edge-colourings of a complete graph. *J. Comb. Inf. Syst. Sci.* **1** (1976), 17–20.
121. W. D. Wallis, The smallest regular graphs without one-factors. *Ars Combinatoria* **11** (1981), 295–300.
122. W. D. Wallis, One-factorizations of graphs: Tournament applications. *College Math. J.* **18** (1987), 116–123.
123. W. D. Wallis, *Combinatorial Designs*, Marcel Dekker, New York, 1988.
124. W. D. Wallis, One-factorizations of complete graphs. In *Contemporary Design Theory*, Wiley, New York, 1992, 593–631.
125. W. D. Wallis, *One-Factorizations*. Kluwer Academic Publishers, Dordrecht, 1997.
126. W. D. Wallis *Magic Graphs*. Birkhäuser, Boston, 2001.
127. D. J. Watts, *Small Worlds*. Princeton, 1999.
128. D. J. Watts, *Six Degrees*. Norton, New York, 2004.
129. D. J. Watts and S. H. Strogatz, Collective dynamics of 'small-world' networks. *Nature* **393**(1998), 440–442.
130. D. de Werra, Scheduling in sports. In *Studies on Graphs and Discrete Programming*, North-Holland, Amsterdam, 1981, 381–395.
131. D. B. West, *Introduction to Graph Theory*, Prentice-Hall, Upper Saddle River, 1996.
132. E. G. Whitehead Jr., Algebraic structure of chromatic graphs associated with the Ramsey number $N(3, 3, 3; 2)$. *Discrete Math.* **1** (1971), 113–114.
133. H. Whitney, Congruent graphs and the connectivity of graphs. *Amer. J. Math.* **54** (1932), 150–168.
134. Wikipedia, `http://www.en.wikipedia.org/wiki/Six_Degrees_of_Kevin_Bacon`.
135. H. P. Yap, *Some Topics in Graph Theory*, Cambridge University Press, Cambridge, 1986.

Hints

2 Show that any two inadjacent vertices have a common neighbor.

6 The graphs need not both be connected.

8 Use the results of Exercise 1.2.3.

8 (iii) By definition, for any $x \in V(G)$ there is a y such that $D(x, y) = R(G)$. Write a and b for the vertices of K_2. Then $D((a, x), (b, y)) = R(G) + 1$. (Note: you must also show that there is no path from $(a.x)$ to (b, y) that is shorter than $R(G) + 1$.)

9 Say G has diameter d. Choose a and t of distance d. Say $a, \ldots, s, t$ is a path of length d. Show that if $G - \{s, t\}$ is disconnected, then $G - \{x, y\}$ is connected for some other pair of vertices.

5 It is easy to express the degree $d(a, b)$ as a function of $d(a)$ and $d(b)$.

8 Consider a complete bipartite graph with its two parts equal or nearly equal.

10 (i) From Theorem 2.7, it suffices to show that there do not exist nonadjacent vertices x and y with $d(x) + d(y) < v$.

5 Use Exercise 1.2.5.

5 Suppose G contains r cutpoints. Use induction on r. Define an *endblock* in G to be a block containing exactly one cutpoint y. Observe that G contains an endblock; consider endblocks.

7 Write c for the vertex of degree 4 and x, y, z, t for the four of degree 1. There are unique paths cx, cy, cz, ct. Show that every vertex other than c lies on exactly one of those paths. If any of those vertices has degree > 2, prove there is another vertex of degree 1.

8 consider a vertex of degree 1, and its unique neighbour. Now work by induction on v.

11 (ii) Use induction on the number of vertices. Given a tree T, look at the tree derived by deleting vertices of degree 1 from T.

4.2.10 Find a one-to-one map from the set of pairs (T_1, T_2), where T_i is a spanning tree of G_i, onto the set of spanning trees of G.

4.3.5 Think about trees in which the weights are the negative of those given.

6.1.8 Select a vertex x of degree 1. If yz is any edge of the tree, define the *distance* from x to yz to be the smaller of $D(x, y)$ and $D(x, z)$. Prove that every one-factor of T must contain preciselt the edges of even distance from x. If these form a one-factor, T has one; otherwise it has none. (There are other proofs.)

6.1.10 Proceed by induction on the number of vertices. Use Exercise 2.1.9.

6.1.13 (ii) Finding a two-factorization consisting of two factors of type [3, 3] is equivalent to factoring K_6 into a one-factor two factors each consisting of two disjoint triangles. But if you delete two disjoint triangles from K_6, the result is $K_{3,3}$, which is triangle-free.

6.3.3 Assuming G has 1 or 2 bridges, it is useful to notice that the proof of Theorem 6.11 works just as well if there were 2 edges joining the vertices x and y instead of just one. Proceed by induction on the number of vertices of G.

6.4.4 Generalize Exercise 6.4.3.

7.1.10 One half of the statement is just the definition of a proper coloring.

7.3.5 Use Theorem 7.8.

7.4.6 Suppose G is a graph with km edges, $k \geq \chi'(G)$. Write $\mathcal{C}$ for the set of all edge-colorings of G in k colors. If $\pi \in \mathcal{C}$, define $n(\pi) = \sum |e_i - m|$, where e_i is the number of edges receiving color c_i under,π, and the sum is over all colors. Then define $n_0 = \min\{n(\pi) : \pi \in \mathcal{C}\}$. Assume $n_0 > 0$ and derive a contradiction. Then a coloring achieving n_0 has the required property.

7.5.1 Verify this exhaustively. But: (i) to prove that whenever an edge is deleted the result can be 3-edge-colored, notice that there are only 3 different sorts of edge (chord, outside edge with both endpoints degree 3, outside edge with one of degree 2) (in fact, the first two are equivalent, but proving this is just as hard as checking one more case); to prove the graph requires 4 colors, notice that there are only three different ways to 3-color the top 5-cycle, and none can be completed.

8.1.1 Suppose a graph consists of two planar subgraphs with a common vertex x which is a cutpoint. It is clea that this graph is planar (draw both subgraphs; make x the leftmost point in one and the rightmost point on the other). Use induction on the number of blocks.

8.2.2 Use Theorems 8.6 and 8.7.

9.1.4 Take the answer to Exercise 9.1.3; try to add one edge and delete another.

9.3.4 (i) Suppose λ is an edge-magic total labeling of K_v with magic sum k. Let V_e denote the set of all vertices x such that $\lambda(x)$ is even, and V_o the set of vertices x with $\lambda(x)$ odd; define p to be the number of elements of V_e. Write E_1 for the set

of edges with both endpoints in the same set, either V_o or V_e, and E_2 for the set of edges joining the two vertex-sets, so that $|E_1| = \binom{p}{2} + \binom{v-p}{2}$ and $|E_2| = p(v-p)$. If k is even, then $\lambda(yz)$ is even whenever yz is an edge in E_1 and odd when yz is in E_2, so there are precisely $p + \binom{p}{2} + \binom{v-p}{2}$ even labels. But these labels must be the even integers from 1 to $\binom{v+1}{2}$, taken once each. A corresponding deduction can be made if k is odd.
Use the fact that $\binom{v+1}{2}$ is even when $v \equiv 0$ or $3 (\mathrm{mod}\ 4)$ and odd otherwise.
(ii) p is an integer, so the functions whose roots are taken must always be perfect squares.

.9 Form a graph whose vertices are the rows $M_1, M_2, \ldots, M_s$ of M If $i < j$, then allocate a color to the edge M_iM_j corresponding to the ordered pair (m_{ij}, m_{ji}).

.3 Follow the proof of Theorem 2.1.

.4 The maximum flow cannot exceed 14 because of the cut $[s, abcdt]$.

.7 Use a function f that oscillates finitely.

Answers and Solutions

Exercises 1.1

.2 (i) S; (ii) RS; (iii) A; (iv) RS.

.6 (i) S; (ii) R; (iii) S; (iv) AS.

Exercises 1.2

.1 $$G : I = \begin{bmatrix} 1 & 0 \\ 1 & 1 \\ 0 & 1 \end{bmatrix}; \; A = \begin{bmatrix} 0 & 1 & 0 \\ 1 & 0 & 1 \\ 0 & 1 & 0 \end{bmatrix}.$$

$$H : I = \begin{bmatrix} 1 & 0 & 0 & 0 \\ 1 & 1 & 1 & 0 \\ 0 & 1 & 0 & 1 \\ 0 & 0 & 1 & 1 \end{bmatrix}; \; A = \begin{bmatrix} 0 & 1 & 0 & 0 \\ 1 & 0 & 1 & 1 \\ 0 & 1 & 0 & 1 \\ 0 & 1 & 1 & 0 \end{bmatrix}.$$

.5 Say V_1, V_2 are non-empty disjoint sets of vertices of G such that there is no edge joining any vertex of V_1 to any vertex of V_2. Concider vertices x, y of G. If one is in V_1 and the other is in V_2, then they are adjacent in $\overline{G}$. If both are in the same set, say V_1, then select any vertex z in V_2; xz and zy are edges in $\overline{G}$.

.9 Say the two parts of G contain p and q vertices respectively. Then G has at most pq edges (it will have fewer unless G is complete bipartite). Moreover, $p+q = v$. Say $p = \frac{v}{2} + \pi, q = \frac{v}{2} - \pi$. Then $pq = (\frac{v}{2} + \pi)(\frac{v}{2} - \pi) = \frac{v^2}{4} - \pi^2 \leq \frac{v^2}{4}$.

.12 $$2n; \; A = \begin{bmatrix} 0 & 1 & 1 & 1 & 1 & 1 \\ 1 & 0 & 1 & 0 & 0 & 1 \\ 1 & 1 & 0 & 1 & 0 & 0 \\ 1 & 0 & 1 & 0 & 1 & 0 \\ 1 & 0 & 0 & 1 & 0 & 1 \\ 1 & 1 & 0 & 0 & 1 & 0 \end{bmatrix}$$

$$I = \begin{bmatrix} 1 & 1 & 1 & 1 & 1 & 0 & 0 & 0 & 0 & 0 \\ 1 & 0 & 0 & 0 & 0 & 1 & 0 & 0 & 0 & 1 \\ 0 & 1 & 0 & 0 & 0 & 1 & 1 & 0 & 0 & 0 \\ 0 & 0 & 1 & 0 & 0 & 0 & 1 & 1 & 0 & 0 \\ 0 & 0 & 0 & 1 & 0 & 0 & 0 & 1 & 1 & 0 \\ 0 & 0 & 0 & 0 & 1 & 0 & 0 & 0 & 1 & 1 \end{bmatrix}.$$

Exercises 1.3

1.3.1 (i) 222242222 (ii) 332233 (iii)1232222 (iv) 23122231

1.3.2 Suppose x and y are not adjacent. Then each of them is adjacent to at least $\frac{v-1}{2}$ of the remaining $v-2$ vertices. So they have a common neighbor, say z, and xzy is a walk in G.

1.3.4 $\{3, 2, 2, 2, 1\}$ is valid iff $\{1, 1, 1, 1\}$ is valid. The latter corresponds to two disjoint edges, so it is valid. Two examples are

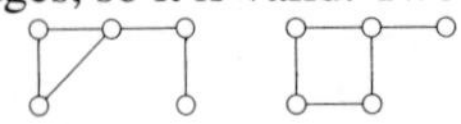

1.3.10 (i) no (you get $\{2, 2, 0, 0\}$); (ii) yes; (iii) no (sum is odd); (iv) yes.

1.3.12 Say x_1 is a vertex of degree 1, x_2 has degree 2, ..., x_5 has degree 5 (and one other vertex, x_0 say, has as its degree one of the numbers already used). Then x_5 must be adjacent to all of x_1, x_2, x_3, x_4, x_0. We now know the vertex adjacent to x_1; it isn't x_4. To have degree 4, x_4 must be adjacent to x_2, x_3, x_5, x_0. We now know x_2's neighbors (x_4, x_5). So x_3 is joined to x_4, x_5, x_0. This defines a graph with degrees 5, 4, 3, 3, 2, 1.

1.3.15 Consider a K_3 with vertices a, b, c. Duplicate edge ac and add edge cd.

1.3.17 Consider a graph with vertices $x_1, x_2, \ldots, x_v$, where the subscripts are integers modulo v. If d is even, say $d = 2n$, the edges are

$$x_i x_j : 1 \le i \le v, i+1 \le j \le i+n.$$

This will yield the required graph provided $2n < v$. If d is odd, $d = 2n+1$, then v must be even. Use the same construction and add an edge $x_i x_{\frac{v}{2}+i}$ for each i.

1.3.18 (i) yes; yes. (ii) Here is one construction. Assign a vertex to each member of the sequence. Arbitrarily pair up the vertices corresponding to odd integers and join the pairs. Then, to a vertex corresponding to integer d, assign $\lfloor d/2 \rfloor$ loops.

Exercises 2.1

2.1.1 $sacbt, sacbdt, sact, sbct, sbdt, sbt, scbt, scbdt, sct$; 2.

2.1.2 $sa:1, sb:1, sc:1, sd:2, st:2, ab:2, ac:1, ad:3, at:2, bc:1, bd:1, bt:1, cd:2, ct:1, dt:1$.

2.1.8 (i) Suppose i is the smallest integer such that $u_i = v_j$ for some j. Then $x, u_1, \ldots, u_i, v_{j-1}, \ldots, v_1, x$ is a cycle unless $i = j = 1$. (ii) Consider $u_1 = v_1 = y$.

Exercises 2.2

.2.2 (i) $g = 3, c = 8, D = 3, R = 2$ (ii) $g = 5, c = 8, D = 4, R = 3$ (iii) $g = 4, c = 8, D = 4, R = 3$ (iv) $g = 4, c = 5, D = 5, R = 3$ (v) $g = 6, c = 6, D = 5, R = 3$

.2.4 (i) 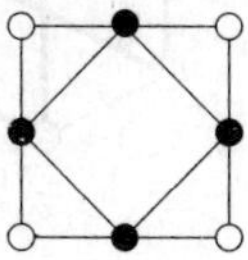 (ii) 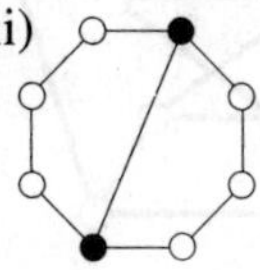 (iii) 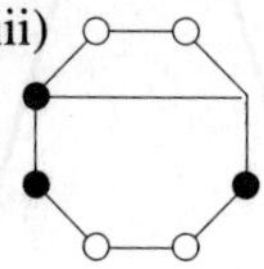

(iv) 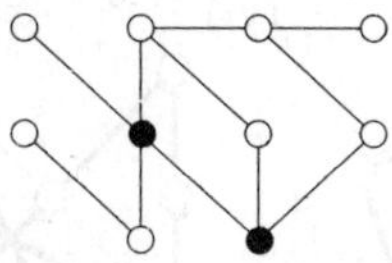 (v) 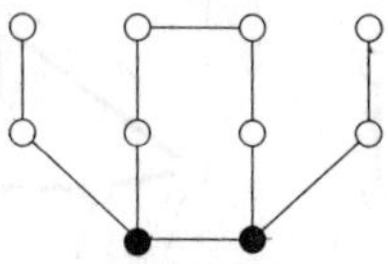

.2.9 The result is clearly true if G is complete. Suppose not. If G has diameter D ($D > 1$), choose two vertices a and t whose distance is D. Let $a, \ldots, r, s, t$ be a path of length D from a to t in G. $a = r$ is possible. $r \not\sim t$.
Suppose $G^* = G - \{s, t\}$ is not connected; say A is the component of G^* that contains a. Every vertex in $G^* - A$ must be adjacent in G to s. (If not, suppose z were a vertex in $G^* - A$ whose distance from s is at least 2. Then the shortest path from a to z must be of length at least $D + 1$, which is impossible.) If $G^* - A$ contains any edge, say bc, then s is still connected to every vertex of $G^* - A - \{b, c\}$; moreover s is adjacent to t and connected to every vertex in A (since A is connected and s is connected to a). So $G - \{b, c\}$ is connected and we could take $\{x, y\} = \{b, c\}$.
We need only consider the case where $G^* - A$ consists of isolated vertices, all adjacent to s. If A has two elements, they together with r and s form an induced $K_{1,3}$. So $|A| = 1$. Say $A = \{w\}$. If $w \sim t$ take $\{x, y\} = \{w, t\}$, and if $b \not\sim t$ then r, s, t, w form an induced $K_{1,3}$.

Exercises 2.3

.3.2 (i) $seft$ (length 9); (ii) $sebt$ (length 11).

Exercises 2.4

.4.1 (i) No Euler walk, as there are 4 odd vertices; 2 edges are needed. (ii) There is a closed Euler walk. (iii) There are two odd vertices, so there is an Euler walk, but not a closed one. Two edges are needed.

Exercises 2.5

2.5.3

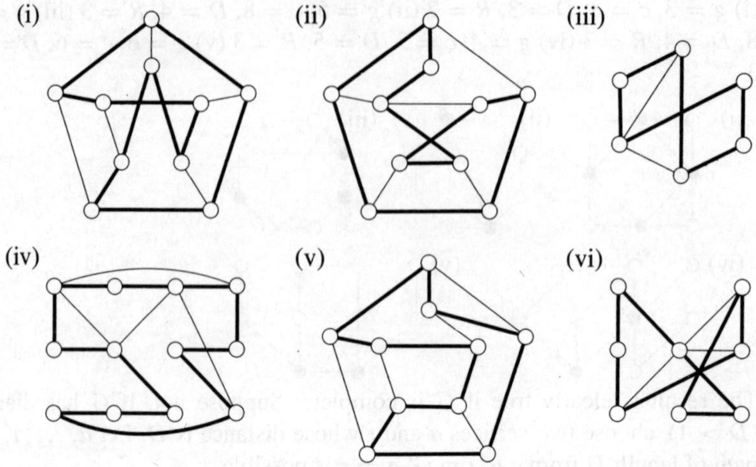

2.5.8 (i) From Theorem 2.7, it suffices to show that there do not exist nonadjacent vertices x and y with $d(x) + d(y) < v$. So, of the $2v - 3$ pairs xz and yz, with $z \in V(G)$, at most $v - 1$ are edges. So G has at most $\binom{v}{2} - (v - 2) = \frac{v^2-3v+4}{2}$ edges.
(ii) If G is formed from K_{v-1} by adding one vertex and one edge connecting it to one of the original vertices, then G has $\frac{v^2-3v+4}{2}$ edges and is not Hamiltonian.

2.5.10 (i) One solution is to seat the people in the following sequences, where the labels are treated as integers mod 11: (1) 1, 2, 3, ...; (2) 1, 3, 5, ...; (3) 1, 4, 7, ...; (4) 1, 5, 9, ...; (5) 1, 6, 11, ... In other words, on day i, the labels increase by i (mod 11). Over 5 days, x sits next to $x \pm 1$, $x \pm 2$, $x \pm 3$, $x \pm 4$ and $x \pm 5$, giving every possible neighbor once.

Exercises 2.6

2.6.1 $(v - 1)!/2$.

2.6.4 (i) SE: 115. NN: 118.
(ii) SE: 286. NN: 286.
(iii) SE: 117. NN: 117.
(iv) SE: 113. NN: 112.

2.6.5 (i) a 118, b 115, c 115, d 115, e 121.
(ii) a 286, b 286, c 286, d 319, e 286.
(iii) a 117, b 117, c 122, d 117, e 117.
(iv) a 112, b 113, c 112, d 113, e 116.

2.6.8 A directed graph model must be used. Replace each edge xy by two arcs xy and yx, with the cost of travel shown on each. In the nearest neighbor algorithm, one considers all arcs with *tail* x when choosing the continuation from vertex x.

Exercises 3.1

.1.1 (i)  (ii) none (iii) 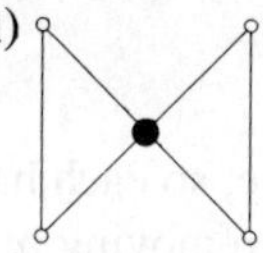

.1.4 If it did, deleting the bridge would yield components with exactly one odd vertex.

.1.6 (i)

$$[a, bcde] = \{ac, ad\} \qquad [abde, c] = \{ac, cd\}$$
$$[ac, bde] = \{ad, cd\} \qquad [acd, be] = \{de\}$$
$$[acde, b] = \{be\}$$

(ii)

$$[a, bcd] = \{ab\} \qquad [ab, cd] = \{bc, bd\}$$
$$[abc, d] = \{bd, cd\} \qquad [abd, c] = \{bc, cd\}$$

.1.9 Suppose G is a connected self-centered graph with a cutpoint x. Select a vertex y such that $D(x, y) = \varepsilon(x)$. Let P be a shortest xy-path. Then P lies completely within some component of $G - x$. Select z, a vertex in some other component of $G - x$. Clearly $\varepsilon(z) \geq D(z, y) > D(x, y) = \varepsilon(x)$, contradicting the centrality of x.

.1.10 (i) Yes; (ii) no; (iii) yes; (iv) no; (v) no; (vi) yes. (Cases (ii) and (iv) do not disconnect the graph; case (v) is not minimal.)

Exercises 3.2

2.1 (i) 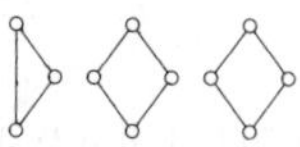 (ii) 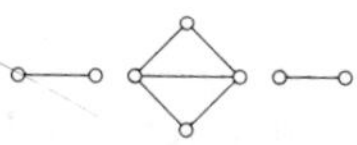

2.3 Suppose G is a connected graph with at least two edges.
(i) G is connected and is not K_2, so each edge is adjacent to some other edge. So "any two adjacent edges lie on a cycle" implies that each edge lies on a cycle. So each point lies on a cycle, and there are no cutpoints.
(ii) Suppose xy and yz are adjacent edges that do not lie on any common cycle. There can be no path from x to z that does not contain y (if there were, that path plus xy and yz would be a cycle containing the two edges). So y is a cutpoint.

2.4 Suppose G contains r cutpoints. We proceed by induction on r. The case $r = 0$ is trivially true; the equation becomes $-1 = -1$. Assume the result is true for graphs with r or fewer cutpoints, $r \geq 0$, and suppose G has $r + 1$ cutpoints. We define an *endblock* in G to be a block containing exactly one cutpoint y. Clearly G contains an endblock. Select an endblock E of G, and form a graph H by deleting from G all vertices and edges of E except for the unique cutpoint. Then $b(H) = b(G) - 1$. For each vertex x of H, $b_H(x) = b_G(x)$, except $b_H(y) = b_G(y) - 1$. The $|V(E)| - 1$ deleted vertices each belonged to 1 block of G. By induction, $b(H) - 1 = \sum_{x \in V(H)} [b_H(x) - 1] = \sum_{x \in V(H), x \neq y} [b_H(x) - 1] + b_H(y) - 1 = \sum_{x \in V(H), x \neq y} [b_G(x) - 1] + b_G(y)$. So $b(G) - 1 = \sum_{x \in V(H)} [b(x) - 1] =$

$\sum_{x \in V(G)}[b(x) - 1]$ (the vertices of G not in H all contribute 0 to the sum, because they were all in one block of G).

Exercises 3.3

3.3.1 Each graph contains a spanning cycle, so each has $\kappa \geq 2$. The third has $\delta = 2$, so by Theorem 3.5 $\kappa' = 2$. In the first, removing of any vertex leaves a Hamiltonian graph, so at least two more must be deleted to disconnect, and $\kappa = 3$, whence $\kappa' = 3$. For the second graph, the preceding argument shows that the only candidates for two vertices whose removal would disconnect it are the top two in the diagram, but they do not work, so $\kappa = \kappa' = 3$. The answers are (i) 3,3, (ii) 3,3, (iii) 2,2.

3.3.3 $\kappa, \kappa', \delta =$ (i) 1,1,1 (ii) 1,2,2 (iii) 1,1,2 :

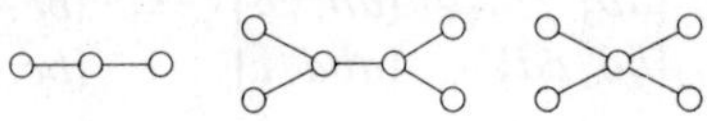

3.3.6 Suppose G has $\delta(G) \geq \frac{1}{2}v(G)$ but $\kappa'(G) < \delta(G)$. Select a set S of $\kappa'(G)$ edges whose removal disconnects G; say $G - S$ consists of disjoint parts with vertex-sets X and Y. Every vertex of X has degree at least δ, and there are fewer than δ edges of G with exactly one endpoint in X (only the members of S fit this description), so there is at least one vertex in X with all its neighbors in X. So $|X| > \delta$; similarly $|Y| > \delta$; so $v(G) = |X| + |Y| > 2 \cdot \delta$, a contradiction.
An example with $v = 6$, $\delta = 2$, $\kappa' = 1$ consists of two disjoint triangles plus an edge joining a vertex of one to a vertex of the other.

Exercises 4.1

4.1.3 Suppose G is a finite acyclic graph with v vertices. If G is connected it is a tree, so it has $v - 1$ edges by Theorem treesize. Now assume G has $v - 1$ edges. Suppose G consists of c components $G_1, G_2, \ldots, G_c$, where G_i has v_i vertices; $\sum v_i = v$. Each G_i is a tree, so it has $v_i - 1$ edges, and G has $\sum (v_i - 1) = v - c$. So $c = 1$ and G is connected.

4.1.5 If G contains edges xy and yz then G^2 contains triangle xyz. So, if G^2 is a tree then G consists of disjoint K_1's and K_2's $\Rightarrow G^2$ consists of disjoint K_1's and K_2's. The only trees like this are K_1 and K_2.

4.1.6 One example: vertices are integers, $x \sim x + 1$.

4.1.9 Suppose x has degree k. The longest path in T contains at most two edges incident with x, so there are $k-2$ edges known not to be on the path. At most $(v-1)-(k-2)$ edges are available.

Exercises 4.2

4.2.3 The "only if" is obvious. When $v \geq 4$ there are many constructions. One example: take the vertices as $1, 2, \ldots, v \pmod v$. One tree is the path $1, 2, \ldots, v$. If v is even, take as the second tree the path $1, 3, \ldots, v, 2, \ldots, v-1$. If v is odd, take the path $1, 3, \ldots, v-1, 2, 4, \ldots, v$. Another example: select four different vertices x, y, z, w. One tree consists of all edges from x to another vertex other than xy, plus zy. The other consists of all edges from y to another vertex other than yz, plus wz.

.2.4 Use

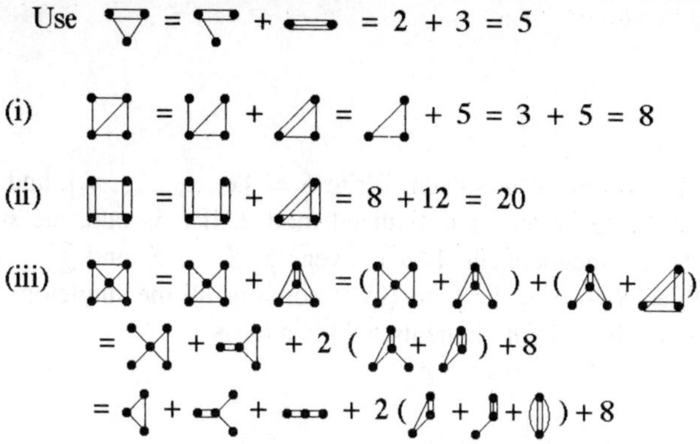

= 3+ 2 + 4 + 2 (5 + 3 + 4) + 8 = 41

.2.7 If G is 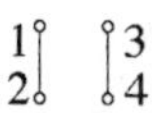

one solution is 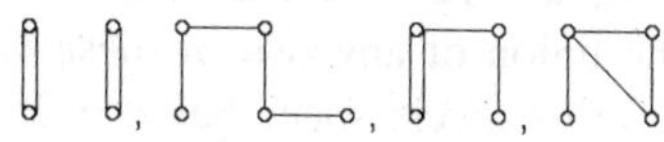

.2.9 16; 125.

.2.14 It is clearly necessary that H have no cycles, and if $H = G$ the result is immediate. So suppose H is acyclic and $H < G$. Say H has disjoint components $H_1, H_2, \ldots, H_n$. Since G is connected, there is in each H_i some vertex x_i that is adjacent to some vertex, y_i say, that is in G but not in H. Write $S = V(G)\backslash V(H)$, and select a spanning tree T in $\langle S\rangle$. Then

$$T \cup H \cup \{x_i y_i : 1 \le i \le n\}$$

is a spanning tree in G.

Exercises 4.3

.3.2 There are several solutions, but the minimum weight is (i) 55, (ii) 28.

.3.6 (i) 27, (ii) (a) 38 (b) 41 (c) 43.

Exercises 5.1

.1.1 (i)

+	0	1	2
0	0	1	2
1	1	0	2
2	2	0	1

×	0	1	2
0	0	0	0
1	0	1	2
2	0	2	1

.1.3 There is exactly one of dimension 0 (∅), one of dimension 4 (V), and none of dimension 5. For dimension 1, the subspaces are $0, x$ where $x \neq 0$, so there are $|V| - 1 = 15$ of them. For dimension 2, any ordered pair x, y of distinct nonzero elements determine the subspace $0, x, y, x+y$. Each of these ordered bases arises

6 times if all ordered pairs are listed, so there are $15 \cdot 14/6 = 35$. For dimension 3, there are $15 \cdot 14 \cdot 12$ ordered bases. Each subspace has 8 elements, so by (5.1) it has $7 \cdot 6 \cdot 4$ ordered bases. So the number of subspaces is $15 \cdot 14 \cdot 12/(7 \cdot 6 \cdot 4) = 15$. (Those who know a little more linear algebra will see from perpendicularity that the number of 3-dimensional subspaces must equal the number of 1-dimensional subspaces.)

Exercises 5.2

5.2.2 They form a basis if and only if n is even. Write $S = \{x_1, x_2, \ldots, x_n\}$, and $S_i = S \backslash \{x_i\}$. $\sum S_i = (n-1)S$, where n is reduced mod 2. If n is odd, the sum is zero, and the S_i are not independent. If n is even, $\sum S_i = S$, and $\sum_{i \neq j} S_i = S_j + S_j + \sum_{i \neq j} S_i = S_j + S = \{x_j\}$, so $\langle \{S_i\} \rangle$ contains all the singletons, so it contains all of S. Since there are n elements, $\{S_i\}$ is a basis.

Exercises 5.3

5.3.1 The cycles of K_4 are 123, 145, 256, 364, 1264, 1563, 2345. The union of any two of these is another of them. So the cycle space has $8 = 2^3$ elements (don't forget $\emptyset$), so it has dimension 3.

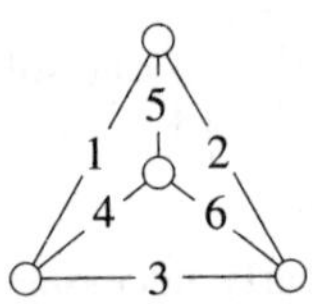

Exercises 5.4

5.4.1 The cycle of length 4 belongs to both. A necessary (not sufficient!) property is that the graph must have a cycle of even length.

5.4.2 (i) (a) Cycle subspace $\{\emptyset, 123, 456, 123456\}$, cutset subspace $\{\emptyset$, 13, 23, 12, 4, 134, 234, 124, 56, 1356, 2356, 1256, 456, 13456, 13456, 23456, 12456, 57, 1357, 2357, 1257, 457, 13457, 23457, 12457, 67, 1367, 2367, 1267, 467, 13467, 23467, 12467$\}$.

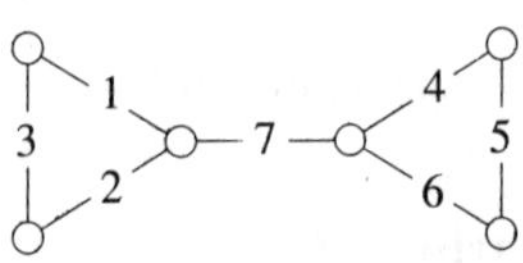

(b) Tree, so cycle subspace = $\emptyset$. Cutset subspace contains all 2^5 subsets of the edges.

(ii) (a) Cycle subspace has $4 = 2^2$ elements, dimension 2. Cutset subspace has $32 = 2^5$ elements, dimension 5. $2 + 5 = 7$. (b) Cycle subspace has $1 = 2^0$ elements, dimension 0. Cutset subspace has $32 = 2^5$ elements, dimension 5. $0 + 5 = 5$.

Exercises 5.5

.1 Choose i such that $2 \le i \le k$ and let L_i be the fundamental cycle corresponding to the edge a_i. Now a_1 is the only edge of T in C and a_i is the only edge of $\overline{T}$ in L_i. So $\{a_i\} \subseteq C \cap L_i \subseteq \{a_1, a_i\}$. By Lemma 5.4, $|L \cap C_i|$ is even, so $L \cap C_i = \{a_1, a_i\}$ whence $a_1 \in L_i$. Now let a_{k+j}, $j \ge 1$, be an edge of $\overline{T}$, and L_{k+j} the corresponding cycle. Since L_{k+j} contains no other edge of $\overline{T}$, $\emptyset \subseteq C \cap L_{k+j} \subseteq \{a_1\}$. Again by Lemma 5.4, $|L \cap C_{k+j}|$ is even, so $L \cap C_{k+j} = \emptyset$, so $a+1 \notin C_{k+j}$.

.4 (i) 2, 3, 4, 5, 6, 7, 8, 9, T.
(ii) (12345), (2347TA), (2379B), (1268C), (348TD), (12369E)
(iii) (15CE), (25$ABCE$), (35$ABDE$), (45AD), (6CE), (7AB), (8CD), (9BE), (TAD)
(iv) Cycles of length 8.

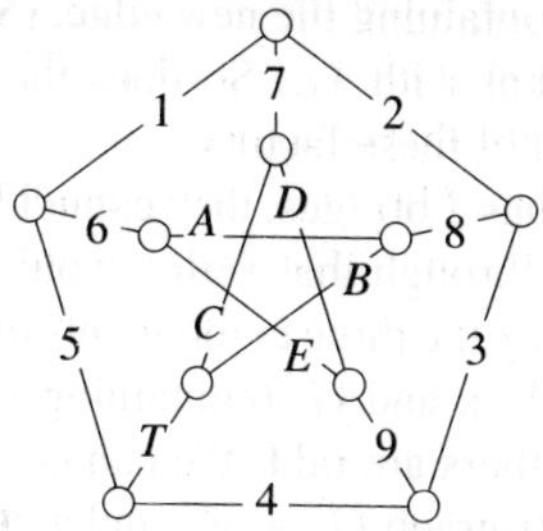

Exercises 6.1

.2 Suppose N has a one-factor. One edge from the center vertex must be chosen; say it is the vertical one. Then the remaining edges of the factor must form a one-factor in the following graph, which has two odd components:

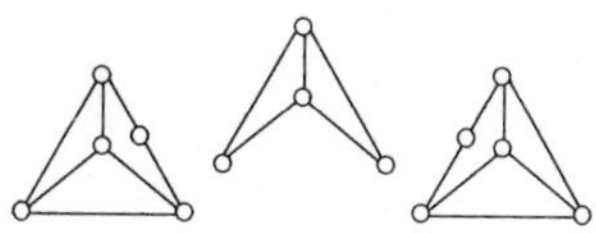

5 (i) $ab\ cd\ ef\ gh$; $ae\ bc\ dh\ fg$.
(ii) $ab\ cd\ ef\ gh$; $ab\ cg\ dh\ ef$; $ae\ bf\ cd\ fg$; $ae\ bf\ cg\ dh$

10 Suppose G has $2n$ vertices. We proceed by induction on n. The result is true for $n = 2$ (see Exercise 6.1.1). Say it is true for $n \le N$. Suppose $v(G) = 2N + 2$. By Exercise 2.1.9, G has an edge xy such that $G - \{x, y\}$ is connected. Now $G - \{x, y\}$ contains no induced $K_{1,3}$, and has $2N$ vertices. So by the induction hypothesis it has a one-factor. Append xy to that factor to construct a one-factor in G.

Exercises 6.2

1 Use a one-factorization of $K_{n,n}$. An example for $n = 4$ is
$1a\ 2b\ 3c\ 4d$, $1b\ 2a\ 3d\ 4c$, $1c\ 2d\ 3a\ 4b$, $1d\ 2c\ 3b\ 4a$.

2 (i) It is required to find the largest possible set of edge-disjoint factors of K_v, each of which consists of $v/3$ triangles.

Exercises 6.3

No. For example, consider $3K_3 \cup 3K_5$.

6.3.3 If G has no bridge, Theorem 6.11 gives the result. for the cases where G has 1 or 2 bridges, it is useful to notice that the proof of Theorem 6.11 works just as well if there were 2 edges joining the vertices x and y instead of just one. We proceed by induction on the number of vertices of G. The result is trivial for 4 vertices.
If G has 1 bridge, xy, write G_x and G_y for the components of $G - xy$, with $x \in G_x$. Say the vertices adjacent to x in $G - x$ are x_1 and x_2. The (multi)graph defined by adding edge x_1x_2 to $G_x - x$ is cubic has no bridge, so it has a 1-factor not containing the new edge. (Simply insist that it contains one of the other edges incident with x_1.) So does the graph similarly derived from G_y. Add xy to the union of these factors.
If G has 2 bridges, they cannot have a common endpoint (if they did, then the third edge through that vertex would also be a bridge.) Say the bridges are xy and zt, and say the three components of $G-xy-zt$ are G_x (containing x), G_y (containing y and z), and G_t (containing t). Then G_y has an even number of vertices, while the others are odd. We can construct a one-factor containing yz in the bridgeless (multi)graph $G_y + yz$, and a one-factor including the bridge xt in $G_x \cup G_t + xt$. Their union is the required factor.

Exercises 6.4

6.4.2 There is no example for $s = 1$. For $s = 2$, $K_3 \cup P_2$ is a $1 - (1, 1, 2)$ graph.

6.4.3 The degrees are clearly correct. But the new vertex is a cutpoint, so G is not Hamiltonian.

Exercises 7.1

7.1.2 3. ($\chi > 2$, because there is an odd cycle. 3 is realized in Figure 7.1.)

7.1.4 Write χ for $\chi(G)$, β for $\beta(G)$.
(i) Select a χ-coloring of G. Write V_i for the color classes. Each V_i is an endependent set, so $|V_i| \leq \beta$, so $v = \sum |V_i| \leq \chi \cdot \beta$.
(ii)Select a maximal independent set S; $|S| = \beta$. G can be colored in $\chi(G-S)+1$ colors (just color all points of S in a new color). $G - S$ has $v - \beta$ vertices, so obviously $\chi(G - S) \leq v - \beta$. so $\chi \leq \chi(G - S) + 1 \leq v - \beta + 1$.

7.1.7 $x_1, x_6, x_2, x_3, x_4, x_5$ works.

7.1.9 Select one edge in the cycle, say xy. By Theorem 7.1, $\chi(G - xy) = 2$. Select a 2-coloring of $G - xy$. Necessarily x and y will receive the same color, as they have a common neighbor. Apply a third color to x.

7.1.12 Color $G - x$ in n colors. There must be a color not on any vertex adjacent to x in G. Apply that color to x. (The "only if" part is obvious.)

Exercises 7.3

3.3 (i) Only one has a vertex of degree 2.
(ii) Neither graph has any coloring in 0, 1, 2 or 3 colors (each contains a K_4), so each has polynomial divisible by $x(x-1)(x-2)(x-3)$. For 4 colors there are 48 colorings: if colors 1, 2, 3, 4 are applied to the K_4 made up of the four left-most vertices, as shown, then the other colors are determined as shown. In the first graph, x can be 2 or 3, and in the second graph, (y, z) can be $(2, 3)$ or $(3, 2)$. This gives 2 colorings each, and $\{1, 2, 3, 4\}$ can be permuted in 24 ways. So each has a polynomial of the form $p(x) = x^6 - 11x^5 + \ldots = x(x-1)(x-2)(x-3)(x^2+ax+b)$ $= x^6 + (a-6)x^5 + \ldots$. Comparing coefficients of x^5, $a - 6 = -11$, $a = -5$. Then $p(4) = 48$ reduces to $(4^2 + 4 \cdot 5 + b) = 2$, or $b = 6$. So the polynomial is $x(x-1)(x-2)(x-3)(x^2 - 5x + 6) = x(x-1)(x-2)^2(x-3)^2$, the same for both graphs.

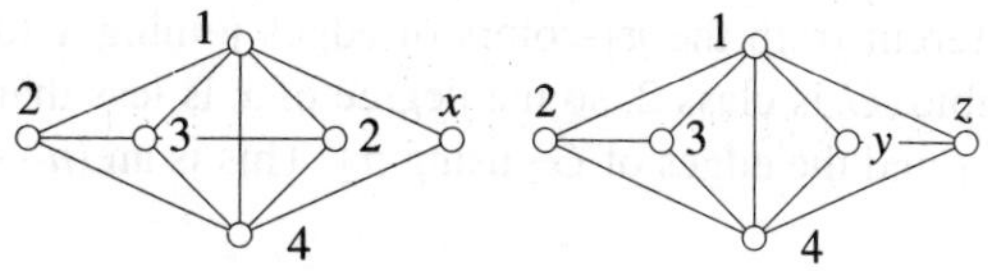

.5 From Theorem 7.8, such a graph would have 4 vertices, 4 edges and 2 components. There is no such graph.

Exercises 7.4

.2 Any 8-edge graph on 5 vertices has $\Delta = 4$ (sum of degrees = 16). There are two such graphs, the complements of $2K_2$ and P_3. For the former, take a one-factorization of K_6, delete the edges of one factor and then delete one vertex; the remaining (partial) factors are the color classes in a 4-edge-coloring. In the latter, consider the K_5 on vertices 1, 2, 3, 4, 5 with edges 15 and 25 deleted. Suitable color classes are $\{12, 34\}$, $\{13, 24\}$, $\{14, 35\}$, $\{23, 45\}$. So the graphs both have edge-chromatic number 4, and both are class 1.

.3 First, observe that any 7-edge graph on 5 vertices can be edge-colored in 4 colors, because it can be embedded in an 8-edge graph on 5 vertices (and use the preceding exercise). Now if a 7-edge graph can be edge-colored in 3 colors, one color would appear on 3 edges. But you can't have 3 disjoint edges on only 5 vertices.

.6 Suppose G is a graph with km edges, $k \geq \chi'(G)$. Write $\mathcal{C}$ for the set of all edge-colorings of G in k colors. If $\pi \in \mathcal{C}$, define $n(\pi) = \sum |e_i - m|$, where e_i is the number of edges receiving color c_i under,π, and the sum is over all colors. Then define $n_0 = \min\{n(\pi) : \pi \in \mathcal{C}\}$. We prove that $n_0 = 0$. Then a coloring achieving n_0 has the required property.
Suppose $n_0 > 0$. Let π_0 be a coloring with $n(\pi_0) = n_0 > 0$. Since G has km edges, there exist color classes M_1 and M_2 under π such that $e_1 = |M_1| < m$ and $e_2 = |M_2| > m$. Say the other color classes have sizes $c_3, c_4, \ldots, c_k$. Now $M_1 \cup M_2$ is a union of paths and cycles. $e_2 > e_1 \Rightarrow$ the union includes at least one

path P with its first and last edges from M_2. Exchange the colors of edges in P. The resulting edge-coloring π' has one more edge in color C_1 and one fewer in color c_2, so its color classes are of sizes $c_1-1, c_2-1, c_3, \ldots, c_k$, and $n(\pi') < n(\pi)$, a contradiction.

7.4.9 (i) By Exercise 6.1.4, $\chi'(P) > 3$, so by Theorem 7.13 $\chi'(P) = 4$. (ii) 4. (iii) 4.

Exercises 7.5

7.5.3 Suppose G has cutpoint x and is edge critical with edge-chromatic number n. Say $G - x$ consists of two subgraphs G_1 and G_2 with common vertex x. Select vertices y in G_1 and z in G_2 adjacent to x. Choose edge-colorings π_1 of $G - xy$ and π_2 of $G - xz$ in the $n - 1$ colors $c_1, c_2, \ldots, c_{n-1}$ (possible by criticality). Permute the names of the colors in π_2 so that the π_2-colors of edges joining x to vertices of G_2 are different from the π_1-colors of edges joining x to vertices of G_1 (this must be possible: G is class 2, so the degree of x is less than n). Color the edges of G_1 using π_1 and the edges of G_2 using π_2. This is an $(n - 1)$-edge-coloring — contradiction.

Exercises 8.1

8.1.3 First, convince yourself that the drawing shown of $K_{2,3}$ is quite general. Now $K_{3,3}$ can be constructed from $K_{2,3}$ by adding one vertex adjacent to all the black vertices. Whichever face it is placed inside, one crossing can be achieved and is unavoidable.

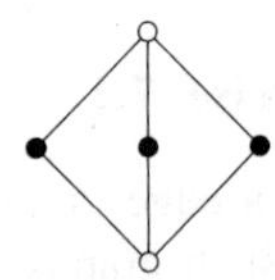

8.1.5 To see that P is not planar, delete the two "horizontal" edges from the representation in figure 2.4. When the vertices of degree 2 in this subgraph are elided, the result is $K_{3,3}$. The crossing number is 2 (this can be shown exhaustively, starting from a representation of $K_{3,3}$ with 1 crossing).

Exercises 8.2

8.2.4 From Theorem 1.1, $2e = \sum(\text{degrees}) \geq 6v$, so $e \geq 3v$. Suppose G is connected: by Theorem 8.7, G is not planar, and the result follows. If G is not connected, the above argument applies to its connected components (in fact, you could prove that every component contains a vertex of degree at most five).

Exercises 8.3

8.3.2 Suppose there are connected planar graphs that cannot be colored in six colors, and let G one with the minimum number of vertices. Let x be a vertex of G of degree less than 6. $G - x$ is 6-colorable; choose a 6-coloring ξ of $G - x$. There will be some color, say c, that is not represented among the vertices adjacent to x in G. Define $\eta(x) = c$, and $\eta(y) = \xi(y)$ if $y \in V(G - x)$. Then η is a 5-coloring of G — contradiction.

Exercises 9.1

.1.3 One solution is 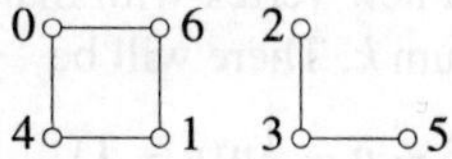

.1.6 There are not enough numbers available for the labeling to be one-to-one.

.1.8 Say the path has e edges.Call the vertices $x_0, x_1, \ldots, x_e$, where $x_i \sim x_{i+1}$. Set $\lambda(x_{2k}) = k, \lambda(x_{2k+1}) = e - k$. This is graceful: differences are $e, e-2, \ldots$ on edges of form x_{2k}, x_{2k+1} and $e-1, e-3, \ldots$ on edges of form x_{2k+1}, x_{2k+2}.

Exercises 9.2

.2.1 The vertex labels are $\{1, 2, 3\}, \{1, 3, 5\}, \{2, 4, 6\}, \{4, 5, 6\}$

.2.2 Write $v = 2n + 1$. Consider the cyclic vertex labeling $(1, 2n+1, 4n+1, 2n-1, \ldots, 2n+3)$; in this case each label is derived from the preceding one by adding $2n (\text{mod } 4n+2)$. The construction is such that the second, fourth, $\ldots$, $2n$-th vertices receive labels between 2 and $2n+1$ inclusive, while the third, fifth, $\ldots$, $(2n+1)$-th receive labels between $2n+2$ and $4n+1$. The successive pairs of vertices have sums $2n+2, 6n+2, 6n, 6n-2, \ldots, 2n+4$; if $k = 3v+1 = 6n+4$, the edge labels are $4n+2, 2, 4, \ldots, 4n$. We have an edge-magic total labeling with $k = 3v+1$ and $s = v^2$ (the case $i = \frac{1}{2}(v+1)$ in the list).

.2.5 Label the endpoints of the common edge with 1 and $n+2$, and the other vertices $2, 3, \ldots, n+1$.

.2.7 P is regular of degree 3 with $v = 10$, $e = 15$. So $2\sigma_0^{10} + \sigma_0^{25} \leq 15k \leq 2\sigma_{15}^{25} + \sigma_0^{25}$, or $435 \leq 15k \leq 735$. So $29 \leq k \leq 49$. To construct an example with $k = 29$, consider the representation as a star shape within a cycle; label the outside vertices 1, 3, 6, 9, 4 and the inside vertices 10, 5, 8, 7, 2 (starting from the top, clockwise, in both cases). Edge-labels can be found by subtraction.

Exercises 9.3

.3.1 From Theorem 9.20, $v(K_n + tK_1) + e(K_n + tK_1) \geq \rho^*(n)$. But $v(K_n + tK_1) = t + n$ and $e(K_n + tK_1) = \binom{n}{2}$.

.3.2 We use the Fibonacci numbers

$$f_2 = 1, f_3 = 2, \ldots, f_{n+1}.$$

Write $k = f_{n+3} + 1$ and define a labeling λ of the K_n with vertices $\{x_1, x_2, \ldots, x_n\}$ by

$$\lambda(x_i) = f_{i-1}$$

for $i = 1, 2, \ldots, n$ and

$$\lambda(x_i x_j) = k - \lambda(x_i) - \lambda(x_j).$$

The smallest edge label is the label on $x_{n-1}x_n$, namely $f_{n+3} + 1 - f_n - f_{n+1} = f_{n+1} + f_{n+2} + 1 - f_n - f_{n+1} = f_{n+1} + 1$, and this is greater than any vertex label.

The largest label used will be at most $k - f_2 - f_3 = k - 3$. So the labels are all positive, and no label is repeated. For each number between 1 and $k - 3$ that has not been used as a label, introduce a new vertex with that number as a label. The result is a magic graph with magic sum k. There will be

$$t = k - 3 - n - \tfrac{1}{2}n(n-1)$$

new vertices, so

$$m(G) \le f_{n+3} - 2 - n - \tfrac{1}{2}n(n-1).$$

Exercises 10.1

10.1.1 (i) Clearly $R(P_3, K_3) \le R(K_3, K_3) = 6$.
(ii) G contains no $P_3 \Leftrightarrow G$ contains no vertex of degree 2. So the components of G are disjoint vertices (degree 0) and edges (degree 1).
(iii) If G contains an isolated vertex and 4 or more components then it has 3 or more components, so $\overline{G}$ has a triangle.
(iii) suppose K_5 is colored so as to contain no red P_3 and no blue K_3. Let G be the subgraph of red edges. By (ii), (iii) $\overline{G}$ contains a K_3 unless $v \le 4$. So $R(P_3, K_3) \le 5$. But The K_4 with edges ab and cd red and the others blue is suitable. So $R(P_3, K_3) = 5$.

10.1.5 Say K_v contains no red or blue K_4. Select a vertex x. R_x (B_x) is the set of vertices joined to x by red (blue) edges. Then $\langle R - x\rangle$ can contain no red K_3 or blue K_4 and $|R_x| < R(3, 4) = 9$. Similarly $|B_x| < 9$. So $|V(x)| \le 1 + (9-1) + (9-1) = 17$, and $R(4, 4) \le 18$.

10.1.7 Suppose the edges of K_{m+n} are colored in red and blue. Any vertex x has degree $m + n - 1$, so if there are less than m red edges incident with x, there must be at least n blue edges. So $R(K_{1,m}, K_{1,n}) \le m + n$.
If m or n is odd, then there exists a regular graph G of degree $m - 1$ on $m + n - 1$ vertices (see Exercise 1.3.17. Its complement $\overline{G}$ is regular of degree $n - 1$. Color the edges of G red and those of $\overline{G}$ blue. This painting avoids any red $K_{1,m}$ and any blue $K_{1,n}$. So m or n odd $\rightarrow R(K_{1,m}, K_{1,n}) = m + n$. In any painting of K_{m+n-1} that avoids both red ($K_{1,m}$ and blue $K_{1,n}$, no vertex can have more than $m - 1$ red and $n - 1$ blue incident edges, so each vertex has exactly $m - 1$ red and $n - 1$ blue, so the red chromatic subgraph is regular of degree $m - 1$. This is impossible if m and n are both even (degree and order can't both be odd — Corollary 1.3). So m and n even $\rightarrow R(K_{1,m}, K_{1,n}) < m + n$. But a painting of K_{m+n-2} is easy to find — $n - 1$ is odd, so we can do it with no red $K_m - 1$ or blue K_{n-1}, let alone K_n. So m and n even $\rightarrow R(K_{1,m}, K_{1,n}) = m + n - 1$.

Exercises 10.2

10.2.3 (i) Suppose n is odd. Suppose the edges of K_{2n} are colored red and blue, and vertex x is incident with r red and b blue edges. If $r \ge n$, x will be the center of at least one red ($K_{1,n}$, and if $r < n$ then $b \ge n$, and x is the center of at least one blue ($K_{1,n}$.

So each vertex is the center of a monochromatic star, and $N_{2,2n}(K_{1,n}) \geq 2n - 1$. But if we select a regular graph of degree n on $2n$ vertices (possible by Exercise 1.3.17), and color all its edges red and insert blue edges between all inadjacent pairs, the result has exactly $2n - 1$ monochromatic (red) n-stars.
(ii) Suppose n is even. Take a K_n with vertices $x_1, x_2, \ldots, x_n$ and a K_{n_1} with vertices $y_1, y_2, \ldots, y_{n-1}$ disjoint from it. Color the following edges red: all the edges of the K_n except x_1x_2 x_3x_4, ..., $x_{n-1}x_n$, all the edges of the K_{n_1} and the edges x_1y_1 x_2y_2, ..., $x_{n-1}y_{n-1}$. The other edges of K_{2n-1}r are colored blue. Every vertex of this graph has red and blue degree $n - 1$ except for x_n, which has n red and $n - 2$ blue edges. So there is exaclty one monochromatic K_1, n, namely $x_n - x_{n-1}y_1y_2 \ldots y_{n-1}$.

Exercises 10.4

4.1 If a graph is to contain no red K_2, it has no red edges, so it is a blue K_v. There is no blue K_q iff $v < q$. So $R(2, q) = q$. Similarly $R(p, 2) = p$.

.4.3 Use Theorem 10.11 with $s = t = 3$. This gives

$$R_2(5) \geq (R_2(3) - 1)(R_2(3) - 1) + 1 = 26.$$

Exercises 11.1

.1.1 (a) (i) $sa, st, as, at, bs, bt, tb$.
(ii) $A(s) = \{a, t\}$, $B(s) = \{a, b\}$, $A(a) = \{s, t\}$, $B(a) = \{s\}$, $A(b) = \{s, t\}$, $B(b) = \{t\}$, $A(t) = \{b\}$, $B(t) = \{a, b, s\}$.
(iii) sat, st. (iv) $satb$. (v) $\{st, at, bt\}$.
(b) (i) $sb, as, bc, ca, ce, dc, et, td$.
(ii) $A(s) = \{b\}$, $B(s) = \{a\}$, $A(a) = \{s\}$, $B(a) = \{c\}$, $A(b) = \{c\}$, $B(b) = \{s\}$, $A(c) = \{a, e\}$, $B(c) = \{b, d\}$, $A(d) = \{c\}$, $B(d) = \{t\}$, $A(e) = \{t\}$, $B(e) = \{c\}$, $A(t) = \{d\}$, $B(t) = \{e\}$.
(iii) $sbcet$. (iv) $sbca$ (not unique). (v) $\{bc\}$.
(c) (i) $sa, sc, se, ab, ac, bd, ce, dc, dt, et$.
(ii) $A(s) = \{a, c, e\}$, $B(s) = \emptyset$, $A(a) = \{b, c\}$, $B(a) = \{s\}$, $A(b) = \{d\}$, $B(b) = \{a\}$, $A(c) = \{d, e\}$, $B(c) = \{s, a\}$, $A(d) = \{t\}$, $B(d) = \{b, c\}$, $A(e) = \{t\}$, $B(e) = \{s, c\}$, $A(t) = \emptyset$, $B(t) = \{d, e\}$.
(iii) $sabdt, sacdt, sacet, scdt, scet, set$. (iv) No cycles. (v) $\{sc, se, ac, bd\}$.

..1.5 (a) (i) DK_4, (ii) one component.
(b) (i) DK_7, (ii) one component.
(c) (i) See the graph; in addition to arcs shown, every vertex is joined to every lower vertex, (ii) each vertex a different component.

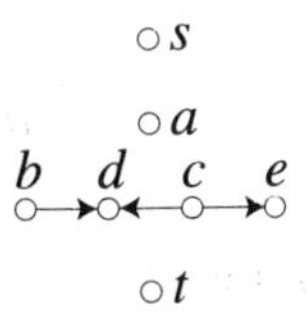

.1.8 No, it has loops.

Exercises 11.2

11.2.1 (i) Suppose the vertices are $x_1, x_2, \ldots, x_v$. Use the orientations $x_1 \to x_2, x_2 \to x_3, \ldots, x_{v-1} \to x_v, x_v \to x_1$. The other edges may be oriented in any way.

11.2.5 (i) 12223, (ii) 11233.

11.2.9 (i) (xcb), $(xcda)$, $(xcbda)$. (ii) (xdb), $(xcdb)$, $(xcdba)$.

11.2.12 (i) vs = sum of the scores = sum of outdegrees. On the other hand, the sum of the outdegrees is $\binom{v}{2}$. So $vs = v(v-1)/2$ and $v = 2s+1$.
(ii) One example: decompose K_{2s+1} into s cycles (see Theorem 6.3), and in each cycle orient each edge so as to form a directed cycle.

Exercises 11.3

11.3.2 Select an Euler walk in G. Orient each edge in the direction of the walk.

Exercises 12.1

12.1.2

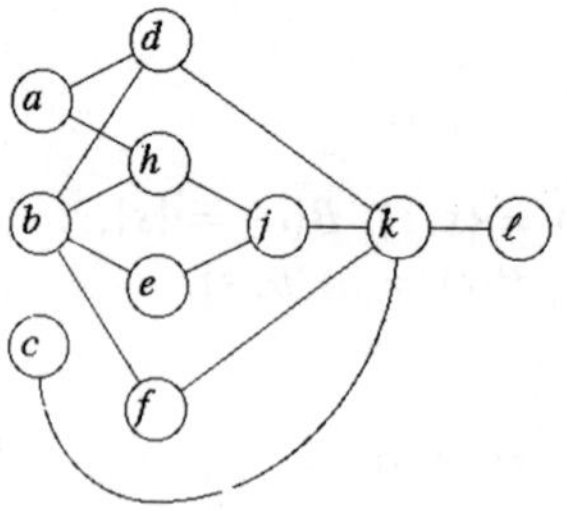

All arcs are directed from left to right.

Exercises 12.2

12.2.2 (ii) 29;

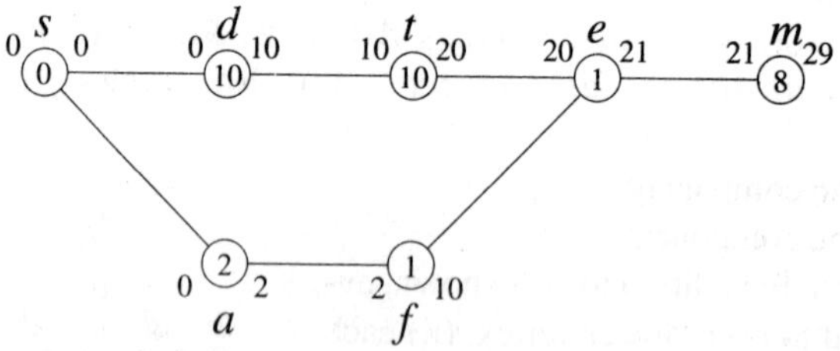

s is start; d is dictate; t is type; a is address; f is stamp; m is mail (and is finish)
All arcs are directed from left to right.

12.2.6 All arcs are directed from left to right. t is an added finish node. Critical path $abdehk\ell t$, duration 46.

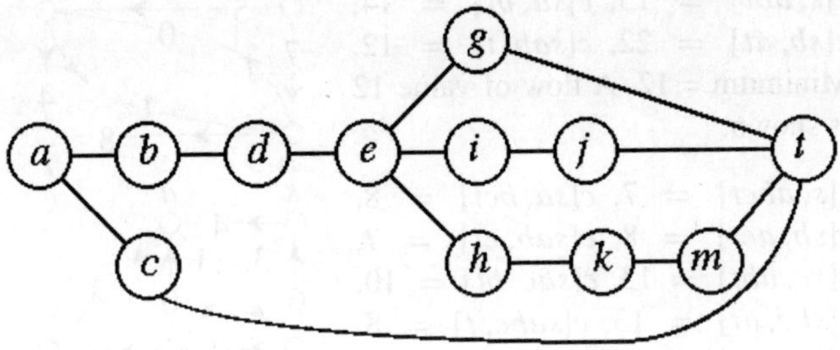

Exercises 12.3

12.3.3 Say the duration of a task in Exercise 12.2.6 was t. Then the expected time in this problem is $4t/3$ and its variance is $(t/6)^2$. The critical path is unchanged, $abdehk\ell t$, and the expected duration is $4 \cdot 46/3 = 61.33$ days. The variance is $336/6^2$, so the probability of completion within 65 days is $P(N(61.33, 3.055) \leq 65 = P(N(0,1) \leq \frac{3.67}{3.055}$ $P(N(0,1) \leq 1.20 = .88$.

12.3.5 Expected times: $a : 16$, $b : 13.5$, $c : 18$, $d : 8$, $e : 16$, $f : 27$, $g : 8.5$, $h : 10$, $i : 17$, $j : 9.5$. Critical path $sbfjt$, length 50. Variances $b : (\frac{5}{6})^2$, $f : 3^2$, $j : (\frac{3}{2})^2$, overall $11.9444 = 3.38^2$.
$P(N(50, 3.38) \leq 52$
$= P(N(0,1) \leq \frac{2}{3.38}$
$= P(N(0,1) \leq .59 = .72$.

All arcs are directed from left to right.

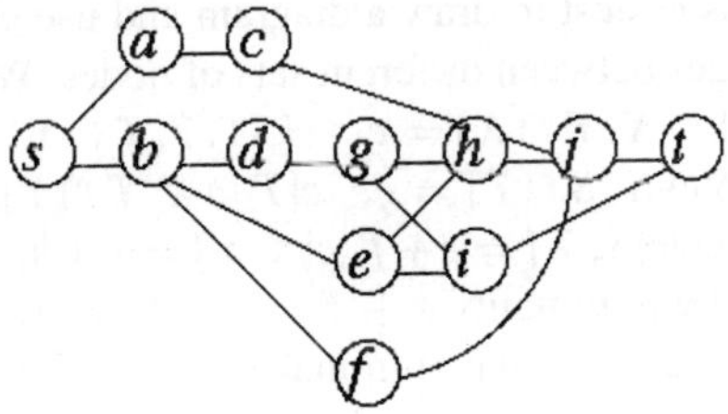

Exercises 13.1

13.1.1 (i) $sadt$. (ii) $sbdt$.
(iii) (a) $\{af, df\}$, (b) $\emptyset$, (c) 11, (d) 8, (e) 5, (f) 14.

13.1.3 (i) No: imbalance at b, g. (ii) Yes.

13.1.5 (i) $sact$, $sbact$, $sbct$, $sbdct$, $sbdt$. (ii) No (flow into t would exceed flow out of s). (iii) $x = 3 - z$, $y = 8 - z$, $0 \leq z \leq 3$.

Exercises 13.2

13.2.2 (ii) 6. (iii) $sabt$. (iv) Change to $f(sa) = 7$, $f(ab) = 5$, $f(bt) = 5$, other flows unchanged. This has value 10. (v) Augment along $suxyzt$: $f(us) = 0$, $f(xu) = 0$, $f(xy) = 4$, $f(yz) = 4$, $f(zt) = 3$. Value is 11. (vi) 11 is maximal because $[saux, bvwyzt]$ is a cut of capacity 11.

13.2.5 (i) $c[s, abt] = 15$, $c[sa, bt] = 14$, $c[sb, at] = 22$, $c[sab, t] = 12$. Minimum = 12. A flow of value 12 is shown.

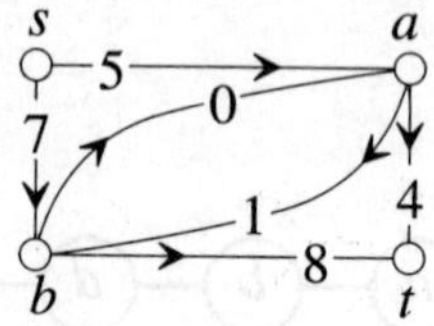

(ii) $c[s, abct] = 7$, $c[sa, bct] = 8$, $c[sb, act] = 8$, $c[sab, ct] = 7$, $c[sc, abt] = 13$, $c[sac, bt] = 10$, $c[sbc, at] = 13$, $c[sabc, t] = 8$. Minimum = 7. A flow of value 7 is shown.

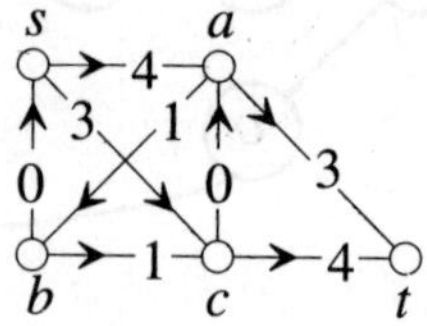

13.2.7 Replace x by two vertices, x_1 and x_2. Every arc into x becomes an arc into x_1; every arc out of x becomes an arc out of x_2; and there is an arcx_1x_2 of capacity d.

Exercises 13.3

13.3.2 First, observe that both are separating cuts:
$T \cap Y = \overline{S} \cap \overline{T} = \overline{S \cup T}$; $T \cup Y = \overline{S} \cup \overline{T} = \overline{S \cap T}$,
$s \in S, X \Rightarrow s \in S \cup X, S \cap X$; $t \in T, Y \Rightarrow t \in T \cap Y, T \cap Y$.
It is easiest to draw a diagram and use single letters to represent the capacities of edges between different sets of nodes. Write:
$c[S \cap X, T \cap X] = e$, $c[S \cap Y, T \cap Y] = f$,
$c[S \cap X, S \cap Y] = g$, $c[T \cap X, T \cap Y] = h$.
Then $c[S, T] = e + f$, $c[X, Y] = g + h$, so by minimality $e + f = g + h = m$, where m is the minimal cut size. So $e + g + f + h = 2m$. Now $c[S \cup X, T \cap Y] = f + h \leq m$, by minimality, and also $c[S \cap X, T \cup Y] = e + g \leq m$. The only possibility is that both capacities equal m.

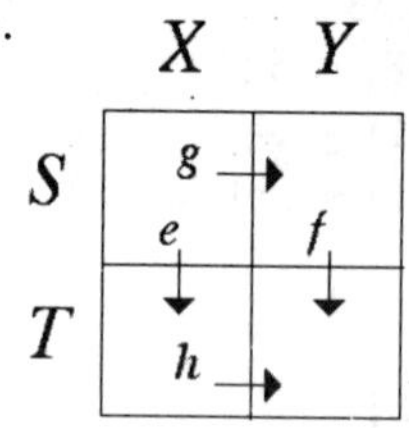

Exercises 13.4

13.4.1 There is a cut, $[sabde, cft]$, of capacity 8.

13.4.2 8 ($[s, abcdeft]$ is a cut of capacity 8).

13.4.5 Max flow values are 9 and 16. Examples of flows realizing these (unmarked arcs receive flow 0):

(i)

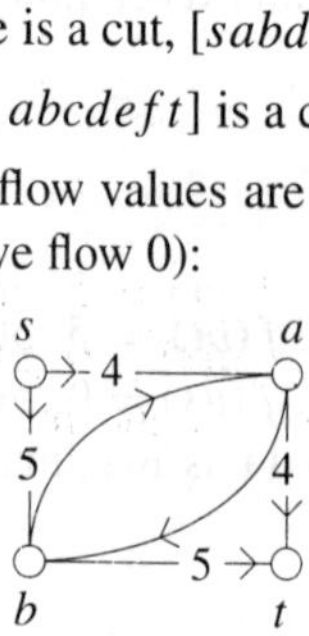

(ii)

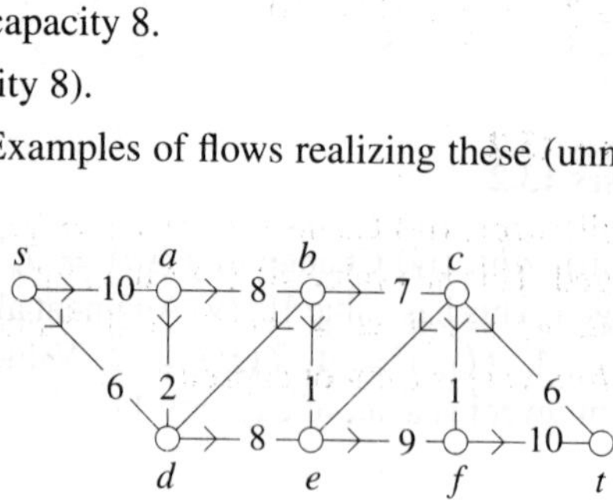

Exercises 13.5

3.5.2 Since there is no restriction on production or sales, add vertices s and t and put infinite capacity on all arcs sFi and Mit. Then carry out the algorithm. The maximum flow is 115; an example is shown. To see that this is maximum, observe the cut of capacity 115 shown by the heavy line.

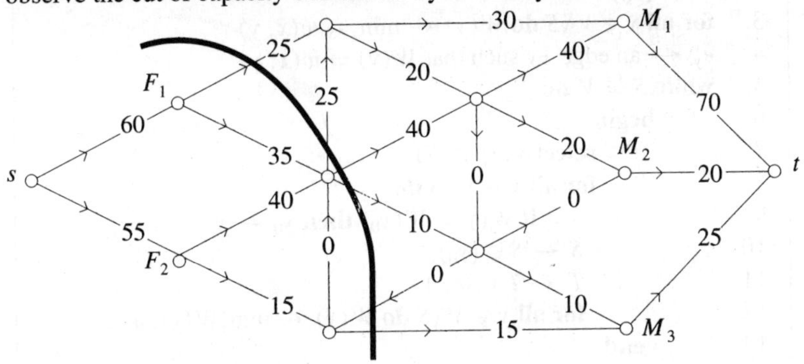

13.5.3 Yes. A suitable flow is shown in the Figure. (Directions are assumed to be as in the original.)

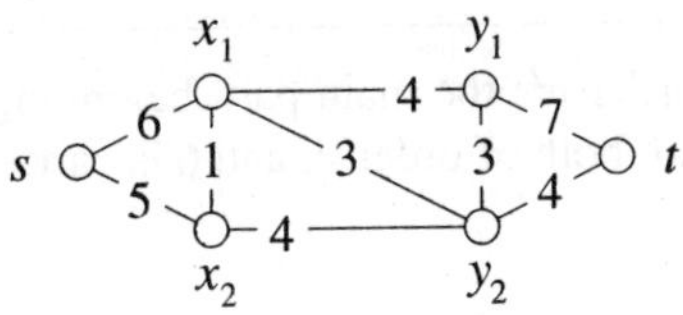

Exercises 14.1

14.1.1 If $f = O(g)$ and $g = O(h)$ then there exist a values n_{01} and n_{02} and positive constants K_1 and K_2 such that $f(n) \le K_1 g(n)$ whenever $n \ge n_{01}$ and $g(n) \le K_2 h(n)$ whenever $n \ge n_{02}$. So, if $n \ge \max\{n_{01}, n_{02}\}$, $f(n) \le K_1 g(n) \le g(n) \le K_1 K_2 h(n)$. So $f = O(h)$ (using $n_0 = \max\{n_{01}, n_{02}\}$, $K = K_1 K_2$).

14.1.8 In testing whether n is prime, one is answering the decision problem: *Is n in the set P_n? where* $P_n = \{x : x \le n, x \text{ is prime}\}$. Since $\mathcal{P}_n$ is asymptotically equal to $\sqrt{n}$, the input size of the problem is $\log n$, not n. If we write $t = \log n$ then $\sqrt{n} = e^{t/2}$, so the problem is actually exponential in the input size.

Exercises 14.3

14.3.2 (i) It is easy to see that $w_{k;ij}$ is the length of the shortest path from x_i to x_j among all paths that contain at most k edges, as required (this can be written formally as an induction). As no path can contain more than $v - 1$ edges, W_v is the matrix of shortest paths.

(ii) In the algorithm, replace line 4. by:

4. **for** $k = 1$ **to** $v - 1$ **do**

and replace line 7. by:

7. **for** $h = 1$ **to** v **do**

8. $w_{k;ij} \leftarrow \min\{w_{k-1;ij}, \min_h\{w_{k-1;ih} + w_{hj}\}\}$.

(iii) Complexity is v^4.

14.3.5 x_0 is the arbitrarily chosen starting vertex. At any stage, S is the set of vertices and T is the set of edges already selected for the tree. For each vertex $y \in V \backslash S$, $W(y)$ is the minimum weight of edges joining y to S.

```
1.  T ← ∅
2.  S ← {x0}
3.  for all y ∈ V\S do W(y) ← min_{x∈S} w(x, y)
4.  e_y ← an edge xy such that W(y) = w(x, y)
5.  while S ≠ V do
6.      begin
7.          select y0 ∈ V\S)
8.          for all y ∈ V\S do
9.              if W(y < W(y0) then y0 ← y
10.         S ← S ∪ {y0}
11.         T ← T ∪ {e_{y0}}
12.         for all y ∈ V\S do W(y) ← min{W(y), w(y, y0))
13.     end
```

This is order v^2: the main part, beginning with step 6, is of complexity v (steps 8 and 12 are both of order v, and 6 is carried out $v - 1$ times.

Exercises 15.2

15.2.5 (i) $n = 2$: mpl $= \frac{29}{14}$; $n = 3$: mpl $= \frac{29}{14}$; $n = 4$: mpl $= \frac{55}{28}$
(ii) cpl $= 2$ in each case
(iii) cc $= \frac{1}{8}$ when $n = 2$, 0 otherwise.

Index